DEVELOPMENTS IN MATHEMATICAL AND EXPERIMENTAL PHYSICS

Volume A: Cosmology and Gravitation

DEVELOPMENTS IN MATHEMATICAL AND EXPERIMENTAL PHYSICS

Volume A: Cosmology and Gravitation

Edited by

Alfredo Macias,
Francisco Uribe, and
Enrique Diaz

Universidad Autónoma Metropolitana–Iztapalapa
Mexico City, Mexico

Kluwer Academic / Plenum Publishers
New York, Boston, Dordrecht, London, Moscow

Library of Congress Cataloging-in-Publication Data

Mexican Meeting on Mathematical and Experimental Physics (1st: 2001: Mexico City, Mexico)
    Developments in mathematical and experimental physics/edited by Alfredo Macias, Francisco Uribe, and Enrique Diaz-Herrera.
      p. cm.
    Proceedings of the first Mexican Meeting on Mathematical and Experimental Physics, held September 10–14, 2001, in Mexico City, Mexico—t.p. verso.
    Includes bibliographical references and index.
    Contents: v. A. Cosmology and gravitation.
    ISBN 0-306-47293-7
      1. Mathematical physics—Congresses. 2. Physics—Congresses. I. Macías, A. (Alfredo) II. Uribe P., Francisco Javier (Uribe Patiño) III. Diaz-Herrera, Enrique, 1957– IV. Title.

QC19.2 .M48 2001
530.15—dc21

2001075482

Proceedings of the First Mexican Meeting on Mathematical and Experimental Physics, held September 10–14, 2001, in Mexico City, Mexico

ISBN 0-306-47293-7

©2002 Kluwer Academic / Plenum Publishers, New York
233 Spring Street, New York, N.Y. 10013

http://www.wkap.com

10 9 8 7 6 5 4 3 2 1

A C.I.P. record for this book is available from the Library of Congress

All rights reserved

No part of this book may be reproduced, stored in a retrieval system, or transmitted in any form or by any means, electronic, mechanical, photocopying, microfilming, recording, or otherwise, without written permission from the Publisher, with the exception of any material supplied specifically for the purpose of being entered and executed on a computer system, for exclusive use by the purchaser of the work

Printed in the United States of America

This book is dedicated to the 70th birthday of Leopoldo García–Colín
and to the 80th birthday of Nicholas van Kampen

Contributing authors

- Dharambir V. Ahluwalia
  Facultad de Fisica, UAZ.
  A. P. C–600, Zacatecas, ZAC 98062, México.
  E–mail: ahluwalia@phases.reduaz.mx

- Eloy Ayón–Beato
  Departamento de Física
  Centro de Investigación y de Estudios Avanzados del IPN.
  Apdo. Postal 14–740, C.P. 07000, México, D.F., México.
  E–mail: ayon@fis.cinvestav.mx

- Abel Camacho
  Departamento de Física,
  Instituto Nacional de Investigaciones Nucleares.
  Apartado Postal 18–1027, México, D. F., México.
  E–mail: acamacho@nuclear.inin.mx

- Cuauhtemoc Campuzano
  Departamento de Física
  Centro de Investigación y de Estudios Avanzados del IPN.
  Apdo. Postal 14–740, C.P. 07000, México, D.F., México.
  E–mail: ccvargas@fis.cinvestav.mx

- Iliana Carrillo–Ibarra
  Departamento de Matemáticas
  Centro de Investigación y de Estudios Avanzados del IPN.
  Apdo. Postal 14-740, 07000, México D.F., México.
  E–mail: iliana@math.cinvestav.mx

- Jorge L. Cervantes–Cota
  Departamento de Física,
  Instituto Nacional de Investigaciones Nucleares (ININ).
  P.O. Box 18–1027, México D.F. 11801, México.
  E–mail:jorge@nuclear.inin.mx

- Pablo Chauvet–Alducin
  Departamento de Física , Universidad Autónoma Metropolitana–Iztapalapa.
  P. O. Box. 55–534, México D. F., C.P. 09340 México.
  E–mail: pcha@xanum.uam.mx

- Sergio del Campo
  Instituto de Física, Facultad de Ciencias Básicas y Matemáticas, Universidad Católica de Valparaíso. Avenida Brasil 2950, Valparaíso, Chile.
  Depart. de Física, Facultad de Ciencia, Universidad de Santiago de Chile.
  E-mail: sdelcamp@ucv.cl

- Jerónimo Cortez
  Instituto de Ciencias Nucleares, Universidad Nacional Autónoma de México.
  A. P. 70–543 México D.F. 04510, Méxoco.
  E-mail: cortez@nuclecu.unam.mx

- Hansjoerg Dittus
  ZARM, University of Bremen. 28359 Bremen, Germany.
  E-mail: dittus@zarm.uni-bremen.de

- S. Formański
  Institute of Physics,Technical University of Łódź.
  Wólczańska 219. 93-005 Łódź, Poland.
  E-mail: sforman@ck-sg.p.lodz.pl

- Rodolfo Gambini
  Instituto de Física, Facultad de Ciencias, Universidad de la República. Iguá 4225, CP 11400 Montevideo, Uruguay.
  E-mail: rgambini@fisica.edu.uy

- Alberto García
  Departamento de Física, CINVESTAV–IPN.
  Apartado Postal 14–740, C.P. 07000, México, D.F., México.
  E-mail: aagarcia@fis.cinvestav.mx

- Hector García–Compeán
  Departamento de Física, Centro de Investigación y Estudios Avanzados del IPN.
  P.O. Box 14-740, 07000, México D.F., México.
  E-mail: compean@fis.cinvestav.mx

- Astrid Haibel
  Universität zu Köln, II. Physikalisches Institut.
  Zülpicher Str. 77, D–50937 Köln, Germany.
  E-mail: ah@uni-koeln.de

- Mustapha Ishak
  Department of Physics, Queens University.

Kingston, Ontario, Canada.
E-mail: ishak@hera.phy.qeensu.ca

- Mariana Kirchbach
  Facultad de Fisica, UAZ. A. P. C-600, Zacatecas, ZAC 98062, México
  E-mail: kirchbach@chiral.reduaz.mx

- Jaime Klapp
  Instituto Nacional de Investigaciones Nucleares.
  Ocoyoacac 52045, Estado de México, México.
  E-mail: klapp@nuclear.inin.mx

- Claus Laemmerzahl
  Institut für Experimentalphysik
  Heinrich–Heine–Universität Duesseldorf.
  Universitaetsstrasse 1, Gebaeude 25.42, Raum O1.44 40225 Duesseldorf, Germany.
  E-mail: claus.laemmerzahl@uni-duesseldorf.de

- Samuel Lepe
  Instituto de Física, Facultad de Ciencias Básicas y Matemáticas
  Universidad Católica de Valparaíso. Avenida Brasil 2950, Valparaíso, Chile.
  Depart. de Física, Facultad de Ciencia, Universidad de Santiago de Chile.
  Avda. Ecuador 3493, Santiago, Chile.
  E-mail: slepe@lauca.usach.cl

- James E. Lidsey
  Astronomy Unit, School of Mathematical Sciences
  Queen Mary, University of London. Mile End Road, LONDON E1 4NS, UK.
  E-mail: J.E.Lidsey@qmul.ac.uk

- Alfredo Macías
  Departamento de Fisica, Universidad Autónoma Metropolitana–Iztapalapa.
  Apartado Postal 55-534, C.P. 09340, México, D.F., México.
  E-mail: amac@xanum.uam.mx

- Tonatiuh Matos
  Departamento de Física, Centro de Investigación y de Estudios Avanzados del IPN.

AP 14-740, 07000 México D.F., México.
E-mail: tmatos@fis.cinvestav.mx

- Eckehard W. Mielke
  Departamento de Física, Universidad Autónoma Metropolitana-Iztapalapa.
  Apartado Postal 55-534, C.P. 09340, México, D.F., México.
  E-mail: ekke@xanum.uam.mx

- Marcos Nahmad
  Instituto de Física, Universidad Autónoma de Puebla.
  P.O. Box J-48, Puebla 72570, México.

- Günter Nimtz
  Universität zu Köln, II. Physikalisches Institut.
  Zülpicher Str. 77, D-50937 Köln, Germany.
  E-mail: g.nimtz@uni-koeln.de

- Shin'ichi Nojiri
  Department of Applied Physics
  National Defence Academy. Hashirimizu Yokosuka 239-8686, Japan.
  E-mail: nojiri@cc.nda.ac.jp, snojiri@yukawa.kyoto-u.ac.jp

- Yuri N. Obukhov
  Institute for Theoretical Physics, University of Cologne.
  50923 Köln, Germany.
  Department of Theoretical Physics, Moscow State University. 117234 Moscow, Russia.
  E-mail: yo@thp.uni-koeln.de

- Octavio Obregón
  Instituto de Física de la Universidad de Guanajuato.
  P.O. Box E-143, 37150, León Gto., México.
  E-mail: octavio@ifug3.ugto.mx

- Sergei D. Odintsov
  Tomsk State Pedagogical University.
  Tomsk, RUSSIA and Instituto de Fisica de la Universidad de Guanajuato.
  Lomas del Bosque 103, Apdo. Postal E-143, 37150 León,Gto., México.
  E-mail: odintsov@ifug5.ugto.mx, odintsov@mail.tomsknet.ru

- Leonardo Patiño
  Instituto de Física, Universidad Nacional Autónoma de México.

A. P. 20-364, México D. F. 01000, México.
E-mail: leonardo@ft.ifisicacu.unam.mx

- Humberto H. Peralta
  Departamento de Física
  Universidad Autónoma Metropolitana-Iztapalapa.
  Apartado Postal 55-534, C.P. 09340, México, D.F., México.
  E-mail: estocastico@msn.com

- Luis O. Pimentel
  Departamento de Física
  Universidad Autónoma Metropolitana-Iztapalapa.
  P.O. Box 55-534, CP 09340, México D. F., México.
  E-mail: lopr@xanum.uam.mx

- Jerzy F. Plebański
  Departamento de Fisica, CINVESTAV. 07000 México D.F., México.
  E-mail: pleban@fis.cinvestav.mx

- Maciej Przanowski
  Departamento de Fisica, CINVESTAV. 07000 México D.F., México.
  Institute of Physics,Technical University of Lódź.
  Wólczańska 219. 93-005 Lódź, Poland.
  E-mail: przan@fis.cinvestav.mx

- Jorge Pullin
  Department of Physics and Astronomy, Louisiana State University.
  202 Nicholson Hall, Baton Rouge, LA 70803-4001, USA.
  E-mail: pullin@phys.lsu.edu

- Hernando Quevedo
  Instituto de Ciencias Nucleares
  Universidad Nacional Autónoma de México.
  A.P. 70-543, 04510 México D.F., México.
  E-mail: quevedo@nuclecu.unam.mx

- Cupatitzio Ramirez
  Facultad de Ciencias Físico Matemáticas
  Universidad Autónoma de Puebla.
  P.O. Box 1364, 72000, Puebla, México.
  E-mail: cramirez@fcfm.buap.mx

- Mario A. Rodriguez-Meza
  Departamento de Física,
  Instituto Nacional de Investigaciones Nucleares (ININ).

P.O. Box 18–1027, México D.F. 11801, México.
E–mail: mar@nuclear.inin.mx

- Guillermo F. Rubilar
  Institute for Theoretical Physics, University of Cologne.
  50923 Köln, Germany.
  E–mail: gr@thp.uni-koeln.de

- Miguel Sabido
  Instituto de Física de la Universidad de Guanajuato.
  P.O. Box E–143, 37150, León Gto., México.
  E–mail: msabido@ifug3.ugto.mx

- Franz E. Schunck
  Institut für Theoretische Physik, Universität zu Köln.
  50923 Köln, Germany.
  E–mail: fs@thp.uni-koeln.de

- Leonardo Di G. Sigalotti
  Instituto Venezolano de Investigaciones Científicas (IVIC).
  Carretera Panamericana Km. 11, Altos de Pipe, Estado Miranda, Venezuela.

- Daniel Sudarsky
  Instituto de Ciencias Nucleares
  Universidad Nacional Autónoma de México.
  A. Postal 70–543, México D.F. 04510, México.
  E–mail: sudarsky@nuclecu.unam.mx

- Roberto A Sussman
  Instituto de Ciencias Nucleares, UNAM.
  Apartado Postal 70–543, México D.F., 04510, México.
  E–mail: sussman@nuclecu.unam.mx

- César A. Terrero–Escalante
  Departamento de Física,
  Centro de Investigación y de Estudios Avanzados del IPN.
  Apdo. Postal 14–740, 07000, México, D.F., México.

Preface

The FIRST MEXICAN MEETING ON MATHEMATICAL AND EXPERIMENTAL PHYSICS was held at EL COLEGIO NACIONAL in Mexico City, Mexico, from September 10 to 14, 2001. This event consisted of the LEOPOLDO GARCÍA–COLÍN SCHERER Medal Lecture, delivered by Prof. Nicholas G. van Kampen, a series of plenary talks by Leopoldo García-Colín, Günter Nimtz, Luis F. Rodríguez, Rubén Barrera, and Donald Saari, and of three parallel symposia, namely, Cosmology and Gravitation, Statistical Physics and Beyond, and Hydrodynamics and Dynamical Systems. The response from the Physics community was enthusiastic, with over 200 participants and around 80 speakers, from all over the world: USA, Canada, Mexico, Germany, France, Holland, United Kingdom, Switzerland, Spain, and Hungary.

The main aim of the conference is to provide a scenario to Mexican researchers on the topics of Mathematical and Experimental Physics in order to keep them in contact with work going on in other parts of the world and at the same time to motivate and support the young and mid-career researchers from our country. To achieve this goal, we decided to invite as lecturers the most distinguished experts in the subjects of the conference and to give the opportunity to young scientist to communicate the results of their work. The plan is to celebrate this international endeavor every three years.

The most outstanding researcher at Universidad Autonoma Metropolitana is Leopoldo García–Colín. His devotion to science and high level of energy and enthusiasm that he brings to his research and teaching are very much appreciated by his students and collaborators. Therefore, the Universidad Autónoma Metropolitana (UAM) instituted in 2001 the Leopoldo García–Colín Medal, which will be awarded every three years. The medal is given in his honor regarded as one of the greatest mexican physicist of the 20th century and a Distinguished Professor of our University. The Leopoldo García–Colín Medal award recognizes outstanding scientists, in the developing and advanced countries, who have made outstanding contributions to theoretical physics. It is the highest recognition accorded by the UAM for excellence in scientific research to whom has significantly contributed to the advancement of science. The main goal of this award is to promote the academic activities at UAM by means of the wisdom of the medal winners.

Preface

An international committee of distinguished scientist selects the winners from a list of nominated candidates. The committee invites nominations from scientists working in the field of Theoretical Physics.

The award consists of a gold medal, which displays Leopoldo García–Colín in a side pose, as well as a plaque on which major contributions of the award winner are mentioned. The selection of the awardees is made solely on scientific merit.

The first awardee is Prof. Nicholas van Kampen from Utrecht University in Netherlands, for his outstanding work which has significantly contributed to the advancement of science, for the inspirational quality of his research, his thoughtful guidance of graduated and undergraduate students, his graciousness as a colleague and his service to the scientific community.

The proceedings of the FIRST MEXICAN MEETING ON MATHEMATICAL AND EXPERIMENTAL PHYSICS consist of three volumes, namely, Volume A: Cosmology and Gravitation, Volume B: Statistical Physics and Beyond, and Volume C: Hydrodynamics and Dynamical Systems.

These three volumes contain lecture notes on the topics covered in each of the three symposia at this conference. Additionally, the Proceedings are dedicated to honor the outstanding contributions to science of Professor Leopoldo García–Colín on the occasion of his 70th birthday, and of Professor Nicholas G. van Kampen on the occasion of his 80th birthday.

We would like to thank everyone who contributed to the success of the 1st MEXICAN MEETING ON MATHEMATICAL AND EXPERIMENTAL PHYSICS. Very special thanks are due to the invited speakers and lecturers who delivered a very interesting set of talks and shared their knowledge and time with participants. We also thank young people just starting out on their careers.

The Meeting could not have been realized without the financial support of El Colegio Nacional, of CONACyT (Mexico), of the Germany–Mexico exchange program of the DLR (Bonn)–CONACYT (Mexico City), and of Silicon Graphics. We wish to thank Dr. José Luis Gázquez, Dr. Luis Mier y Terán, Dr. María José Arroyo, General Rector, Rector of the Campus Iztapalapa, and Dean at Iztapalapa of the Faculty of Basic Sciences and Engineering of the Universidad Autónoma Metropolitana, respectively, for sponsoring this international and multidisciplinary endeavor. We also thank Juan Azorín from the Mexican Physical Society for his help and support.

We thank specially Prof. Leopoldo García–Colín and all the staff of EL COLEGIO NACIONAL for the warm hospitality that was extended to all participants.

We hope this Proceedings will be of interest to all the participants, and indeed to mathematicians and physicists in general, specially to young people just starting their scientific careers.

ALFREDO MACIAS, ENRIQUE DIAZ, FRANCISCO J. URIBE

Preface to the Volume A: Cosmology and Gravitation

The FIRST MEXICAN MEETING ON MATHEMATICAL AND EXPERIMENTAL PHYSICS was held at EL COLEGIO NACIONAL in Mexico City, Mexico from September 10 to 14, 2001.

These Proceedings contain lecture notes on the topics covered during the SYMPOSIUM ON COSMOLOGY AND GRAVITATION, the contributions cover the invited talks as well as the short talks. The short talks provided a pleasant and efficient way to learn about research work being developed by the participants and bring people together.

We would like to thank everyone who contributed to the success of the Symposium on Cosmology and Gravitation, in particular to Prof. Alberto García, Dean of the Physics Department at CINVESTAV–IPN for his valuable contribution to the organization of the academic program. Very special thanks are due to the invited speakers and lecturers who gave a very interesting and high quality set of talks and shared their knowledge and time with participants.

The Symposium on Cosmology and Gravitation could not have been realized without the financial support of the CONACyT (Mexico), of the Germany–Mexico exchange program of the DLR (Bonn)–CONACYT (Mexico City), and of Silicon Graphics. We wish to thank Dr. José Luis Gázquez, Dr. Luis Mier y Terán, Dr. María José Arroyo, General Rector, Rector of the campus Iztapalapa, Dean at Iztapalapa of the Faculty of Basic Sciences and Engineering of the Universidad Autónoma Metropolitana, respectively, for sponsoring this international and multidisciplinary endeavor. The financial support of ININ is gratefully acknowledged. We also thank Dr. Juan Azorín from the Mexican Physical Society for his help and support.

We thank specially Prof. Leopoldo García–Colín and all the stuff at EL COLEGIO NACIONAL for the warm hospitality that was extended to all participants.

We hope that these Proceedings will serve to further the impressive growth of Cosmology and Relativity in the region and reinforce the already strong ties between Mexican scientists and scientist and relativists in other parts of the world.

ALFREDO MACIAS

Contents

Part I Quantum Gravity and String Theories

On the Berezin Description of Kähler Quotients		3
Iliana Carrillo-Ibarra, Hugo Garcia-Compeán		
1	Motivation	3
2	Geometric quantization of Kähler quotients	5
3	Berezin quantization of Kähler quotients	6
(2+1)–dimensional supergravity		13
Alfredo Macias		
1	Introduction	13
2	Geometry in (2+1) dimensions	14
3	Standard supergravity in (2+1) dimensions	16
	3.1 Einstein–Cartan gravity in vacuum	16
	3.2 Rarita–Schwinger field	17
	3.3 Standard simple supergravity	18
	3.4 Cosmological term	20
4	Chern–Simons supergravities	21
5	Supergravity with translational Chern–Simons term	24
	5.1 Three–dimensional supergravity	24
	5.2 Energy–momentum and spin	24
	5.3 Supersymmetric transformations	25
	5.4 Supersymmetric dynamical symmetries	27
6	Concluding remarks	28
Noncommutative gravity and quantum cosmology		31
H. Garcia-Compeán, O. Obregón, C. Ramirez		
1	Introduction	32
2	String theory and noncommutative gauge theory	33
3	Chamseddine's Moyal deformation of Einstein gravity	36
4	Noncommutative quantum cosmology	37
5	Concluding remarks	41
Consistent discretizations in classical and quantum general relativity		47
Rodolfo Gambini, Jorge Pullin		
1	Introduction	47
2	Numerical relativity	48
3	Canonical quantization	49
4	The path integral	51
5	Conclusions	53

The field–to–particle transition problem		55
Jerónimo Cortez, Leonardo Patiño, Hernando Quevedo		
1	Introduction	56
2	The holographic principle	57
3	Nonlinear sigma models	60
4	The field–to–particle transition problem in gravity	63
5	Conclusions	67

Towards non–commutative topological gauge theory of gravity		71
H. Garcia-Compeán, O. Obregón, C. Ramirez, M. Sabido		
1	Introduction	72
2	Non commutative gauge symmetry and the Seiberg–Witten map	73
3	Topological gravity	74
4	Non–commutative topological gravity	75
5	Conclusions	77

Part II Cosmology and Black Holes

Improving the "no–hair" theorem for the Proca field		81
Eloy Ayón-Beato		
1	Introduction	81
2	Proca fields on static spacetimes	82
3	The "no–Proca–hair" theorem	83
4	Conclusions	86
Appendix: On the suitable integration measure of the horizon		87

New model calculations of protostellar collapse and fragmentation			89
Jaime Klapp, Leonardo Di G. Sigalotti			
1	Introduction		89
2	Observational constraints		91
3	New protostellar collapse calculations		92
	3.1	Adaptive calculations: Jeans resolution constraint	93
	3.2	Binary vs filament formation	95
4	Concluding remarks		96

Inflationary cosmology and the braneworld scenario	99
James E. Lidsey	

Galaxies formation from the scalar field dark matter model			111
Tonatiuh Matos			
1	Introduction		111
2	Scalar field as cosmological dark matter		112
3	Scalar field dark matter in galaxies		114
4	Scalar field collapse		115
5	The galactic model with scalar field dark matter		116
	5.1	A long exposition photograph of a galaxy	116
	5.2	The analytical solution	117
	5.3	Physical features of the model	117

Scalar soliton modelling dark matter halos		123
Eckehard W. Mielke, Humberto H. Peralta, Franz E. Schunck		
1	Introduction		123

2	Approximate NTS solution	125
3	Rotation curves	127
4	Comparison with observations	128
5	Outlook: Towards generally relativistic rotation curves	130

The de Sitter/Anti– de Sitter black holes phase transition?   135
Shin'ichi Nojiri, Sergei D. Odintsov

Understanding hairy black holes with the isolated horizon formalism   143
Daniel Sudarsky
1	Introduction	143
	1.1 Isolated horizon and horizon mass	144
	The main relation between black holes and solitons	146
	More complicated theories and crossing phenomena	148
2	Discussion	152

Sketching the inflaton potential   155
César A. Terrero-Escalante
1	Introduction	155
2	Stewart–Lyth inverse problem	157
3	Constraining the inflaton potential	159
4	Conclusions	162

Part III   Exact Solutions

Cosmological solutions and their stability in scalar–tensor theories   167
Jorge L. Cervantes-Cota, M. A. Rodriguez-Meza, Marcos Nahmad
1	Introduction	167
2	Equations for FRW and Bianchi models	168
3	Solutions for the potential $V = V_o \psi^2$	170
4	Stability analysis	171
5	Conclusions	174

Considerations on accelerated expansion in a scalar–tensor cosmology   177
Pablo Chauvet-Alducin
1	Introduction	177
2	Scalar–tensor field equations	180
3	The vacuum	181
4	Radiation	183

On conformally flat stationary axisymmetric spacetimes   187
Cuauhtemoc Campuzano and Alberto Garcia
1	Introduction	187
2	Conformally flat axisymmetric stationary spacetimes, $\epsilon = -1$	190
3	Conclusions	191

Quantum cosmology in some extended scalar–tensor theories   193
Sergio del Campo, Samuel Lepe, Luis O. Pimentel
1	Introduction	193
2	Scalar–tensor theories	194
3	Non–linear gravity	196
4	Multidimensional gravity	198

Deformation quantization of sdiff(Σ_2) SDYM equation — 201
M. Przanowski, J. F. Plebański, S. Formański
1. Introduction — 201
2. Conservation laws and hidden symmetries — 203
3. Linear systems for ME and dressing operators — 206
4. Infinite algebra of hidden symmetries — 209
5. Twistor construction. — 211

Part IV Experiments and Other Topics

Nonlocality and superluminal signal–velocity in photonic tunnelling — 217
Günter Nimtz and Astrid Haibel
1. Introduction — 217
2. Experimental set–up — 218
3. Partial reflection by photonic barriers — 219
4. Tunnelling time and superluminal velocity — 220
5. Conclusions — 221

At the interface of quantum and gravitational realms — 225
D. V. Ahluwalia
1. Introduction — 225
2. Non–commutative nature of spacetime and gravitationally–modified wave particle duality — 226
3. Quantum test particles in classical sources of gravity — 228
4. Quantum test particles in quantum sources of gravity — 229
5. Spatial and temporal fluctuations in spacetime foam — 229
6. Concluding remarks — 230

Non–Newtonian gravity and coherence properties of light — 233
A. Camacho
1. Introduction — 233
2. Young's experiment and non–Newtonian gravity — 234
3. Interference patterns — 236
   3.1 Time independent interference pattern — 236
   3.2 Time dependent interference pattern — 237
4. Hanbury–Brown–Twiss effect — 238

On a possible new type of a T odd skewon field linked to electromagnetism 241
Friedrich W. Hehl, Yuri N. Obukhov, Guillermo F. Rubilar
1. The constitutive tensor density χ of vacuum spacetime and its irreducible decomposition — 242
   1.1 Principal piece $^{(1)}\chi$ and light cone structure — 243
   1.2 Abelian axion α — 243
   1.3 Skewon piece $\rlap{/}{S}_i{}^j$ and dissipation — 243
2. The skewon field $\rlap{/}{S}_i{}^j$ — 245
3. Decomposing the "dual" constitutive tensor density $\kappa_{ij}{}^{kl}$ and recovering the skewon field — 247
4. The skewon field as 6×6 matrix — 248
5. Spatially isotropic skewon field and the ansatz of Nieves and Pal — 250

Contents

6	On the four electromagnetic constants for vacuum spacetime with spatial isotropy	250
7	How does the skewon field affect light propagation?	251
8	Discussion	253
Appendix: Two identities		254

A proposal for testing the weak equivalence principle for charged particles in space   257
Hansjoerg Dittus, Claus Laemmerzahl

1	Introduction	257
2	An Eötvös coefficient for charged particles	260
3	The Witteborn–Fairbank setup	261
4	Side effects in WEP experiments for charged particles	265
	4.1 Gravity induced side–effects	265
	The Schiff–Barnhill effect	265
	The DMRT–field	266
	4.2 Other side effects and errors	267
	Patch effects	267
	Electric field gradients	268
	Thermoelectric fields	268
	Magnetic fields and field gradients	268
	Radiation pressure	269
	Gas scattering in the tube	269
5	Experiments in space	270
	5.1 The idealized case	270
	5.2 Errors from microgravity conditions	273
	5.3 Total estimate of errors in space–born Witteborn–Fairbank experiment	274
6	Outlook	274

Massive $(1/2,1/2)$ bosons   277
D. V. Ahluwalia, M. Kirchbach

1	Introduction	277
2	Constructing $(1/2, 1/2)$ representation space	278
3	Conclusions and discussion	282

Inhomogeneous cosmologies with adiabatic evolution   285
Roberto A Sussman, Mustapha Ishak

1	Introduction	285
2	Integration of the field equations	286
3	Interpretation of the solutions	288
4	Initial conditions	289
5	Conserved quantities and scaling laws	290
6	Adiabatic initial conditions and evolution	291

Leopoldo Garcia–Colin Scherer: Brief biography   293
Eduardo Piña

Nicholas G. van Kampen: Brief biography   295
Rosalio F. Rodriguez

Index   299

I

QUANTUM GRAVITY AND STRING THEORIES

ON THE BEREZIN DESCRIPTION OF KÄHLER QUOTIENTS

Iliana Carrillo–Ibarra
Departamento de Matemáticas
Centro de Investigación y de Estudios Avanzados del IPN
Apdo. Postal 14-740, 07000, México D.F., México
iliana@math.cinvestav.mx

Hugo Garcia–Compeán
Departamento de Física
Centro de Investigación y de Estudios Avanzados del IPN
Apdo. Postal 14-740, 07000, México D.F., México
compean@fis.cinvestav.mx

Abstract We survey geometric quantization of finite dimensional affine Kähler manifolds. Its corresponding prequantization and the Berezin's deformation quantization, as proposed by Cahen *et al.*, is used to quantize their corresponding Kähler quotients. Equivariant formalism greatly facilitates the description.

Keywords: Berezin's Deformation Quantization, Kähler quotients, Equivariant Formalism

1. Motivation

Chern–Simons (CS) gauge theory in 2+1 dimensions is a very interesting quantum field theory which has been very useful to describe diverse sorts of physical and mathematical systems. On the physical side, we have the fractional statistics of quasi-particles in the fractional quantum Hall effect [1], Einstein gravity in 2+1 dimensions with nonzero cosmological constant [2]. On the mathematical side it is related to beautiful mathematics like knot and link invariants [3] and to quantum groups [4]. There is also a nice relation with conformal field theory in two dimensions [3, 5]. From the quantization of this theory we also have learned a lot of things like a very non–trivial generalization of the original Jones representations of the braid group [6].

CANONICAL QUANTIZATION

In 2+1 dimensions CS gauge theory is based in the Lagrangian

$$L_{CS} = \frac{k}{4\pi} \int_M Tr\left(A \wedge dA + \frac{2}{3} A \wedge A \wedge A\right), \quad (1)$$

where A is a Lie algebra ($Lie(G)$) valued gauge connection on the G–bundle E over M and where G is a compact and simple finite dimensional Lie group, e.g. $SU(2)$. Thus $A = \sum_{a=1}^{dim\ G} A_I^a T_a dx^I$ (with $I, J = 0, 1, 2$) where T_a are the generators of $Lie(G)$ and $k \in \mathbf{Z}$ is the level of the theory.

In the classical theory the equations of motion are given by the "flat connection" condition

$$F(A) = dA + A \wedge A = 0 \iff F_{IJ}^a = 0. \quad (2)$$

While the constraints are given by the Gauss law:

$$\frac{\delta L}{\delta A_0^a} = \varepsilon_{IJ} F_{IJ}^a = 0. \quad (3)$$

Canonical quantization on $M = \Sigma \times \mathbf{R}$ consists in the construction of a Hilbert space $\mathcal{H}_\Sigma$ associated to the two–dimensional surface Σ. The construction of $\mathcal{H}_\Sigma$ is as follows. First, the decomposition of M allows to fix the temporal gauge $A_0 = 0$. In this gauge, the Poisson bracket is given by

$$\{A_I^a(x), \pi_J^b(y)\} = \frac{4\pi}{k} \varepsilon_{IJ} \delta^{ab} \delta^2(x - y), \quad (4)$$

where $\pi_I^a(x) = \frac{\partial L}{\partial \dot{A}_I^a} = A_I^a(x)$.

The phase space $\mathcal{A}$ consist of the solutions of Eq. (2). That means all flat connections on Σ. The incorporation of the constraints (3) leads to the modulus space of certain families of vector bundles over Σ. This can be identified with the reduced phase space $\mathcal{M} = \mathcal{A}/\mathcal{G}$. Pick a complex structure J on Σ leads to consider to $\mathcal{M}$ as a compact Kähler manifold $\mathcal{M}_J$ of finite dimension. This space is precisely the modulus space of certain family of holomorphic vector bundles over Σ. The quantization of the manifold $\mathcal{M}_J$ leads to the quantization of the Chern–Simons theory given by the Lagrangian (1). This can be constructed with the help of the Quillen determinant line bundle $\mathcal{L}$. This is a line bundle over $\mathcal{M}_J$ whose first Chern class $c_1(\mathcal{L})$ coincides with the simplectic form ω_0 which determines the Poisson brackets $\{A_I^a(x), A_J^b(y)\} = \omega_0^{-1}(dA, dA)$.

For the arbitrary level k the symplectic form $\omega = k\omega_0$ is given by the first Chern class of the k-th power of the line bundle $\mathcal{L}^{\otimes k}$. Finally the Hilbert space $\mathcal{H}_\Sigma$ is constructed from the space of L^2-completion of holomorphic sections $H^0_{L^2}(\mathcal{M}_J, \mathcal{L}^{\otimes k})$ of the determinant line bundle $\mathcal{L}^{\otimes k}$.

The quantization of the CS theory is given by the geometric quantization of the physical reduced phase space $\mathcal{M}_J$. This was given by Axelrod, Della Pietra and Witten [6].

In the present note we use the Berezin's deformation quantization formalism [7, 8, 9, 10, 11] to quantize the reduced phase space $(\mathcal{M}_J, \tilde{\omega})$ which can be regarded as a finite dimensional Kähler quotient space. This is a preliminary step to apply the Berezin's formalism to quantize the more involved infinite dimensional case of Chern–Simons gauge theory in three dimensions with compact groups. One possible guide to address the infinite dimensional case would be the case of the quantization of $\mathbf{CP}^\infty$ [12]. This will be reported in a forthcoming paper [13]. Recent interesting applications of Berezin's quantization are found in [14, 15].

2. Geometric quantization of Kähler quotients

UP–STAIRS GEOMETRIC QUANTIZATION

We first consider the finite dimensional affine symplectic manifold $(\mathcal{A}, \omega)$ with a chosen complex structure J on $\mathcal{A}$, invariant under affine translations [6]. This induces a Kähler structure on $\mathcal{A}$. In order to quantize the affine Kähler space $\mathcal{A}_J$ we consider the prequantum line bundle $\mathcal{L}$ over $\mathcal{A}$ with connection ∇. This connection has curvature $[\nabla, \nabla] = -i\omega$ since symplectic connection ω is of the $(1,1)$ type and type $(0,2)$ component of the curvature vanishes. This induces a holomorphic structure on $(\mathcal{L}, \nabla)$. Consider the L^2-completion of the subset of holomorphic sections $\mathcal{H}_Q|_J \equiv H^0_{L^2}(\mathcal{A}_J, \mathcal{L})$ of the prequantum Hilbert space $H^0_{L^2}(\mathcal{A}, \mathcal{L})$. This constitutes the Hilbert space of the quantization of $(\mathcal{A}_J, \omega)$. The Kähler quantization depends on the choice of J. In the theory of geometric quantization it is impossible to choice in a natural way a Kähler polarization. That is, to choice a complex structure J for which ω has the properties of a Kähler form. Thus the Kähler polarization is not unique. In the procedure of quantization we should make sure that the final result will be independent on this complex structure J and it depends only on the underlying symplectic geometry. Thus the idea is to find a canonical identification of the $\mathcal{H}_Q|_J$ as J varies (for further details see [6]).

DOWN–STAIRS GEOMETRIC QUANTIZATION

Now we will consider symplectic quotients of finite-dimensional affine symplectic spaces. The idea is geometric quantize the reduced phase space $(\mathcal{M}_J, \widetilde{\omega})$ where $\mathcal{M}_J = \mathcal{A}_J//\mathcal{G}$ is the *Marsden-Weinstein quotient*.

We start from $(\mathcal{A}_J, \omega)$ with the action of the group $\mathcal{G}$ acting on $\mathcal{A}_J$ by symplectic diffeomorphisms.

Let **g** be the Lie algebra of $\mathcal{G}$ and consider the map $T : \mathbf{g} \to Vect(\mathcal{A}_J)$. Since $\mathcal{G}$ preserves the symplectic form, the image of T is a subspace of $Vect(\mathcal{A}_J)$ consisting in the symplectic vector fields on $\mathcal{A}_J$. The co-moment map is given by $F : \mathbf{g} \to Ham(\mathcal{A}_J)$ where $Ham(\mathcal{A}_J)$ is the space of Hamiltonian functions on $\mathcal{A}_J$. Take a basis of **g** to be $\{L_a\}$ and we have $T_a = T(L_a)$. T is a Lie algebra representation since $[T_a, T_b] = f_{ab}^c T_c$ with $a, b, c = 1, \ldots, dim\mathcal{G}$.

For each $x \in \mathcal{A}_J$, $F_a(x)$ are the components of a vector in the dual space $\mathbf{g}^V$. That means that there is a mapping $F : \mathcal{A}_J \to \mathbf{g}^V$ called the *moment map*. $F^{-1}(0)$ is $\mathcal{G}$-invariant so one can define the symplectic quotient of $\mathcal{A}_J$ and $\mathcal{G}$ as $\mathcal{M}_J = F^{-1}(0)/\mathcal{G}$. Thus one have $\pi : F^{-1}(0) \to F^{-1}(0)/\mathcal{G} \equiv \mathcal{M}_J$ where $x \mapsto \widetilde{x}$. $\mathcal{M}_J$ also have structure of a symplectic manifold whose symplectic structure $\widetilde{\omega}$ is given by $\widetilde{\omega}_{\widetilde{x}}(\widetilde{V}, \widetilde{W}) = \omega_x(V, W)$ for $\widetilde{V}, \widetilde{W} \in T_{\widetilde{x}}\mathcal{A}_J$.

We consider on $\mathcal{A}_J$ only $\mathcal{G}$-invariant quantities so that when restricted to $F^{-1}(0) \subset \mathcal{A}_J$ they are pushed-down to the corresponding objects in $\mathcal{M}_J$.

The prequantum line bundle can be pushed-down as follows. The symplectic action of $\mathcal{G}$ on $\mathcal{A}_J$ can be lifted to $\mathcal{L}$ in such a way that it preserves the connection and Hermitian structure on $\mathcal{L}$. One may define the pushdown bundle $\widetilde{\mathcal{L}}$ by stating that its sections $\Gamma(\mathcal{M}_J, \widetilde{\mathcal{L}}) \equiv \Gamma(F^{-1}(0), \mathcal{L})^{\mathcal{G}}$ constitutes a $\mathcal{G}$-invariant subspace of the space of sections $\Gamma(\mathcal{A}_J, \mathcal{L})$. The connection also can be pushed-down and it satisfies $\widetilde{\nabla}_{\widetilde{V}}\psi = \nabla_V \psi$. Meanwhile the curvature of the connection $\widetilde{\nabla}$ is $-i\widetilde{\omega}$. Thus the prequantization is given by $(\widetilde{\mathcal{L}}, \widetilde{\nabla}, \langle \cdot | \cdot \rangle_{\widetilde{\mathcal{L}}})$. Here $\langle \cdot | \cdot \rangle_{\widetilde{\mathcal{L}}}$ is the $\mathcal{G}$-invariant inner product $\langle \cdot | \cdot \rangle_{\mathcal{L}}^{\mathcal{G}}$.

3. Berezin quantization of Kähler quotients

The main goal of this section is to describe the Berezin's quantization of the Kähler manifold $(\mathcal{M}_J, \widetilde{\omega})$ where $\mathcal{M}_J$ is the Marsden–Weinstein quotient. That means we find an associative and noncommutative family of algebras $(\widetilde{\mathcal{S}_B}, \widetilde{*_B})$ with $\widetilde{\mathcal{S}_B} \subset C^\infty(\mathcal{M}_J)$ which is indexed with a real and positive parameter $\hbar$ which helps to recover the classic limit when $\hbar \to 0$. Here we set $\hbar = 1$.

In order to do that we follows the same strategy that for the geometric quantization case of the previous section. We first Berezin quantize

$(\mathcal{A}_J, \omega)$, i.e. we find an associative and noncommutative family of algebras $(\mathcal{S}_B, *_B)$ with $\mathcal{S}_B \subset C^\infty(\mathcal{A}_J)$ (see [8, 9]). After that we project out all relevant quantities to be $\mathcal{G}$-invariant i.e. $\widetilde{\mathcal{S}_B} \subset C^\infty(\mathcal{A}_J)^{\mathcal{G}} \equiv C^\infty(F^{-1}(0)/\mathcal{G})$ with $F^{-1}(0) \subset \mathcal{A}_J$.

UP-STAIRS BEREZIN'S QUANTIZATION

Let $(\mathcal{L}, \nabla, \langle \cdot | \cdot \rangle_\mathcal{L})$ be a prequantization of the affine Kähler manifold $\mathcal{A}_J$. The inner product $\langle \cdot | \cdot \rangle_\mathcal{L}$ is compatible with the connection ∇ and it is defined as

$$\langle \chi | \psi \rangle_\mathcal{L} \equiv \int_{\mathcal{A}_J} \langle \chi | \psi \rangle \frac{\omega^n}{n!}, \tag{5}$$

for all $\chi, \psi \in H^0_{L^2}(\mathcal{A}_J, \mathcal{L})$ where $\langle \chi | \psi \rangle = \exp(-\Phi)\overline{\chi}\psi$ and Φ is the Kähler potential $\Phi(Z, \overline{Z}) = \sum_i Z^i \overline{Z}^i$. The norm of an element ψ of $H^0_{L^2}(\mathcal{A}_J, \mathcal{L})$ is given by $\langle \psi | \psi \rangle_\mathcal{L} \equiv \|\psi\|^2_\mathcal{L}$. As $\mathcal{A}_J$ is topologically trivial, the line bundle $\mathcal{L}$ can be identified with the *trivial* holomorphic line bundle whose holomorphic sections are holomorphic functions ψ with Hermitian structure i.e. $\langle \psi | \psi \rangle = \exp(-\Phi)|\psi|^2$. The curvature of the connection compatible with the Hermitian structure is given by $\bar{\partial}\partial(-\Phi) = \sum_i dZ^i \wedge d\overline{Z}^i = -i\omega$. Of course the existence of a prequantization bundle implies that $\frac{\omega}{2\pi} \in H^2(\mathcal{A}_J, \mathbf{Z})$.

Take $\mathcal{Q} \in \mathcal{L}_0$ and $\pi[\mathcal{Q}] = x \in \mathcal{A}_J$ with local complex coordinates $\{Z^i, \overline{Z}^i\}$. Here $\mathcal{L}_0$ is the line bundle $\mathcal{L}$ without the zero section. Now consider $\psi \in H^0_{L^2}(\mathcal{A}_J, \mathcal{L})$ a holomorphic section such that $\psi(x) = \psi[\pi(\mathcal{Q})] = L_\mathcal{Q}(\psi)\mathcal{Q}$, where $L_\mathcal{Q}(\psi)$ is a linear continuous functional of ψ.

By the Riesz theorem there is a section $e_\mathcal{Q} \in H^0_{L^2}(\mathcal{A}_J, \mathcal{L})$ such that

$$L_\mathcal{Q}(\psi) = \langle e_\mathcal{Q} | \psi \rangle_\mathcal{L} \tag{6}$$

with $\mathcal{Q} \in \mathcal{L}_0$. $e_\mathcal{Q}$ is known in the literature as a *generalized coherent state*.

Now consider a bounded operator $\hat{\mathcal{O}} : H^0_{L^2}(\mathcal{A}_J, \mathcal{L}) \to H^0_{L^2}(\mathcal{A}_J, \mathcal{L})$. Define the *covariant* symbol of this operator as

$$\mathcal{O}_B(x) = \frac{\langle e_\mathcal{Q} | \hat{\mathcal{O}} | e_\mathcal{Q} \rangle_\mathcal{L}}{\|e_\mathcal{Q}\|^2_\mathcal{L}}, \tag{7}$$

where $\mathcal{Q} \in \mathcal{L}_0$ and $\pi(\mathcal{Q}) = x$. Define the space of covariant symbols $\mathcal{S}_B = \{\mathcal{O}_B(x),$ which are covariant symbols of operators $\hat{\mathcal{O}}\}$. Each covariant symbol can be analytically continued to the open dense subset of $\mathcal{A}_J \times \mathcal{A}_J$ in such a way $\langle e_\mathcal{Q} | e_{\mathcal{Q}'} \rangle_\mathcal{L} \neq 0$ with $\pi(\mathcal{Q}) = x$ and $\pi(\mathcal{Q}') = y$ (with local coordinates $\{W^i, \overline{W}^i\}$). The obtained symbol is holomorphic in the first entry and anti–holomorphic in the second entry. This analytic continuation is reflected in the covariant symbol in the form

$$\mathcal{O}_B(Z,\overline{W}) = \frac{\langle e_\mathcal{Q}|\widehat{\mathcal{O}}|e_{\mathcal{Q}'}\rangle_\mathcal{L}}{\langle e_\mathcal{Q}|e_{\mathcal{Q}'}\rangle_\mathcal{L}}. \tag{8}$$

The operator $\widehat{\mathcal{O}}$ can be recovered from its symbol in the form

$$\widehat{\mathcal{O}}\psi(Z) = \langle e_\mathcal{Q}|\widehat{\mathcal{O}}|\psi\rangle_\mathcal{L} \mathcal{Q}. \tag{9}$$

The consideration of completeness condition, which can be written as follows, $1 = \int_{\mathcal{A}_J} |e_\mathcal{Q}\rangle\langle e_\mathcal{Q}| \exp\left(-\Phi(Z,\overline{Z})\right) \frac{\omega^n}{n!}(Z,\overline{Z})$ leads to

$$\widehat{\mathcal{O}}\psi(Z) = \int_{\mathcal{A}_J} \mathcal{O}_B(Z,\overline{W}) \mathcal{B}_\mathcal{Q}(Z,\overline{W}) \psi(W) \exp\left(-\Phi(W,\overline{W})\right) \frac{\omega^n}{n!}(W,\overline{W}) \mathcal{Q}, \tag{10}$$

where $\psi(W) = \langle e_{\mathcal{Q}'}|\psi\rangle_\mathcal{L}$, $\pi[\mathcal{Q}] = x$, $\mathcal{Q} \in \mathcal{L}_0$ and $\mathcal{B}_\mathcal{Q}(Z,\overline{W}) \equiv \langle e_\mathcal{Q}|e_{\mathcal{Q}'}\rangle_\mathcal{L}$. $\mathcal{B}_\mathcal{Q}(Z,\overline{W})$ is the generalized Bergman kernel.

In order to connect this global description with the (local) standard Berezin formalism is important to set a dense open subset $\mathcal{U}_J$ of the affine space $\mathcal{A}_J$. Then there is a holomorphic section $\psi_0 : \mathcal{U}_J \to \mathcal{L}_0$ and a holomorphic function $\phi : \mathcal{U}_J \to \mathbf{C}$ such that $\psi(x) = \phi(x)\psi_0(x)$ with $x \in \mathcal{U}_J \subset \mathcal{A}_J$. Define the map $\mathcal{U}_J \to H^0_{L^2}(\mathcal{U}_J,\mathcal{L})$ such that $x \mapsto \phi_x$. Let ϕ_x and ϕ' be two elements of $H^0_{L^2}(\mathcal{U}_J,\mathcal{L})$ then

$$\phi'(x) = \langle \phi'|\phi_x\rangle_\mathcal{U} = \int_{\mathcal{U}_J} \phi'(y)\overline{\phi}_x(y)|\psi_0|^2(y) \exp\left(-\Phi(y)\right) \frac{\omega^n}{n!}(y), \tag{11}$$

where $\langle\cdot|\cdot\rangle_\mathcal{U}$ is the inner product in $H^0_{L^2}(\mathcal{U}_J,\mathcal{L})$.

Let $\widehat{\mathcal{O}}$ be a bounded operator on $H^0_{L^2}(\mathcal{A}_J,\mathcal{L})$. Define the corresponding operator $\widehat{\mathcal{O}}_0$ acting on $H^0_{L^2}(\mathcal{U}_J,\mathcal{L})$ by $\widehat{\mathcal{O}}\psi = [\widehat{\mathcal{O}}_0\phi]\psi_0$ where $\psi \in H^0_{L^2}(\mathcal{A}_J,\mathcal{L})$ and $\psi = \phi\psi_0$ on $\mathcal{U}_J$.

The analytic continuation of the covariant symbol when restricted to $\mathcal{U}_J \times \mathcal{U}_J$ is given by

$$\mathcal{O}_{B(0)}(Z,\overline{W}) = \frac{\langle \phi_x\psi_0|\widehat{\mathcal{O}}_0|\phi_y\psi_0\rangle_{\mathcal{U}_J}}{\langle \phi_x\psi_0|\phi_y\psi_0\rangle_{\mathcal{U}_J}} = \frac{\langle \phi_x|\widehat{\mathcal{O}}_0|\phi_y\rangle_{\mathcal{U}_J}}{\langle \phi_x|\phi_y\rangle_{\mathcal{U}_J}}. \tag{12}$$

The function $\mathcal{O}_{B(0)}(Z,\overline{Z}) \in C^\infty(\mathcal{U}_J)$ is called the *covariant symbol* of the operator $\widehat{\mathcal{O}}_0$. Now if $\mathcal{O}_{B(0)}(Z,\overline{W})$ and $\mathcal{O}'_{B(0)}(Z,\overline{W})$ are two covariant symbols of $\widehat{\mathcal{O}}_0$ and $\widehat{\mathcal{O}'}_0$, respectively, then the covariant symbol of $\widehat{\mathcal{O}}_0\widehat{\mathcal{O}'}_0$

is given by the *Berezin–Wick star product* $\mathcal{O}_{B(0)} *_B \mathcal{O}'_{B(0)}$ given by

$$(\mathcal{O}_{B(0)} *_B \mathcal{O}'_{B(0)})(Z,\overline{Z}) = \int_{\mathcal{U}_J} \mathcal{O}_{B(0)}(Z,\overline{W})\mathcal{O}'_{B(0)}(W,\overline{Z})$$

$$\times \frac{\mathcal{B}(Z,\overline{W})\mathcal{B}(W,\overline{Z})}{\mathcal{B}(Z,\overline{Z})} \exp\left\{-\Phi(W,\overline{W})\right\} \frac{\omega^n}{n!}(W,\overline{W})$$

$$= \int_{\mathcal{U}_J} \mathcal{O}_{B(0)}(Z,\overline{W})\mathcal{O}'_{B(0)}(W,\overline{Z}) \exp\left\{\mathcal{K}(Z,\overline{Z}|W,\overline{W})\right\} \frac{\omega^n}{n!}(W,\overline{W}), \quad (13)$$

where $\mathcal{K}(Z,\overline{Z}|W,\overline{W}) := \Phi(Z,\overline{W}) + \Phi(W,\overline{Z}) - \Phi(Z,\overline{Z}) - \Phi(W,\overline{W})$ is called the *Calabi diastatic function* and $\mathcal{B}(Z,\overline{Z})$ is the usual Bergman kernel.

Thus we have find the pair $(\mathcal{S}_B, *_B)$ which constitutes the Berezin's quantization of $(\mathcal{A}_J, \omega)$.

Down–Stairs Berezin's Quantization

Finally we are in position to get the desired Berezin quantization of the Kähler quotient $(\mathcal{M}_J, \widetilde{\omega})$. That means to find the family of algebras $(\widetilde{\mathcal{S}}_B, \widetilde{*}_B)$. Having the Berezin's quantization $(\mathcal{S}_B, *_B)$ of $(\mathcal{A}_J, \omega)$ and following the description of the pushed–down prequantization bundle, we restrict ourselves to $F^{-1}(0) \subset \mathcal{A}_J$ and consider only $\mathcal{G}$–invariant quantities.

Consider $(\widetilde{\mathcal{L}}, \widetilde{\nabla}, \langle\cdot|\cdot\rangle_{\widetilde{\mathcal{L}}})$ the pushed–down prequantization with $\widetilde{\mathcal{L}} = \mathcal{L}^{\mathcal{G}}$ being the $\mathcal{G}$–complex line bundle over $\mathcal{M}_J$. The inner product $\langle\cdot|\cdot\rangle_{\widetilde{\mathcal{L}}}$ is the $\mathcal{G}$–invariant product $\langle\cdot|\cdot\rangle_{\mathcal{L}}$ given by

$$\langle\widetilde{\chi}|\widetilde{\psi}\rangle_{\widetilde{\mathcal{L}}} = \langle\chi|\psi\rangle_{\mathcal{L}}^{\mathcal{G}} = \int_{\mathcal{M}_J} \langle\widetilde{\chi}|\widetilde{\psi}\rangle \frac{\widetilde{\omega}}{n!} = \langle\chi|\psi\rangle_{\mathcal{L}} \quad (14)$$

for all $\widetilde{\chi}, \widetilde{\psi} \in H^0_{L^2}(\mathcal{M}_J, \widetilde{\mathcal{L}}) = H^0_{L^2}(F^{-1}(0), \mathcal{L})^{\mathcal{G}}_J$ where $\langle\widetilde{\chi}|\widetilde{\psi}\rangle = \langle\chi|\psi\rangle$ and $\widetilde{\omega}$ is preserved by the action of $\mathcal{G}$, i.e. ω is $\mathcal{G}$–invariant. The norm of an element $\widetilde{\psi}$ of $H^0_{L^2}(F^{-1}(0), \mathcal{L})^{\mathcal{G}}_J$ is given by $\langle\widetilde{\psi}|\widetilde{\psi}\rangle_{\widetilde{\mathcal{L}}} \equiv [||\widetilde{\psi}||^2]_{\widetilde{\mathcal{L}}}$.

Now take $\widetilde{\mathcal{Q}} \in \widetilde{\mathcal{L}}_0$, $\pi[\widetilde{\mathcal{Q}}] = \widetilde{x} \in \mathcal{M}_J$ with local complex coordinates $\{z^i, \overline{z}^i\}$ and $\pi[\widetilde{\mathcal{Q}}'] = \widetilde{y} \in \mathcal{M}_J$ with local complex coordinates $\{w^i, \overline{w}^i\}$. Here $\widetilde{\mathcal{L}}_0$ is the line bundle $\widetilde{\mathcal{L}}$ without the zero section. Now consider $\widetilde{\psi}(\widetilde{x}) = \widetilde{\psi}[\pi(\widetilde{\mathcal{Q}})] = \widetilde{L}_{\widetilde{\mathcal{Q}}}[\widetilde{\psi}]\widetilde{\mathcal{Q}}$ with $\widetilde{L}_{\widetilde{\mathcal{Q}}}[\widetilde{\psi}]$ being a linear functional of $\widetilde{\psi}$. The group $\mathcal{G}$ acts on $H^0_{L^2}(F^{-1}(0), \mathcal{L})_J$ in the form

$$(\widetilde{\Gamma}\widetilde{\psi})(\widetilde{x}) \equiv \widetilde{\Gamma}\widetilde{\psi}(\widetilde{\Gamma}^{-1}\widetilde{x}), \quad (15)$$

where $\widetilde{\Gamma} \in \mathcal{G}$, $\widetilde{x} \in \mathcal{M}_J$ and $\widetilde{\psi} \in H^0_{L^2}(F^{-1}(0), \mathcal{L})^{\mathcal{G}}_J$.

Again the Riesz theorem ensures the existence of a section $\widetilde{e_Q} \in H^0_{L^2}(F^{-1}(0), \mathcal{L})^{\mathcal{G}}_J$ such that $\widetilde{L_Q}[\tilde{\psi}] = \langle \widetilde{e_Q}|\tilde{\psi}\rangle_{\widetilde{\mathcal{L}}}$ with $\tilde{Q} \in \widetilde{\mathcal{L}}_0$. $\widetilde{e_Q}$ is the push–down of the generalized coherent state e_Q.

Let $\widehat{\mathcal{O}}^{\mathcal{G}} : H^0_{L^2}(F^{-1}(0), \mathcal{L})^{\mathcal{G}}_J \to H^0_{L^2}(F^{-1}(0), \mathcal{L})^{\mathcal{G}}_J$ be a bounded operator. The *covariant* symbol of this operator is defined as

$$\mathcal{O}^{\mathcal{G}}_B(\tilde{x}) = \frac{\langle \widetilde{e_Q}|\widehat{\mathcal{O}}^{\mathcal{G}}|\widetilde{e_Q}\rangle_{\widetilde{\mathcal{L}}}}{[\|\widetilde{e_Q}\|^2]_{\widetilde{\mathcal{L}}}} \equiv \frac{\langle e_Q|\widehat{\mathcal{O}}|e_Q\rangle^{\mathcal{G}}_{\mathcal{L}}}{[\|e_Q\|^2]^{\mathcal{G}}_{\mathcal{L}}}, \tag{16}$$

where $\tilde{Q} \in \widetilde{\mathcal{L}}_0$ and $\pi(\tilde{Q}) = \tilde{x}$. Now the space of covariant symbols $\widetilde{\mathcal{S}}_B$ is defined as the pushing–down of $\mathcal{S}_B$, i.e. $\mathcal{S}^{\mathcal{G}}_B$.

Similarly to the case of the quantization of $(\mathcal{A}_J, \omega)$, each covariant symbol can be analytically continued to the open dense subset of $\mathcal{M}_J \times \mathcal{M}_J$ in such a way $\langle \widetilde{e_Q}|\widetilde{e_{Q'}}\rangle_{\widetilde{\mathcal{L}}} \neq 0$ with $\pi(\tilde{Q}) = \tilde{x}$ and $\pi(\widetilde{Q'}) = \tilde{y}$ which is holomorphic in the first entry and anti–holomorphic in the second entry. This analytic continuation is written as

$$\mathcal{O}^{\mathcal{G}}_B(z, \overline{w}) = \frac{\langle e_Q|\widehat{\mathcal{O}}|e_{Q'}\rangle^{\mathcal{G}}_{\mathcal{L}}}{\langle e_Q|e_{Q'}\rangle^{\mathcal{G}}_{\mathcal{L}}}. \tag{17}$$

Similar considerations apply to other formulas. But an essential difference with respect to the quantization of $(\mathcal{A}_J, \omega)$ is that, in the present case, the Kähler quotient is topologically nontrivial and therefore the line bundle $\widetilde{\mathcal{L}}$ is *non-trivial*. It is only locally trivial i.e. $\widetilde{\mathcal{L}}_{(j)} = \mathcal{W}^{(j)} \times \mathbf{C}$ for each dense open subset $\mathcal{W}^{(j)}_J \subset \mathcal{M}_J$ with $j = 1, 2, \ldots, N$. Analogous global formulas found on $\mathcal{L}$, can be applied only on each local trivialization of $\widetilde{\mathcal{L}}$. Of course, transition functions on $\mathcal{W}^{(i)}_J \cap \mathcal{W}^{(j)}_J$ with $i \neq j$ are very important and *sections* and other relevant quantities like the *Bergman kernel, Kähler potential, covariant symbols*, etc., transform nicely under the change of the open set (see [12]). Thus in a particular trivialization $\widetilde{\mathcal{L}}_{(j)}$ and in the local description, the function $\mathcal{O}^{(j)}_{B(0)}(z, \overline{z}) \in C^{\infty}(\mathcal{W}^{(j)}_J)$ is called the *covariant symbol* of the operator $\widehat{\mathcal{O}}^{(j)}_0$. Now if $\mathcal{O}^{(j)}_{B(0)}(z, \overline{z})$ and $\mathcal{O}'^{(j)}_{B(0)}(z, \overline{z})$ are two covariant symbols of $\widehat{\mathcal{O}}^{(j)}_0$ and $\widehat{\mathcal{O}}'^{(j)}_0$, respectively, then the covariant symbol of $\widehat{\mathcal{O}}^{(j)}_0 \widehat{\mathcal{O}}'^{(j)}_0$ is given by the *Berezin–Wick star product* $\mathcal{O}^{(j)}_{B(0)} \widetilde{*}_B \mathcal{O}'^{(j)}_{B(0)}$

$$(\mathcal{O}^{(j)}_{B(0)} \widetilde{*}_B \mathcal{O}'^{(j)}_{B(0)})(z, \overline{z}) = \int_{\mathcal{W}^{(j)}_J} \mathcal{O}^{(j)}_{B(0)}(z, \overline{w}) \mathcal{O}'^{(j)}_{B(0)}(w, \overline{z})$$

$$\times \frac{\mathcal{B}^{(j)}(z, \overline{w}) \mathcal{B}^{(j)}(w, \overline{z})}{\mathcal{B}^{(j)}(z, \overline{z})} \exp\left\{ -\Phi^{(j)}(w, \overline{w}) \right\} \frac{\widetilde{\omega}}{n!}(w, \overline{w})$$

$$= \int_{\mathcal{W}_J^{(j)}} \mathcal{O}_{B(0)}^{(j)}(z,\overline{w}) \mathcal{O}_{B(0)}^{\prime(j)}(w,\overline{z}) \exp\left\{\mathcal{K}^{(j)}(z,\overline{z}|w,\overline{w})\right\} \frac{\widetilde{\omega}}{n!}(w,\overline{w}) \quad (18)$$

where $\mathcal{K}^{(j)}(z,\overline{z}|w,\overline{w}) := \Phi^{(j)}(z,\overline{w}) + \Phi^{(j)}(w,\overline{z}) - \Phi^{(j)}(z,\overline{z}) - \Phi^{(j)}(w,\overline{w})$ is called the *Calabi diastatic function* on $\mathcal{W}_J^{(j)}$. This construction is valid for all local prequantization $(\widetilde{\mathcal{L}}_{(j)}, \widetilde{\nabla}^{(j)}, \langle \cdot | \cdot \rangle_{\widetilde{\mathcal{L}}_{(j)}})$. Finally, this structure leads to the pair $(\widetilde{\mathcal{S}}_B, \widetilde{*_B})$ which constitutes the Berezin quantization of $(\mathcal{M}_J, \widetilde{\omega})$.

Acknowledgments

H. G.-C. wish to thank the organizers of the First Mexican Meeting on Mathematical and Experimental Physics, for invitation, he also thanks M. Przanowski and F. Turrubiates for very useful discussions. I. C.-I. is supported by a CONACyT graduate fellowship. This work was partially supported by the CONACyT grant No. 33951E.

References

[1] F. Wilczek (ed.), *Fractional Statistics and Anyon Superconductivity*, Singapore, (World Scientific, 1990).

[2] A. Achúcarro and P.K. Townsend, *Phys. Lett.* **B180** (1986) 89; E. Witten, *Nucl. Phys.* **B311** (1988) 46.

[3] E. Witten, *Commun. Math. Phys.* **121** (1989) 351.

[4] E. Witten, *Nucl. Phys.* **B330** (1990) 285.

[5] G. Moore and N. Seiberg, *Phys. Lett.* **B220** (1989) 422; S. Elitzur, G. Moore, A. Schwimmer and N. Seiberg, *Nucl. Phys.* **B326** (1989) 108.

[6] S. Axelrod, S. Della Pietra and E. Witten, *J. Diff. Geom.* **33** (1991) 787.

[7] E.A. Berezin, *Math. USSR–Izv.* **6** (1972), 1117; *Soviet Math. Dokl.* **14** (1973) 1209; *Math. USSR–Izv.* **8** (1974) 1109; *Math. USSR–Izv.* **9** (1975) 341; *Commun. Math. Phys.* **40** (1975) 153.

[8] M. Cahen, S. Gutt and J. Rawnsley, *J. Geom. Phys.* **7** (1990) 45.

[9] M. Cahen, S. Gutt and J. Rawnsley, *Trans. Amer. Math. Soc.* **337** (1993) 73.

[10] N. Reshetikhin and L.A. Takhtajan, math.QA/9907171.

[11] M. Schlichenmaier, math.QA/9910137.

[12] H. García–Compeán, J.F. Plebański, M. Przanowski and F.J. Turrubiates, hep-th/0112049.

[13] H. García–Compeán, "Berezin's Quantization of Chern–Simons Gauge Theory", to appear (2002).

[14] M. Spradlin and A. Volovich, hep-th/0106180.

[15] J.M. Isidro, quant-ph/0112032.

(2+1)-DIMENSIONAL SUPERGRAVITY

Alfredo Macias
Departamento de Física, Universidad Autónoma Metropolitana
Apartado Postal 55-534, C.P. 09340, México, D.F., México
amac@xanum.uam.mx

Abstract In this review, we revisit the different approaches existing to $(2+1)$-dimensional supergravity, namely, the standard one and the ones based in the introduction of Lorentz and translational Chern–Simons terms. Moreover, we review the conditions for reducing extended supergravities with and without cosmological constant to Chern–Simons theories of the super–Poincaré or super–anti-de Sitter supergroups.

Keywords: Supergravity, Chern–Simons gravity/supergravity.

1. Introduction

Interest in $(2+1)$-dimensional gravity, general relativity, dates back to 1963 when Staruszkiewicz [1] showed that a simple geometrical description for point particles in a $(2+1)$-dimensional space–time could be given. Over the next twenty years occasional papers on classical [2, 3] and quantum mechanical [5, 6] aspects appeared, but until recently the subject remained as a curiosity.

In 1984, Deser, Jackiw, and 't Hooft began a systematic investigation on the behavior of classical and quantum mechanical point sources in $(2+1)$-dimensional gravity [7, 8, 9, 10], showing that such systems exhibit interesting features both as a toy models for $(3+1)$-dimensional quantum gravity and as realistic models of cosmic strings.

Witten, in 1988, showed that $(2+1)$-dimensional general relativity could be rewritten as a Chern–Simons theory, permitting exact computations of topology changing amplitudes [11, 12]. The Chern–Simons terms had been recognized a few year earlier by Achúcarro and Townsend [13], however Witten's rediscovery came at a time that quantum mechanical treatment of Chern–Simons theory was advancing rapidly.

The study of Chern–Simons theories in $(2+1)$-dimensions [14] has recently been of much interest due to a number of reasons. These range from the very mathematical applications related to knot theories [15] to the more physical implications in the description of the quantum Hall effect through the idea of anyons, which correspond to particles with fractional statistic [16]. Another very interesting property of the (2+1)-Chern–Simons theories is their exact solubility in terms of a finite number of degrees of freedom, which make them an adequate laboratory to test many ideas related to the quantization of gauge and gravitational theories [17]. In (2+1)-dimensions, the standard Einstein theory of gravity together with the de Sitter gravity [17], conformal gravity [18], and supergravity [19] turn out to be equivalent to Chern–Simons theories.

Since Chern–Simons theories are of topological nature, their description is most naturally based on the holonomies of the connection, thus providing a convenient realization to study loop quantization methods in gauge theories.

In this paper, we review the different approaches to supergravity in (2+1) dimensions, we begin in Sec. II revisiting the geometry of (2+1)-dimensional space–times. In Sec. III we revisited the standard formulation of supergravity. In Sec. IV we present the conditions for reducing the Chern–Simons formulation of supergravity to the super–Poncaré and anti–de Sitter supergroups. In Sec. V we present the formulation of (2 + 1) supergavity including Lorentz and translational Chern–Simons terms.

2. Geometry in (2+1) dimensions

We are dealing with a three–dimensional C^∞ manifold endowed with a metric g with signature +1 and a connection ω. A general tangent vector field and a general co–vector field are written as follows:

$$a = a^i e_i, \qquad \alpha = \alpha_i \vartheta^i, \tag{1}$$

respectively. ϑ^i is a basis one–form and e_i the corresponding dual basis, i.e., $<\vartheta^i, e_j> = \delta^i_j$. The components a^i are zero–forms which transform under $A \in GL(3, R)$ according to the rule $\tilde{a}^i = A^i{}_j a^j$. The covariant exterior derivative reads

$$Da^i = da^i + \omega^i{}_j a^j, \tag{2}$$

with d the ordinary exterior derivative. The restriction to zero–forms can be removed if the ordinary product is replaced by an exterior product $\omega^i{}_j \wedge a^j$ for a vector-valued p–form a^i. It is possible to extend the definition of exterior covariant derivative to any field transforming under the group $GL(3, R)$. The metricity condition reads $Dg_{ij} = 0$, where

$g = g_{ij}\vartheta^i \otimes \vartheta^j$, i.e., g_{ij} is a tensor–valued zero–form. The presence of a metric allows the introduction of orthonormal frames ϑ^a, hence

$$g(\vartheta^a, \vartheta^b) = o^{ab} = \text{diag}(-1, +1, +1), \tag{3}$$

and

$$Dg_{ab} = Do_{ab} = 0, \quad \Rightarrow \quad \omega_{ab} = -\omega_{ba}. \tag{4}$$

The covariant derivative of the triad $e^a{}_\mu$ reads

$$de^a{}_\mu + \omega^\nu{}_\mu e^a{}_\nu - \omega^a{}_\nu e^\nu{}_b = 0, \tag{5}$$

where $\omega^\mu{}_\nu = \Gamma^\mu{}_{\nu\rho} dx^\rho$ and $\Gamma^\mu{}_{\nu\rho}$ is the usual connection. We remain in an orthonormal basis if it is imposed the restriction that the $GL(3,R)$ transformation to belong to the subgroup $SO(2,1)$. Since the group $SO(2,1)$ is a three parameter group it is convenient to introduce a one index set of one–forms:

$$\omega^a = -\frac{1}{2}\eta^a{}_{bc}\omega^{bc},, \tag{6}$$

the Latin indices $a, b, ...$ are raised and lowered by means of the orthonormal metric o^{ab}, and η^{abc} is the Levi–Civita symbol ($\eta^{012} = +1$).

A p–form field χ transforms under the group $SO(2,1)$ infinitesimally as

$$\delta\chi = l^a S_a \chi, \tag{7}$$

where

$$[S_a, S_b] = -\eta_{ab}{}^c S_c. \tag{8}$$

The exterior covariant derivative then reads

$$D\chi = d\chi + \omega^a \wedge S_a \chi. \tag{9}$$

The eight forms 1, ϑ^a, $\vartheta^a \wedge \vartheta^b$, $\vartheta^a \wedge \vartheta^b \wedge \vartheta^c$ generate the exterior algebra in the cotangent space of the three–dimensional manifold, and their Hodge duals are given as follows:

$$\eta^{abc} = {}^*(\vartheta^a \wedge \vartheta^b \wedge \vartheta^c) \tag{10}$$

$$\eta^{ab} = \vartheta^c \eta^{ab}{}_c = {}^*(\vartheta^a \wedge \vartheta^b), \tag{11}$$

$$\eta^a = \frac{1}{2}\vartheta^b \wedge \eta^a{}_b = {}^*\vartheta^a \tag{12}$$

$$\eta = \frac{1}{3}\vartheta^a \wedge \eta_a = {}^*1. \tag{13}$$

The η forms fulfill the following algebraic identities:

$$\vartheta^d \eta_{abc} = \delta^d_c \eta_{ab} + \delta^d_b \eta_{ac} + \delta^d_a \eta_{bc}, \tag{14}$$

$$\vartheta^c \wedge \eta_{ab} = \delta^c_b \eta_a - \delta^c_a \eta_b \tag{15}$$

$$\vartheta^b \wedge \eta_a = \delta^b_a \eta. \tag{16}$$

The volume element of our space η is given by

$$\eta = (1/3!)\eta_{abc}\vartheta^a \wedge \vartheta^b \wedge \vartheta^c = e\, dx^0 \wedge dx^1 \wedge dx^2, \tag{17}$$

where $e = \det e^a{}_\mu = (-\det g_{\mu\nu})^{1/2}$.

The torsion two–form T^a is defined as follows

$$T^a = D\vartheta^a = d\vartheta^a + \eta^a{}_b \wedge \omega^b = \frac{1}{2}T^a{}_{bc}\vartheta^b \wedge \vartheta^c. \tag{18}$$

The curvature two–form R^a is given by

$$D^2\chi = S_a R^a \wedge \chi \tag{19}$$

$$R^a = d\omega^a - \frac{1}{2}\eta^a{}_{bc}\omega^b \wedge \omega^c = \frac{1}{2}R^a{}_{bc}\vartheta^b \wedge \vartheta^c. \tag{20}$$

R^a is related to the standard two index curvature two–form $R^a{}_b$ by

$$R^a = -\frac{1}{2}\eta^a{}_{bc}R^{bc}, \tag{21}$$

$$R^a{}_b = \frac{1}{2}R^a{}_{bcd}\vartheta^c\vartheta^d. \tag{22}$$

The Bianchi identities read

$$DT^a = -\eta^a{}_b \wedge R^b, \tag{23}$$
$$DR^a = 0. \tag{24}$$

The interior multiplication is defined for a tangent vector field ρ and a p–form χ as follows

$$\rho \rfloor \chi = \rho^a \chi_{bc...}\, e_a \rfloor (\vartheta^b \wedge \vartheta^c \wedge ...) \tag{25}$$

where

$$e_a \rfloor (\vartheta^b \wedge \vartheta^c \wedge ...) = \delta^b_a \vartheta^c \wedge ... - \delta^c_a \vartheta^b \wedge ... + ... \tag{26}$$

3. Standard supergravity in (2+1) dimensions

3.1. Einstein–Cartan gravity in vacuum

The action for the Einstein–Cartan gravity in three dimensions [20] reads

$$S = \int \vartheta^a \wedge R_a \equiv \int L. \tag{27}$$

ϑ^a is a one–form, R^a is a two–form, and L is scalar-valued three–form. In order to compute the field equations we vary with respect to the frame ϑ^a, and to the connection ω^a. The frame variation leads to

$$R^a = 0, \tag{28}$$

while
$$\delta R^a = d(\delta\omega_a) - \eta_{abc}\omega^b \wedge \delta\omega^c. \quad (29)$$

Since $\vartheta^a \wedge d(\delta\omega_a) = d\vartheta^a \wedge \delta\omega_a - d(\vartheta^a \wedge \delta\omega_a)$, one obtains

$$\delta L = \delta\vartheta^a \wedge R^a + \left(d\vartheta^a + \eta^a{}_b \wedge \omega^b\right) \wedge \delta\omega_a - d(\vartheta^a \wedge \delta\omega_a). \quad (30)$$

The last term in (30) is an exact form an could affect only the boundary conditions of the space under consideration. By means of (18), the variation with respect to the connection $\delta\omega_a$ yields $T^a = 0$. Therefore, the only solution to the Einstein–Cartan in vacuum three–space is the Minkowski space. The standard Einstein equations $R_{\mu\nu} = 0$ determine uniquely the six independent components of the curvature tensor [20].

It is interesting to note that Eq. (27) is indeed the standard Einstein–Hilbert action, using (20), (21), (22), and (17) in (27) yield:

$$\begin{aligned} L &= \frac{1}{2}\vartheta^a \wedge \vartheta^b \wedge \vartheta^c R_{abc} = \frac{1}{2}\eta^{abc}R_{abc} = -\frac{1}{4}\eta^{abc}\eta_{adf}R^{df}{}_{bc} \\ &= \frac{1}{2}\eta R^{ab}{}_{ab} = \frac{1}{2}\eta R = \frac{1}{2}eR d^3 x, \end{aligned} \quad (31)$$

which is the standard Einstein–Cartan action in three dimensions, and where also the relation $\eta^{abc}\eta_{adf} = \delta^c_f \delta^b_d - \delta^b_f \delta^c_d$ has been used.

3.2. Rarita–Schwinger field

In this section a brief survey of the spinors is presented. The covering group of $SO(2,1)$ is the group $SL(2,R)$, isomorphic to $SU(1,1)$, hence one can use real component spinor. The generators of $SL(2,R)$ satisfy the relation (8) as well as the Clifford algebra in three dimensions [20]

$$\{\gamma_a, \gamma_b\} = \gamma_a\gamma_b + \gamma_b\gamma_a = 2o_{ab}. \quad (32)$$

A realization of this algebra in terms of real matrices can be written as follows

$$\gamma_0 = \begin{pmatrix} 0 & 1 \\ -1 & 0 \end{pmatrix}, \quad \gamma_1 = \begin{pmatrix} 0 & 1 \\ 1 & 0 \end{pmatrix}, \quad \gamma_2 = \begin{pmatrix} 1 & 0 \\ 0 & -1 \end{pmatrix}. \quad (33)$$

For any two–component spinor ψ, its adjoint $\overline{\psi}$ is given by

$$\overline{\psi} = \psi^T \gamma^0, \quad (34)$$

and there are only two bilinear invariants, i.e., a pseudoscalar $\overline{\psi}\psi$ and a vector $\overline{\psi}\gamma_a\psi$. The exterior covariant derivative of a spinor–valued p–form ψ reads

$$D\psi = d\psi + \frac{1}{2}\gamma_a\omega^a \wedge \psi. \quad (35)$$

Due to the fact that the spinors are odd elements of some Grassmann algebra, any two spinors satisfy

$$\overline{\psi}_1\psi_2 = -\overline{\psi}_2\psi_1, \qquad \overline{\psi}_1\gamma_a\psi_2 = -\overline{\psi}_2\gamma_a\psi_1, \qquad (36)$$

while for any three spinors one has the Fierz formula

$$\left(\overline{\psi}_1\psi_2\right)\psi_3 = -\frac{1}{2}\sum_A \left(\overline{\psi}_1\gamma_A\psi_3\right)\gamma^A\psi_2, \qquad (37)$$

where $\gamma_A = (\mathbf{1}, \gamma_a)$ and $\gamma^A = (\mathbf{1}, \gamma^a)$.

Let us introduce the Rarita–Schwinger type spinor–valued one–form

$$\Psi = \Psi_i dx^i = \Psi_a \vartheta^a, \qquad (38)$$

and the Clifford–algebra–value one–form

$$\gamma = \gamma_a \vartheta^a \quad \text{with} \quad \gamma_a\gamma_b + \gamma_b\gamma_a = 2o_{ab}. \qquad (39)$$

The Lagrangian three–form for the gravitino field reads

$$\mathcal{L}_{RS} = \frac{i}{2}\overline{\Psi} \wedge D\Psi, \qquad (40)$$

where $D\Psi$ is given by (35). The corresponding field equations for the gravitino vector –spinor are

$$D\Psi = 0. \qquad (41)$$

Here, we will us the rules $\Phi \wedge^* \Psi = \Psi \wedge^* \Phi$ for p–forms Φ and Ψ of the same degree and $\overline{AB} = \overline{B}\,\overline{A}$ as well as $\overline{\gamma} = \gamma$, and for the Dirac adjoint, one has $\overline{\Psi} := \gamma^0 \Psi^\dagger$ where $\Psi^\dagger$ is the Hermitian–conjugated form.

3.3. Standard simple supergravity

The basic fields variables of the simple supergravity are the frame ϑ^a and the Rarita–Schwinger vector–spinor Ψ. The action is the sum of the vacuum Einstein–Cartan action (27) and the gravitino action (40). Therefore, the Lagrangian three–form of simple supergravity reads

$$\mathcal{L}_{SG} = \vartheta^a \wedge R^a + \frac{i}{2}\overline{\Psi} \wedge D\Psi. \qquad (42)$$

The supergravity field equations are obtained by the variation with respect to the frame, the connection, and to the gravitino field. The variation of $D\Psi$ reads

$$\begin{aligned}\delta(D\Psi) &= \delta(d\Psi) + \frac{1}{2}\gamma_a \omega^a \wedge \delta\Psi - \gamma_a \Psi \wedge \delta\omega^a \\ &= D(\delta\Psi) - \frac{1}{2}\gamma_a \Psi \wedge \delta\omega^a.\end{aligned} \qquad (43)$$

Therefore, the total variation yields

$$\delta \mathcal{L}_{sg} = \delta\vartheta^a \wedge R_a + i\delta\overline{\Psi} \wedge D\Psi + (T^a - \frac{i}{4}\overline{\Psi} \wedge \Psi) \wedge \delta\omega_a + \text{an exact form}. \quad (44)$$

Hence, the field equation read

$$R^a = 0, \quad (45)$$

$$T^a = \frac{i}{4}\overline{\Psi} \wedge \Psi \quad (46)$$

$$D\Psi = 0. \quad (47)$$

Only two of the field equations (45), (46), and (47) are dynamical, the Cartan relation (46) allows to solve the torsion in terms of the Rarita–Schwinger field. This feature is characteristic of the first order formalism. It is standard to decompose the connection ω in terms of the pure Riemannian part $\widetilde{\omega}$ and the contortion $K^a{}_b$, where

$$T^a = K^a{}_b \wedge \vartheta^b, \quad (48)$$

or solving for the contortion

$$2K_{abc} = T_{cab} - T_{abc} - T_{bca}, \quad (49)$$

from (46) one obtains

$$T_{abc} = \frac{i}{2}\overline{\Psi}_b \gamma_a \Psi_c \quad (50)$$

It is worthwhile to note that the spaces considered have vanishing curvature (c.f. (45)) but non–vanishing torsion. This is not sufficient for the space to be flat. One is free to use $\widetilde{\omega}$, instead of the complete ω, to define a covariant derivative $\widetilde{D}$ in the Lagrangian (42). In this case, the connection $\widetilde{\omega}$ is torsion free but the curvature is not vanishing and the field equation for the gravitino field become

$$\widetilde{D}\Psi + \frac{1}{2}\gamma_a K^a \wedge \Psi = 0, \quad (51)$$

where $K^a = -(1/2)\eta^a{}_{bc}K^{bc}$. The relation (51) corresponds to a non–minimal coupling of a vector–spinor gravitino field to standard torsion free Einstein gravity.

The Lagrangian (42) is indeed supersymmetric: Let us consider the transformations

$$\delta\vartheta^a = i\overline{\alpha}\gamma^a\Psi, \qquad \delta\Psi = 2D\alpha, \quad (52)$$

α is spinor–valued zero–form. Therefore,

$$\delta \mathcal{L}_{SG} = i\overline{\alpha}\gamma^a \wedge R_a + 2iD\overline{\alpha} \wedge D\Psi + (T^a - \frac{1}{4}i\overline{\Psi} \wedge \gamma^a\Psi) \wedge \delta\omega^a + \text{an exact form}. \quad (53)$$

On the other hand,
$$D\overline{\alpha} \wedge D\Psi = d(\overline{\alpha}D\Psi) - 2i\overline{\alpha}D^2\Psi, \tag{54}$$
and $D^2\Psi = (1/2)\gamma^a R_a \wedge \Psi$, from (19). Consequently, the first two terms in (53) give an exact form and only one term is left which vanishes in the second order formalism.

3.4. Cosmological term

As it is well known that in four dimensions, it is possible to extend supergravity to include a cosmological term, this implies the addition of a mass term for the gravitino field. The same can be done in three dimensions. The supergravity Lagrangian now reads

$$\mathcal{L}_{SG} = \vartheta^a \wedge R_a + \frac{i}{2}\overline{\Psi} \wedge D\Psi - \Lambda\eta + \frac{i}{4}m\overline{\Psi} \wedge \gamma \wedge \Psi. \tag{55}$$

Hence, the field equations read

$$R_a = \Lambda\eta_a - T_a, \tag{56}$$

$$D\Psi + \frac{1}{2}m\gamma \wedge \Psi = 0, \tag{57}$$

$$T^a = \frac{i}{4}\overline{\Psi} \wedge \gamma^a \Psi. \tag{58}$$

The inclusion of a cosmological term is enough to provide curvature, making the standard Einstein gravity in three dimensions non-trivial.

The supersymmetry transformations become

$$\delta\vartheta^a = i\overline{\alpha}\gamma^a\Psi, \qquad \delta\Psi = 2D\alpha + m\gamma\alpha, \tag{59}$$

Hence, after using (58)

$$\delta\mathcal{L}_{SG} = i\overline{\alpha}\gamma^a \wedge R_a + 2iD\overline{\alpha} \wedge D\Psi - i\overline{\alpha}\gamma^a\Psi \wedge \left[\Lambda\eta_a - \frac{i}{4}m\overline{\Psi} \wedge \gamma_a\Psi\right]$$
$$+ \frac{i}{2}mD\overline{\alpha} \wedge \gamma \wedge \Psi - im\overline{\alpha}\gamma \wedge \left[D\Psi + \frac{i}{2}m\gamma \wedge \Psi\right]. \tag{60}$$

As before, the first two terms combine to give an exact form. The fifth and sixth terms can be re-arranged to give an exact form plus a torsion term arising from $D\gamma = T^a\gamma_a$. This term, together with the other torsion term are of the form $\overline{\alpha}\gamma^a\Psi \wedge \overline{\Psi} \wedge \gamma_a\Psi$ and vanish by a Fierz re-arrangement (37). Therefore, one is left with the two terms:

$$-i\overline{\alpha}\gamma^a\Psi \wedge \Lambda\eta_a - \frac{i}{2}m^2\overline{\alpha}\gamma \wedge \gamma \wedge \Psi = \left(\frac{m^2}{2} - \Lambda\right)i\overline{\alpha}\Psi \wedge \gamma^a\eta_a. \tag{61}$$

In order that the Lagrangian (55) is supersymmetric, Eq. (61) leads to the condition $\Lambda = m^2/2$.

4. Chern–Simons supergravities

The purpose of this section is to revisit the problem of how to understand extended supergravities in (2+1) dimensions, with and without cosmological constant, as Chern–Simons theories of the super–Poincaré ($\Lambda = 0$) or super–anti–de Sitter ($\Lambda < 0$) super-groups [13].

Let us start by reviewing the bosonic case. The Einstein–Hilbert action for general relativity in (2+1) dimensions is a Chern–Simons action for the three–dimensional Poincaré group $ISO(2,1)$, i.e.,

$$S = \int_M d^3x \frac{1}{2} eR = \frac{1}{2} \int_M \mathrm{tr} \left[\Gamma \wedge d\Gamma + \frac{2}{3} \Gamma \wedge \Gamma \wedge \Gamma \right]. \tag{62}$$

The gauge potential (connection) is as usual a one–form with values in the Lee algebra of $ISO(2,1)$, i.e.,

$$\Gamma = \vartheta^a P_a + \omega^a M_a, \tag{63}$$

where $M_a = -(1/2)\eta_{abc} M^{bc}$ are the Lorentz generators. Therefore, the Poincaré algebra reads

$$[P_a, P_b] = 0, \quad [M_a, M_b] = \eta_{abc} M^c, \quad [M_a, P_b] = \eta_{abc} P^c, \tag{64}$$

$SO(2)$ indices are raised and lowered with the Minkowski metric (3). The quadratic invariant

$$I = 4 M_a P^a = 4 \left[\frac{M_a + P_a}{2} \right] \left[\frac{M^a + P^a}{2} \right] - 4 \left[\frac{M_a - P_a}{2} \right] \left[\frac{M^a - P^a}{2} \right], \tag{65}$$

corresponds to the inverse of the non–degenerate Casimir invariant of the group $ISO(2,1)$.

For the bosonic sector, the relation between the Chern–Simons actions for Poincaré and anti–de Sitter gravity is immediate. The generators M_{AB} ($A = a, 3$) of the anti–de Sitter group $SO(2,2)$ satisfy the algebra

$$[M_{AB}, M_{CD}] = g_{BC} M_{AD} - g_{BD} M_{AC} - g_{AC} M_{BD} + g_{AD} M_{BC}, \tag{66}$$

where $g_{AB} = \mathrm{diag}(-,+,+,-)$. Defining $M_a = (1/2)\eta_{abc} M^{bc}$, and $P_a = M_{a3}$ and re-scaling $P^a \to P^a/\Lambda$, and $M^a \to M^a$, the Poincaré algebra is recovered in the limit of vanishing cosmological constant. This procedure is known as (Wigner–Inönu) group contraction [13]. Moreover, the group $SO(2,2) \cong SO(2,1) \times SO(2,1)$ whose generators read

$$J_{\pm a} = \frac{1}{2}[M_a \pm P_a]. \tag{67}$$

Since each factor has a unique non-degenerate quadratic Casimir invariant $J_a J^a$, any linear combination of them is a quadratic $SO(2,2)$ invariant

$$I = \alpha J_+^a J_{+a} + \beta J_-^a J_{-a}, \tag{68}$$

for $\alpha, \beta \neq 0$. Hence, by using (67), the invariant becomes

$$I = \left[\frac{\alpha+\beta}{4}\right] \frac{P^a P_a}{\Lambda^2} + \left[\frac{\alpha-\beta}{2}\right] \frac{M^a P_a}{\Lambda} + \left[\frac{\alpha+\beta}{2}\right] \frac{M^a M_a}{\Lambda^2}, \tag{69}$$

for $\alpha = -\beta = 4$ the Chern–Simons theory reduces to the anti–de Sitter gravity. Of course, there exist other choices for the coefficients α, β, nevertheless, without Poincaré limit [13].

The generalization of these results to extended Poincaré supergravity is straightforward. The super–Poincaré algebra with N supercharges Q_μ^I ($\mu = 1, 2$ and $I = 1, ..., N$) read

$$\left[M^a, M^b\right] = \eta^{abc} M_c, \tag{70}$$

$$\left[P^a, P^b\right] = 0, \tag{71}$$

$$\left[M^a, P^b\right] = \eta^{abc} P_c, \tag{72}$$

$$\left[M^a, Q_\mu^I\right] = -\frac{1}{2}(\gamma^a)_\mu{}^\nu Q_\nu^I, \tag{73}$$

$$\left[P^a, Q_\mu^I\right] = 0, \tag{74}$$

$$\left[Q_\mu^I, Q_\nu^J\right] = (\gamma^a P_a)_{\mu\nu} \delta^{IJ}, \tag{75}$$

and has a non–degenerate Casimir invariant

$$I = 4 M_a P^a - \overline{Q}_\mu^I Q^{I\mu}. \tag{76}$$

The action for pure supergravity for any N reads

$$S = \int_M \vartheta^a \wedge R^a + \frac{i}{2} \overline{\Psi} \wedge D\Psi, \tag{77}$$

and results as a Chern–Simons action for the gauge potential (connection)

$$\Gamma = \vartheta^a P_a + \omega^a M_a + \overline{\Psi}^I Q^I, \tag{78}$$

where the gravitinos Ψ^I are the gauge fields associated to the supercharges.

The anti–de Sitter case, is more interesting because it is possible to supersymmetrize [13] both factors $SO(2,2) \cong SO(2,1) \times SO(2,1)$ independently. The super-group $OSp_\pm(n|2, R)$ has $Sp(2, R) \cong SO(2, 1)$

generators J_a ($a = 0,1,2$), n supercharges Q_μ^i and if $n > 1$ also $O(n)$ generators t^{ij} ($i,j = 1,...,n$), satisfying the following algebra

$$[J_a, J_b] = \eta_{abc} J^c, \tag{79}$$

$$[t^{ij}, t^{kl}] = \delta^{jk} t^{il} - \delta^{jl} t^{ik} - \delta^{ik} t^{jl} + \delta^{il} t^{jk}, \tag{80}$$

$$[J_a, t^{ij}] = 0, \tag{81}$$

$$[J^a, Q_\mu^i] = \frac{1}{2}(\gamma^a)_\mu{}^\nu Q_\nu^i, \tag{82}$$

$$[t^{ij}, Q_\mu^i] = \delta^{ij} Q_\mu^i - \delta^{ik} Q_\mu^j, \tag{83}$$

$$\{Q_\mu^i, Q_\nu^j\} = \pm \left[2\delta^{ij} (\gamma^a)_{\mu\nu} J_a + \eta_{\mu\nu} t^{ij} \right]. \tag{84}$$

The result is a (p,q) supergravity, where p indicates the number of supercharges in the first factor and q in the second one of (84). (p,q) models based on the supergroups $OSp(p|2,R) \times OSp(p|2,R)$ can be also interpreted as Chern–Simons theories.

The Jacobi identities allow either sign on the right–hand side of the anticommutator (84). Since the supercharges are real the sign cannot be changed by a redefinition of them. Moreover, a change in the sign of J^a and t^{ij} would modify the $Sp(2,R) \times O(n)$ commutation relations (79)–(84). Then, the two algebras are not equivalent and are denoted $OSp_\pm$, respectively. The quadratic Casimir invariants for the two inequivalent supergroups read

$$I_\pm = 4J^a J^b o_{ab} + \frac{1}{2} t^{ij} t_{ij} \mp \overline{Q}^i Q^i. \tag{85}$$

Considering, for example, the simple case $p = q = 1$ for the group $OSp_+(1|2,R) \times OSp_+(1|2,R)$, the super–anti–de Sitter Casimir invariant reads

$$I = \Lambda[I_+ - I_-] = 4P^a M_a - \overline{Q}^I Q^I, \quad (I = 1,2), \tag{86}$$

an reduces to the form of the one of Eq.(76). The limit $\Lambda \to 0$ is well defined in the anti–de–Sitter(1,1) action

$$S = \int_M \mathcal{L}_{SG} = \int_M \left[\vartheta^a \wedge R_a + \frac{i}{2} \overline{\Psi} \wedge D\Psi - \Lambda \eta + \frac{i}{4} m \overline{\Psi} \wedge \gamma \wedge \Psi \right], \tag{87}$$

after the re–scaling $\vartheta^a \to \Lambda \vartheta^a$, $\Psi \to \sqrt{\Lambda} \Psi$, hence, $\mathcal{S} = \Lambda^{-1} S$. The limit $\Lambda \to 0$ is not clear for the case $(p,q) > (1,1)$.

5. Supergravity with translational Chern–Simons term

5.1. Three–dimensional supergravity

The topological supergravity Lagrangian in first order formalism, we will consider, is given by

$$\mathcal{L}_{SG} = \chi \mathcal{L}_{EC} - \Lambda \eta + \Theta_L \mathcal{C}_L + \Theta_T \mathcal{C}_T + \mathcal{L}_\Psi, \tag{88}$$

where

$$\mathcal{L}_{EC} = -\chi \vartheta^a \wedge R_a, \tag{89}$$

$$\Lambda \eta = \frac{\Lambda}{3} \vartheta^a \wedge \eta_a, \tag{90}$$

$$\Theta_L \mathcal{C}_L = \Theta_L \left[\omega^a \wedge R_a - \frac{1}{3!} \eta^{abc} \omega^a \wedge \omega^b \wedge \omega^c \right], \tag{91}$$

$$\Theta_T \mathcal{C}_T = \frac{\Theta_T}{2} \vartheta^a \wedge T_a, \tag{92}$$

$$\mathcal{L}_\Psi = \frac{i}{2} \overline{\Psi} \wedge D\Psi + \frac{i}{4} m \overline{\Psi} \wedge \gamma \wedge \Psi + s_1 \overline{D\Psi} \wedge^\star (D\Psi)$$
$$+ s_2 \overline{\gamma \wedge D\Psi} \wedge^\star (\gamma \wedge D\Psi). \tag{93}$$

$\mathcal{L}_{EC}$ is the standard Einstein–Cartan Lagrangian, $\Lambda \eta$ the cosmological constant term, $\mathcal{C}_L$ and $\mathcal{C}_T$ are the rotational and translational Chern–Simons terms, respectively, and $\mathcal{L}_\Psi$ is a generalization of the Rarita–Schwinger Lagrangian $\mathcal{L}_{RS} = \frac{i}{2} \overline{\Psi} \wedge D\Psi$ in 3 dimensions. χ, Θ_L, and Θ_T are coupling constants. Note that $\overline{\gamma \wedge D\Psi} = \overline{D\Psi} \wedge \gamma$.

5.2. Energy–momentum and spin

The *energy–momentum current* Σ_α is given by

$$\Sigma_a = \frac{\partial \mathcal{L}_\Psi}{\partial \vartheta^a} = e_a \rfloor \mathcal{L}_\Psi - (e_a \rfloor D\Psi) \wedge \frac{\partial \mathcal{L}_\Psi}{\partial D\Psi} - (e_a \rfloor \overline{D\Psi}) \wedge \frac{\partial \mathcal{L}_\Psi}{\partial \overline{D\Psi}}$$
$$- (e_a \rfloor \Psi) \wedge \frac{\partial \mathcal{L}_\Psi}{\partial \Psi} - (e_a \rfloor \overline{\Psi}) \wedge \frac{\partial \mathcal{L}_\Psi}{\partial \overline{\Psi}}$$
$$= \frac{i}{\Psi} m \overline{\Psi} \gamma_a \Psi + s_1 \{e_a \rfloor (\overline{D\Psi} \wedge^\star (D_\Psi))$$
$$- (e_a \rfloor D\Psi) \wedge^\star \overline{D\Psi} - (e_a \rfloor \overline{D\Psi}) \wedge^\star D\Psi\}$$
$$+ s_2 \overline{\gamma_a D\Psi} \wedge^\star (\gamma \wedge D\Psi) + \gamma_a D\Psi \wedge^\star (\overline{\gamma \wedge D\Psi}). \tag{94}$$

Note that $s_2 \overline{\gamma_a D\Psi} \wedge^\star (\gamma \wedge D\Psi) = 2 s_2 \overline{\gamma_a D\Psi} \wedge^\star (\gamma \wedge D\Psi)$. Thus we have

$$\Sigma_a = \frac{i}{\Psi} m \overline{\Psi \gamma \Psi} + s_2 \gamma_a D\Psi \wedge^\star \overline{\gamma \wedge D\Psi}, \tag{95}$$

whereas all the other terms give no contribution.

Note that the Rarita–Schwinger term in $\mathcal{L}_\Psi$ gives no contribution to the energy–momentum current, since it does not depend on the coframe ϑ^α.

The 3–dual of the *spin current* is given by

$$\begin{aligned}
\tau_a &= \frac{1}{2}\eta_{abc}\tau^{bc} = \frac{(-1)^s}{2}\frac{\delta L_\Psi}{\delta\omega_a} \\
&= \frac{(-1)^s}{2}\left[-\frac{i}{4}\overline{\Psi}\gamma_a\Psi + \frac{s_1}{2}\overline{\Psi}\gamma_a\wedge{}^\star(D\Psi) + \frac{s_1}{2}\gamma_a\Psi\wedge{}^\star(\overline{D\Psi})\right. \\
&\quad \left. + \frac{s_2}{2}\overline{\Psi}\gamma_a\wedge\gamma\wedge{}^\star(\gamma\wedge D\Psi) + \frac{s_2}{2}\gamma_a\Psi\wedge\gamma\wedge{}^\star(\overline{\gamma\wedge D\Psi})\right]. \quad (96)
\end{aligned}$$

Therefore, the spin current reads

$$\tau_a = (-1)^s\left[-\frac{i}{8}\overline{\Psi}\gamma_a\Psi + s_1\overline{\Psi}\gamma_a\wedge{}^\star(D\Psi) + s_2\overline{\Psi}\gamma_a\wedge\gamma\wedge{}^\star(\gamma\wedge D\Psi)\right]. \quad (97)$$

5.3. Supersymmetric transformations

Let us consider the first–order Lagrangian

$$\mathcal{L} = \mathcal{L}(\vartheta^a, T^a, \omega^a, R^a{}_,\Psi, D\Psi). \quad (98)$$

The variation of $\mathcal{L}$ with respect to its independent variables ϑ^α, ω_a, and Ψ) fields reads

$$\delta\mathcal{L} = \delta\vartheta^a\wedge\frac{\delta\mathcal{L}}{\delta\vartheta^a} + \delta\omega_a\wedge\frac{\delta\mathcal{L}}{\delta\omega_a} + \delta\overline{\Psi}\wedge\frac{\delta\mathcal{L}}{\delta\overline{\Psi}}, \quad (99)$$

for convenience, we vary with respect to the Dirac–Adjoint $\overline{\Psi}$.

The Lagrangian (98) is invariant under the supersymmetric transformation of Deser [22], which in exterior form notation reads

$$\delta\vartheta^a = i\overline{\alpha}\gamma^a\Psi, \qquad \delta\Psi = 2D\alpha + c\gamma_a, \quad (100)$$

where α is a spinor–valued zero–form. Insertion of (99) into (99) yields

$$\delta L = i\overline{\alpha}\gamma^a\Psi\wedge\frac{\delta\mathcal{L}}{\delta\vartheta^a} + 2\overline{D\alpha}\wedge\frac{\delta\mathcal{L}}{\delta\overline{\Psi}} + c\overline{\alpha}\gamma\wedge\frac{\delta\mathcal{L}}{\delta\overline{\Psi}} + \delta\omega_a\wedge\frac{\delta\mathcal{L}}{\delta\omega_a}. \quad (101)$$

In the following, we assume that

$$\frac{\delta\mathcal{L}}{\delta\omega_a} = \delta\mathcal{L}_{SG_{bosonic}}\delta\omega_a + 2(-1)^s\tau_a \cong 0, \quad (102)$$

is fulfilled on shell. From $(\delta L/\delta \omega_a) \cong 0$ follows that

$$(-1)^s \frac{\chi}{2} T_a - \frac{\Theta_T}{2} \eta_a - \Theta_L R_a = \tau_a, \tag{103}$$

compared with Eq. (6.9) of Ref. [21]. Then we obtain

$$\delta L \cong \overline{\alpha} \left[i\gamma^a \Psi \wedge \frac{\delta L}{\delta \vartheta^a} + c\gamma \wedge \frac{\delta L}{\delta \overline{\Psi}} - 2D \frac{\delta L}{\delta \overline{\Psi}} \right] + 2d \left(\overline{\alpha} \wedge \frac{\delta L}{\delta \overline{\Psi}} \right). \tag{104}$$

Let us restrict for the moment to the Rarita–Schwinger Lagrangian L_Ψ with $s_1 = s_2 = 0$, then the Rarita–Schwinger equation becomes massive

$$\frac{2}{i} \frac{\delta L}{\delta \overline{\Psi}} = D\Psi + \frac{m}{2} m\gamma \wedge \Psi = 0. \tag{105}$$

Hence, the first term in (104) following form supersymmetric transformations reads:

$$\begin{aligned}
i\ \ & \gamma^a \Psi \wedge \frac{\delta \mathcal{L}}{\delta \vartheta^a} + c\gamma \wedge \frac{\delta \mathcal{L}}{\delta \overline{\Psi}} - 2D \frac{\delta \mathcal{L}}{\delta \overline{\Psi}} \\
= &\ i\gamma^a \Psi \left(-\frac{\chi}{\ell} R_a - \frac{\Lambda}{\ell} \eta_a + \frac{\Theta_T}{e^2} T_a + \Sigma_a \right) \\
+ &\ c\gamma \wedge \left(\frac{i}{2} D\Psi + \frac{i}{4} m\gamma \wedge \Psi \right) - D \left(iD\Psi + \frac{i}{2} m\gamma \wedge \Psi \right) \\
= &\ i\gamma^a \Psi \left(-\frac{\chi}{\ell} R_a - \frac{\Lambda}{\ell} \eta_a + \frac{\Theta_T}{\ell^2} T^a \right) + i\gamma^a \Psi \frac{i}{4} m \overline{\Psi} \gamma_a \Psi \\
+ &\ c\gamma \wedge \left(\frac{i}{2} D\Psi + \frac{i}{4} m\gamma \wedge \Psi \right) \\
- &\ iR_\alpha^\star \gamma^\alpha \Psi - \frac{i}{2} m T_a \gamma^\alpha \Psi + \frac{i}{2} m\gamma_\alpha \wedge D\Psi. \tag{106}
\end{aligned}$$

By a Fierz-rearrangement, we have $\gamma^\alpha \Psi \overline{\Psi} \gamma_\alpha \Psi = 0$. Moreover, in our restricted model with $s_1 = s_2 = 0$ we have to put $c = -m$. Using the formula of Howe and Tucker [20], i.e.

$$\gamma \wedge \gamma = -2\gamma^\alpha \eta_\alpha, \tag{107}$$

we find the following two requirements

$$i\gamma^a \Psi \wedge \frac{\delta \mathcal{L}}{\delta \vartheta^a} + c\gamma \wedge \frac{\delta \mathcal{L}}{\delta \overline{\Psi}} - 2D \frac{\delta \mathcal{L}}{\delta \overline{\Psi}} = i \left(-\frac{\chi}{\ell} - 1 \right) R_\alpha^\star \gamma^\alpha \Psi + \left(\frac{\Theta_T}{\ell^2} - \frac{m}{2} \right) T_\alpha \gamma^\alpha \Psi, \tag{108}$$

$$i \left(\frac{m^2}{2} - \frac{\Lambda}{\ell} \right) \gamma^\alpha \eta_\alpha \Psi = 0 \tag{109}$$

in order that our Lagrangian is supersymmetric. This leads to the conditions

$$c = -m, \quad \chi = -\ell, \quad \Theta_T = \frac{m\ell^2}{2}, \quad \Lambda = \frac{m^2}{2}\ell, \qquad (110)$$

for the coupling constants of the geometrical part $\mathcal{L}_{SG_{bosonic}}$ of our Lagrangian.

5.4. Supersymmetric dynamical symmetries

The second field equation $\delta L/\delta \omega_a \cong 0$ yields

$$(-1)^s \frac{\chi}{2} T_a + \frac{\Theta_T}{2\ell} \eta_a - \Theta_L \ell R_a = \ell \tau_a = \ell(-1)^{s+1} \frac{i}{8} \overline{\Psi} \gamma_a \Psi. \qquad (111)$$

If we insert the above condition (110), for the constants, into equation (111) one obtains

$$T_a - (-1)^s \frac{m}{2} \eta_a + (-1)^s 2\Theta_L R_a = \frac{i}{4} \overline{\Psi} \gamma_a \Psi. \qquad (112)$$

Generalizing the peculiar dynamical symmetry of Ref. [21], we try the ansatz [13]

$$\vartheta_a = c \omega_a + e \overline{\alpha} \gamma_a \Psi, \qquad (113)$$

for the solution of (112), where α is again the spinor–valued zero–form.

By exterior differentiation, we find

$$d\vartheta_a = c\, d\omega_a + e\, d\overline{\alpha} \gamma_a \Psi + e\, \overline{\alpha} \gamma_a d\Psi, \qquad (114)$$

or

$$T_a + (-1)^s \eta_{ab} \wedge \omega^\beta = cR_a - \frac{(-1)^s}{2} \eta_{abc} \omega^b \wedge \omega^c$$
$$+ eD\overline{\alpha}\gamma_a\Psi - \frac{e}{2}\overline{\alpha}\gamma_b\omega^b\gamma_a\Psi + e\overline{\alpha}\gamma_a D\Psi - \frac{e}{2}\overline{\alpha}\gamma_a\gamma_b\omega^b\Psi \qquad (115)$$

Let us re–substitute again our ansatz, in order to replace the connection terms ω^b. Then we obtain

$$T_a + \frac{(1)^s}{c}\eta_{ab}\vartheta^b - (-1)^s \frac{e}{c}\eta_{ab}\overline{\alpha}\gamma^b\Psi$$
$$= cR_a - \frac{(-1)^s}{2c^2}\eta_{abc}(\vartheta^b - e\overline{\alpha}\gamma^a\Psi) \wedge (\vartheta^b - e\overline{\alpha}\gamma^b\Psi)$$
$$+ eD\overline{\alpha}\gamma_a\Psi + e\overline{\alpha}\gamma_a D\Psi - \frac{e}{c}\overline{\alpha}(\vartheta_a - e\overline{\alpha}\gamma_a\Psi)\Psi. \qquad (116)$$

Solving (116) one can get rid from the torsion in terms of the gravitino field.

6. Concluding remarks

In this review we presented the different approaches to $(2+1)$-dimensional supergravity, namely, the standard one, and the ones based in the introduction of Lorentz and translational Chern–Simons terms. Also, we review the conditions for reducing extended supergravities with and without cosmological constant to Chern–Simons theories of the super–Poincaré or super–anti–de Sitter supergroups. Moreover, we formulate a supergravity theory including Lorentz as well as translational Chern–Simons terms. Future work will consist in finding solutions to the theory which generalize the existing ones in the literature, in particular the Bañados–Teitelboim–Zanelli solution [25], contained in the Baeckler et al. one [21].

Acknowledgments

We thank Friedrich W. Hehl for useful discussions and literature hints. This research was supported by CONACYT Grant 28339E.

References

[1] A. Staruszkiewicz, *Acta Physics Polonica* **6** (1963) 735.

[2] G. Clement, *Nucl, Phys,* **B114** (1976) 437.

[3] P. Collas, *Am. J. Phys.* **45** (1977) 833.

[4] Y. Choquet-Bruhat, C. DeWitt-Morette, and M. Dillard-Bleick, *Analysis, Manifolds and Physics* (North–Holland, Amsterdam 1977).

[5] H. Leutwyler, *Nuovo Cimento* **42** (1966) 159.

[6] G. Ponzano and T. Regee, in: *Spectrospic and group theoretical methods in physics*, F. Bloch, S.G. cohen, A. De-Shalit, and I. Talmi, eds. (North–Holland, Amsterdam 1968).

[7] S. Deser and R. Jackiw, *Ann. Phys.* **153** (1984) 405.

[8] S. Deser and R. Jackiw, *Comm. Math. Phys.* **118** (1988) 495.

[9] S. Deser, R. Jackiw, and G. 't Hooft, *Ann. Phys.* **152** (1984) 220.

[10] G. 't Hooft, *Comm. Math. Phys.* **117** (1988) 685.

[11] E. Witten, *Nucl. Phys.* **B311** (1988) 46.

[12] E. Witten, *Nucl. Phys.* **B323** (1989) 113.

[13] A. Achúcarro and P.K. Townsend, *Phys. Lett.* **B180** (1986) 89.

[14] D. Birmingham, M. Blau, M. Rakowski, and G. Thompson, *Phys. Rep.* **209** (1991) 129.

[15] E. Witten, *Commun. Math. Phys.* **121** (1989) 351. *Nucl. Phys.* **B322** (1989) 629.

[16] R. Iengo and K. Lechner, *Phys. Rep.* **213** (1992) 1.

[17] E. Witten, *Nucl. Phys.* **B311** (1988/89) 46.

[18] J.H. Horne and E. Witten *Phys. Rev. Lett.* **62** (1989) 501.

REFERENCES

[19] K. Koehler, F. Mansouri, C. Vaz, and L. Witten, *Mod. Phys. Lett.* **A5** (1990) 935. *Nucl. Phys.* **B341** (1990) 167. *Nucl. Phys.* **B348** (1992) 373.

[20] P.S. Howe and R.W. Tucker, *J. Math. Phys.* **19** (1978) 869.

[21] P. Baekler, E. W. Mielke and F.W. Hehl, *Nuovo Cimento* **B107** (1992) 91.

[22] S. Deser: "*Cosmological Topological Supergravity*", in: *Quantum Theory of Gravity, essays in honor of the 60th Birthday of Bryce S. DeWitt*, S.M. Chistensen, ed. (Adam Hilger, Bristol 1984), p. 374.

[23] S. Deser: "*Three topics in three dimensions*", in: *Supermembranes and Physics in 2+1 Dimensions*, Proc. of the Trieste Conference, July 17-21 1989, ICTP, Trieste, M.J. Duff, C.N. Pope, and E. Sezgin, eds. (World Scientific, Singapore 1989), p. 239.

[24] F.W. Hehl, J. Lemke, and E.W. Mielke, *Two lectures on fermions and gravity*, in Proc. of the School on *Geometry and Theoretical Physics*, Bad Honnet, 12 – 16 Feb. 1990, J. Debrus and A.C. Hirshfeld, eds. (Springer, Berlin 1991), p. 56.

[25] M. Bañados, C. Teitelboim, and J. Zanelli, *Phys. Rev. Lett.* **69** (1992) 1849.

NONCOMMUTATIVE GRAVITY AND QUANTUM COSMOLOGY

H. Garcia-Compeán
Departamento de Física
Centro de Investigación y de Estudios Avanzados del IPN
P.O. Box 14-740, 07000, México D.F., México
compean@fis.cinvestav.mx

O. Obregón
Instituto de Física de la Universidad de Guanajuato
P.O. Box E-143, 37150, León Gto., México
octavio@ifug3.ugto.mx

C. Ramirez
Facultad de Ciencias Físico Matemáticas
Universidad Autónoma de Puebla
P.O. Box 1364, 72000, Puebla, México
cramirez@fcfm.buap.mx

Abstract We survey recent developments in noncommutative gravity and noncommutative quantum cosmology. On the side of gravity we overview a proposal by Chamseddine on a deformation of Einstein gravity. For quantum cosmology we survey our proposal on the Moyal deformation of the Wheeler–De Witt equation by making a deformation of the minisuperspace. Within this context we find an exact solution of the deformed Wheeler–De Witt equation for the case of the Kantowski–Sachs metric and review its physical consequences.

Keywords: Noncommutative Field Theory, General Relativity, Quantum Cosmology

1. Introduction

Noncommutativity of space-time [1] has been recently subject of renewed interest (see reviews [2, 3, 4, 5]). This was greatly motivated by developments in M(atrix)–theory in a background B-field [6] and in string theory, where the noncommutativity arises in the low energy effective field theory of open string theory and in the gauge theory on a D–brane in the presence of constant B-field [7, 8].

One of the most exciting recent applications of the idea of a minimal size to field theory, is that concerning the description of Yang–Mills instantons in four dimensional noncommutative spacetimes. It has been shown that in these spaces instantons acquire an effective size in terms of the noncommutativity parameter Θ. As a consequence, the modulus space of noncommutative instantons no longer has the singularities corresponding to small instantons [9]. This effect has a nice stringy interpretation [8].

It is very well known that gauge and gravitational interactions have different nature in string theory. The former one come from the open string sector and for the case of D–branes, gauge fields are living only on the world–volume of the brane. In the low energy limit, the dynamics of these degrees of freedom are described by a supersymmetric Yang–Mills theory. Gravitational field is no captured by the brane in this limit and it can live in the bulk of entire spacetime. This is related to the fact that the gravitational field is described by closed strings and D–branes can emit closed strings. Thus gravity and gauge interactions enter in an asymmetric way in string theory. Thus, at this stage of things there are not "gravitational D–branes" in whose world–volume lives only the gravitational field. At present, there is not evidence that this happens, however one would like to explore this idea and apply the results of Seiberg and Witten [8] to the gravitational field. The application of these ideas to the gravitational field leads to a new notion of *noncommutative gravity*. Noncommutative gravity has been considered in [10, 11, 12]. In particular in reference [12] a deformed Einstein gravity is constructed by using the Seiberg–Witten map [8], by gauging the noncommutative ISO(3,1) group. A different approach of noncommutative gravity has been proposed in Ref. [13]. Some solutions of noncommutative 3–dimensional Chern–Simons gravity have been also obtained for the Euclidean case in [14]. A Lorentzian noncommutative gravity in 2+1 dimensions is proposed in Ref. [15] as a deformation of the corresponding three–dimensional Chern–Simons theory. Recently, the idea of noncommutativity has been lifted to the minisuperspace models of quantum cosmology [16]. In the present paper we will survey some of these ideas.

One of the puzzles in quantum gravity is the measurement of length, which seems to be limited to distances greater than the Planck length L_P, because to locate a particle we would need an energy greater than the Planck mass M_P. The corresponding gravitational field will have an horizon given by the Schwarzschild radius $R = \frac{2GM_P}{c^2} = 2L_P$ and, whatever happens inside, this radius is shielded and therefore a minimal size should exist for quanta of space and time configuration. Consequently, at very early times of the universe, before the Planck time, nontrivial effects of noncommutativity can be expected.

This paper is organized as follows: in Sec. 2 we overview the derivation of the noncommutative Yang–Mills gauge theory from string theory in the presence of a constant background B-field. In Sec. 3 we briefly survey the Chamseddine's proposal of noncommutative Einstein gravity [12]. Sec. 4 is devoted to develop the main part of this paper which is the application of noncommutative geometry to the minisuperspace in quantum cosmology. Finally in Sec. 5 we give our final remarks.

2. String theory and noncommutative gauge theory

In this section we describe briefly some new developments on the relation between string theory and noncommutative Yang–Mills theory. For this we follow Seiberg–Witten paper [8]. We do not intend to be exhaustive but only to point out some key points and notation for future convenience.

Roughly speaking, the idea consists in describing an open string propagating in a background spacetime $\mathcal{M}$ with metric g_{IJ} and a NS constant B-field, B_{IJ}. The action I is given by

$$I = \frac{1}{4\pi\alpha'} \int_D d^2\sigma \left(g_{IJ} \partial_\alpha X^I \partial^\alpha X^J - 2\pi i \alpha' B_{IJ} \varepsilon^{\alpha\beta} \partial_\alpha X^I \partial_\beta X^J \right), \quad (1)$$

where D is the disc. Equivalently we have

$$I = \frac{1}{4\pi\alpha'} \int_D d^2\sigma \, g_{IJ} \partial_\alpha X^I \partial^\alpha X^J - \frac{i}{2} \int_{\partial D} d\tau B_{IJ} X^I \partial_\tau X^J.$$

Equations of motion from this action must satisfy the boundary conditions

$$g_{IJ} \partial_n X^J + 2\pi i \alpha' B_{IJ} \partial_\tau X^J |_{\partial \Sigma} = 0. \quad (2)$$

The propagator of open string vertex operators is given by

$$\langle X^I(\tau) X^J(\tau')\rangle = -\alpha' G^{IJ} \log(\tau - \tau')^2 + \frac{i}{2}\Theta^{IJ}\varepsilon(\tau - \tau') \qquad (3)$$

where

$$G^{IJ} = \left(\frac{1}{g + 2\pi\alpha' B}\right)^{IJ}_S, \qquad \Theta^{IJ} = 2\pi\alpha'\left(\frac{1}{g + 2\pi\alpha' B}\right)^{IJ}_A. \qquad (4)$$

Here S and A stand for the symmetric and antisymmetric part of the corresponding matrix, and the logarithmic term determines the anomalous dimensions as usual. Thus G_{IJ} is the effective metric seen by the open strings. While, as was suggested in [7, 8], the antisymmetric part Θ^{IJ} determines the spacetime *noncommutativity*.

The product of two tachyon vertex operators $exp(ip \cdot X)$ and $exp(iq \cdot X)$ for $\tau > \tau'$ in the short distance singularity is written as

$$\exp\left(ip \cdot X\right)(\tau) \exp\left(iq \cdot X\right)(\tau') \sim |\tau - \tau'|^{2\alpha' G^{IJ} p_I q_J}$$
$$\times\ \exp\left(ip \cdot X\right) * \exp\left(iq \cdot X\right) \qquad (5)$$

where

$$\exp\left(ip \cdot X\right) * \exp\left(iq \cdot X\right) \equiv \exp\left(\frac{i}{2}\Theta^{IJ} p_I q_J\right)\exp\left(i(p+q)X\right). \qquad (6)$$

This last expression is the Moyal $*$-product in the momentum representation. In the coordinate representation it is defined for any two smooth functions f and g over $\mathcal{M}$ and it is given by

$$f * g = exp\left(\frac{i}{2}\Theta^{IJ}\frac{\partial}{\partial U^I}\frac{\partial}{\partial V^J}\right)f(X+U)g(X+V). \qquad (7)$$

Here the operation $*$ is associative: $f * (g * h) = (f * g) * h$ and non-commutative: $f * g \neq g * f$. The above product can be written as $f * g = fg + i\{f,g\} + \ldots$ where $\{f,g\}$ is the Poisson bracket given by $\Theta^{IJ}\partial_I f \partial_J g$. Θ is determined in terms of B (see Eq. (4)). This give an associative and noncommutative algebra $\mathcal{A}_*$. In the limit $\alpha' \to 0$ (ignoring the anomalous dimensions of open string sector) the product of vertex operators turns out to be the Moyal product of functions on the spacetime $\mathcal{M}$.

Now one can consider scattering amplitude (parametrized by G and Θ) of k gauge bosons of momenta p_I, polarizations ε_I and Chan–Paton wave functions λ_I, $I = 1, \ldots, k$

$$A(\lambda_I, \varepsilon_I, p_I)_{G,\Theta} = Tr\left(\lambda_1 \lambda_2 \ldots \lambda_k\right) \int d\tau'_I \langle \prod_{I=1}^{k} \varepsilon_I \cdot \frac{dX}{d\tau}$$
$$\times \exp\left(ip_I \cdot X\right)(\tau'_I)\rangle_{G,\Theta}. \qquad (8)$$

The Θ dependence come from the factor $\exp\left(-\frac{i}{2}\sum_{s>r} p_I^{(s)} p_J^{(r)} \Theta^{IJ}\right)$. Thus amplitude factorizes as

$$A(\lambda_I, \varepsilon_I, p_I)_{G,\Theta=0} \cdot \exp\left(-\frac{i}{2} \sum_{s>r} p_I^{(s)} \Theta^{IJ} p_J^{(r)} \varepsilon(\tau_r - \tau_s)\right) \qquad (9)$$

which depends only on the cycle ordering of the points $\tau_1, \ldots, \tau_k$ on the boundary of the disc ∂D.

For $B = 0$ the effective action is obtained under the assumption that the divergences are regularized through the Pauli–Villars procedure and it is given by

$$S_G = \frac{1}{g_{st}} \int d^n x \sqrt{G} G^{II'} G^{JJ'} \left(Tr F_{IJ} F_{I'J'} + \alpha' \text{ corrections}\right). \qquad (10)$$

The important case of the effective theory when $\Theta \neq 0$ is incorporated through the phase factor and thus one have to replace the ordinary multiplication of wave functions on the brane world–volume, by the $*$–product (effective action is computed by using point splitting regularization)

$$\widehat{S}_G = \frac{1}{g_{st}} \int d^n x \sqrt{G} G^{II'} G^{JJ'} \left(Tr \widehat{F}_{IJ} * \widehat{F}_{I'J'} + \alpha' \text{ corrections}\right), \qquad (11)$$

where $\widehat{F}_{IJ} = \partial_I \widehat{A}_J - \partial_J \widehat{A}_I - i\{\widehat{A}_I, \widehat{A}_J\}_*$ is the noncommutative field strength. Here $\{f, g\}_* \equiv f * g - g * f$. Thus we get a noncommutative Yang–Mills theory as the Θ (or B) dependence of the effective action to all orders in α'. Infinitesimal gauge field transformation is given by $(\widehat{\lambda} * \widehat{A})_{IJ} = \widehat{\lambda}_{IK} * A_J^K$ and $\delta \widehat{A}_I = \partial_I \widehat{\lambda} + i\widehat{\lambda} * \widehat{A}_I - i\widehat{A}_I * \widehat{\lambda}$.

For the low varying fields the effective action is given by the Born–Infeld–Dirac action

$$S = \frac{1}{g_{st}(\alpha')^2} \int d^n x \sqrt{\det(g + \alpha'(F + B))}. \qquad (12)$$

The same effective action is described by noncommutative Yang–Mills theory but also by the standard Yang–Mills theory. They differ only in the regularization prescription. For the standard commutative case it is the Pauli–Villars one, while for the noncommutative case it is the point splitting prescription. The two frameworks are equivalent and thus there is a redefinition of the fields which can be regarded as a transformation connecting standard commutative and noncommutative descriptions. The change of variables known as the *Seiberg–Witten map* is as follows:

$$\widehat{A}_I = A_I - \frac{1}{4}\Theta^{KL}\{A_K, \partial_L A_I + F_{LI}\} + O(\Theta^2),$$

$$\widehat{\lambda} = \lambda + \frac{1}{4}\Theta^{KL}\{\partial_L \lambda, +A_J\} + O(\Theta^2). \tag{13}$$

3. Chamseddine's Moyal deformation of Einstein gravity

As we have seen, noncommutative Yang–Mills theories are very natural in string theory with a constant and nonzero B-field. They are realized as the effective field theory on the D–brane world–volume. However one immediately can ask about the possibility that gravity theory arises in a more symmetric way in string theory. Up to now gravity and gauge theory are realized in a very different ways in string theory. Gravitational interaction is associated with a massless mode of a closed string while Yang–Mills are more naturally described in open strings or in heterotic closed string theory. Motivated by this progress in string theory, Chamseddine proposed a Moyal deformation of the Einstein's gravity [12]. In order to do that he start from a gauge–like theory of gravity with gauge group SO(4,1) and reduce it by symmetry contraction to the subgroup ISO(3,1). After that the remnant noncommutative group ISO(3,1) is gauged out by using the technique of [17]. Finally, the Seiberg–Witten map is used to express the deformed fields in terms of the undeformed ones.

The deformed gauge fields $\widehat{\omega}_\mu^{AB}$ are subjected to the conditions determined in [17]. This is for the case of the gauging a noncommutative orthogonal group from an noncommutative unitary group. These conditions are: $\widehat{\omega}_\mu^{AB\dagger}(x,\Theta) = -\widehat{\omega}_\mu^{BA}(x,\Theta)$ and $\widehat{\omega}_\mu^{AB}(x,\Theta)^r \equiv \widehat{\omega}_\mu^{AB}(x,-\Theta) = -\widehat{\omega}_\mu^{BA}(x,\Theta)$. These gauge fields can be expressed in terms of the usual gauge fields by the expansion $\widehat{\omega}_\mu^{AB}(x,\Theta) = \omega_\mu^{AB} - i\Theta^{\nu\rho}\omega_{\mu\nu\rho}^{AB} + \cdots$.

The basic assumption is that there are no extra degrees of freedom and the deformed and undeformed fields are related trough the Seiberg–Witten map

$$\widehat{\omega}_\mu^{AB}(\omega) + \delta_{\widehat{\lambda}}\widehat{\omega}_\mu^{AB}(\omega) = \widehat{\omega}_\mu^{AB}(\omega + \delta_\lambda \omega), \tag{14}$$

where $\delta_{\widehat{\lambda}}\widehat{\omega}^{AB} = \partial_\mu \widehat{\lambda}^{AB} + \widehat{\omega}_\mu^{AC} * \widehat{\lambda}^{CB} - \widehat{\lambda}^{AC} * \widehat{\omega}_\mu^{CB}$ and $\delta_\lambda \omega^{AB} = \partial_\mu \lambda^{AB} + \omega_\mu^{AC} * \lambda^{CB} - \lambda^{AC} * \omega_\mu^{CB}$.

The solution of the Seiberg–Witten map is given at first order in Θ by

$$\widehat{\omega}_\mu^{AB} = \omega_\mu^{AB} - \frac{i}{4}\Theta^{\nu\rho}\{\omega_\nu, \partial_\rho \omega_\mu + R_{\rho\mu}\}^{AB} + O(\Theta^2),$$

$$\widehat{\lambda}^{AB} = \lambda^{AB} + \frac{i}{4}\Theta^{\nu\rho}\{\partial_\nu \lambda, +\omega_\rho\}^{AB} + O(\Theta^2), \tag{15}$$

where $\{u,v\}^{AB} \equiv u^{AC}v^{CB} + v^{AC}u^{CB}$. Thus in general to all order in Θ we have

$$\delta\widehat{\omega}^{AB}(\Theta) = -\frac{i}{4}\Theta^{\nu\rho}\{\widehat{\omega}_\nu, \partial_\rho \widehat{\omega}_\mu + \widehat{R}_{\rho\mu}\}_*^{AB}, \tag{16}$$

where $\widehat{R}_{\mu\nu}^{AB} = \partial_\mu \widehat{\omega}_\nu^{AB} - \partial_\nu \widehat{\omega}_\mu^{AB} + \{\widehat{\omega}_\mu, \widehat{\omega}_\nu\}_*^{AB}$.

In order to see how the dependence of the deformed vierbein $\widehat{e}_\mu^a$ enters into the game, we consider the contraction of the gauge symmetry group SO(4,1) to ISO(3,1). This is a usual procedure in the context of gauge theories of gravity. Of course this procedure is done at the level of undeformed gauge fields by decomposing capital indices $A, B, \ldots$ into $(a, 5)$. The undeformed field strength is given by $R_{\mu\nu}^{AB} = \partial_\mu \omega_\nu^{AB} - \partial_\nu \omega_\mu^{AB} + [\omega_\mu, \omega_\nu]^{AB}$ and it decomposes into

$$R_{\mu\nu}^{ab} = \partial_\mu \omega_\nu^{ab} - \partial_\nu \omega_\mu^{ab} + \omega_\mu^{ac}\omega_\nu^{cb} - \omega_\nu^{ac}\omega_\mu^{cb} + k^2(e_\mu^a e_\nu^b - e_\nu^a e_\mu^b)$$
$$R_{\mu\nu}^{a5} \equiv kT_{\mu\nu}^a = k(\partial_\mu e_\nu^a - \partial_\nu e_\mu^a + \omega_\mu^{ac}e_\nu^c - \omega_\nu^{ac}e_\mu^c). \tag{17}$$

One can solve ω_μ^{ab} in terms of e_μ^a in the case when $T_{\mu\nu}^a = 0$. Similarly the in the deformed case we take $\widehat{\omega}_\mu^{a5} = k\widehat{e}_\mu^a$ and $\widehat{\omega}_\mu^{55} = k\widehat{\phi}_\mu$. For our purposes it is enough to consider the undeformed torsion $T_{\mu\nu}^a = 0$ but $\widehat{T}_{\mu\nu}^a \neq 0$. The deformed action which give rise to these considerations is given by

$$I_* = \int d^4x \sqrt{\widehat{e}} * \widehat{e}_{*a}^\mu * \widehat{R}_{\mu\nu}^{ab} * (\widehat{e}_{*b}^\nu)^\dagger * (\sqrt{\widehat{e}})^\dagger, \tag{18}$$

where $\widehat{e} \equiv det(\widehat{e}_\mu^a)$. This action is Lorentz invariant but not diffeomorphism invariant. To make this action diffeomorphism invariant one has to replace Moyal $*_M$-product by the Kontsevich $*_K$-product [12].

This action can be obtained in terms of the undeformed fields at second order in Θ.

4. Noncommutative quantum cosmology

In this section we overview the results found in Ref. [16]. In the study of homogeneous universes, the metric depends only on the time parameter. Thus, the space dependence can be integrated out in the action and a model with a finite dimensional configuration space arises, called also *minisuperspace*. The coordinates on the minisuperspace are the three–metric components. These theories have been considered by themselves, and their quantization is performed following the rules of quantum mechanics.

The minisuperspace construction is a procedure to define quantum cosmology models in the search to describe the quantum behavior of the very early stages of the universe [18, 19]. By defining these models one necessarily freezes out degrees of freedom, so that these are only simple and probably approximate models of full quantum gravity at Planckian times. Actually, the validity of this approach remains as an open question to date. It has been argued that one can find conditions that must be satisfied to justify the minisuperspace approximation [20]. In string theory formalism, general relativity, and consequently the Wheeler–DeWitt equation, corresponds to the s–wave approximation [21]. Nevertheless, by considering a more general analysis, [22], it seems that we can expect that the fundamental behavior of the wave function will be preserved.

Recently, attempts to connect M–String theory to cosmology on the brane [23] have been done. The fact that in the former theory noncommutativity has been shown to be present, motivated us to consider it also in models of the universe. In this paper we make a proposal in order to explore the influence of noncommutativity at early times, by the introduction of an effective noncommutativity in quantum cosmology. Instead of a deformation of space-time, we consider a *"deformation of minisuperspace"*.

It is assumed that the minisuperspace variables do not commute, as it has been proposed for the spacetime coordinates. As a consequence of it, the fields do not commute among them in a specific manner dictated by the Moyal product. In our ansatz, we propose a simple and direct noncommutativity among certain components of the gravitational field.

It is well known that noncommutativity is usually defined in the "preferred" frame of Cartesian coordinates, where the noncommutative parameters are taken to be constant. For any other coordinate systems, the corresponding noncommutativity will be, in general, in terms of parameters related in a complicated manner to those of Cartesian coordinates. For the even–dimensional case, the spacetimes can be interpreted as symplectic manifolds with the noncommutativity parameters playing the role of components of the symplectic form. In this case, Darboux's theorem ensures that always, locally, there exists a coordinate system in which these components are constant. In string theory this is assumed when a 'constant' B–field is considered.

On the other hand, the Seiberg–Witten map for gravitation has been proposed in [12], where noncommutative tetrads and connections are computed. In the case of quantum cosmology, the minisuperspace variables play the role of the "coordinates" of the configuration space. Thus, it seems reasonable to propose a kind of noncommutativity among these specific gravitational variables, as it is the case in standard spacetime, when Cartesian coordinates are selected. From the canonical quantization of a cosmological model a quantum mechanical version, with a finite number of degrees of freedom, of the general Wheeler–DeWitt equation for general relativity arises. In this way, quantum cosmology is usually understood as a quantum mechanical model of the universe. We will then further assume, that it can be proceeded as in standard noncommutative quantum mechanics.

The noncommutativity we propose, can be reformulated in terms of a Moyal deformation of the Wheeler–DeWitt equation, similar to the case of the noncommutative Schrödinger equation [24]. Actually, cosmology depends only on time and a Moyal product of functions of only one variable is trivially realized. Other authors have worked out noncommutativity in the early universe, however without considering the gravitational field. In Ref. [25] the noncommutativity of space–time is interpreted as a magnetic field on a horizon scale. In [26, 27], it is argued that noncommutativity affecting matter or gauge fields could have played an important role to produce inflation.

Our proposal will be examined through an specific example. Consider the cosmological model of the Kantowski–Sachs metric. In the parametrization due to Misner, this metric looks like [28]:

$$ds^2 = -N^2 dt^2 + e^{2\sqrt{3}\beta} dr^2 + e^{-2\sqrt{3}\beta} e^{-2\sqrt{3}\Omega}(d\theta^2 + \sin^2\theta d\varphi^2). \quad (19)$$

The Wheeler–DeWitt equation for this metric in a particular factor ordering, is given by

$$\exp(\sqrt{3}\beta + 2\sqrt{3}\Omega)\bigg(-P_\Omega^2 + P_\beta^2 - 48\exp(-2\sqrt{3}\Omega)\bigg)\psi(\Omega,\beta) = 0, \quad (20)$$

where $P_\Omega = -i\frac{\partial}{\partial \Omega}$ and $P_\beta = -i\frac{\partial}{\partial \beta}$. In this parametrization the Wheeler–DeWitt equation has a simple form, which can be formally identified with usual quantum mechanics in Cartesian coordinates. The solutions to this Wheeler–DeWitt equation are given by (see [28])

$$\psi_\nu^\pm(\beta,\Omega) = e^{\pm i\nu\sqrt{3}\beta} K_{i\nu}(4e^{-\sqrt{3}\Omega}), \quad (21)$$

where $K_{i\nu}$ is the modified Bessel function. Packet waves of these solutions have been constructed as superpositions of these solutions. Summing over $e^{i\nu\sqrt{3}\beta}$ and $e^{-i\nu\sqrt{3}\beta}$ to make real trigonometric functions, the

"Gaussian" state

$$\Psi(\beta,\Omega) = 2i\mathcal{N}\int_0^\infty \nu\left[\psi_\nu^+(\beta,\Omega) - \psi_\nu^-(\beta,\Omega)\right]d\nu$$
$$= \mathcal{N}e^{-\sqrt{3}\Omega}\sinh(\sqrt{3}\beta)\exp[-2\sqrt{3}e^{-\sqrt{3}\Omega}\cosh(\sqrt{3}\beta)], \quad (22)$$

has been obtained [28, 29]. A possible connection with quantum black holes [29, 30] and quantum wormholes [31, 32] has been suggested.

For our noncommutative proposal of quantum cosmology, we will follow the procedure outlined above. We will assume that the "Cartesian coordinates" Ω and β of the Kantowski–Sachs minisuperspace obey a kind of commutation relation, like the ones in noncommutative quantum mechanics [24], $[\Omega,\beta] = i\Theta$.

This deformation can be reformulated in terms of a noncommutativity of minisuperspace functions, with the Moyal product $f(\Omega,\beta)*g(\Omega,\beta) = f(\Omega,\beta)e^{i\frac{\Theta}{2}(\overrightarrow{\partial}_\Omega\overrightarrow{\partial}_\beta - \overleftarrow{\partial}_\beta\overrightarrow{\partial}_\Omega)}g(\Omega,\beta)$. Thus, the corresponding noncommutative Wheeler–DeWitt equation is given by

$$\exp(\sqrt{3}\beta + 2\sqrt{3}\Omega) * [-P_\Omega^2 + P_\beta^2 - 48\exp(-2\sqrt{3}\Omega)] * \psi(\Omega.\beta) = 0. \quad (23)$$

It is possible to reformulate noncommutativity of the coordinates in terms of commutative variables and the ordinary product of functions, if new variables are introduced, $\Omega \to \Omega - \frac{1}{2\hbar}\Theta P_\beta$ and $\beta \to \beta - \frac{1}{2\hbar}\Theta P_\Omega$, the momenta remain the same. As a consequence, the original equation changes, with the modified potential

$$V(\Omega,\beta) * \psi(\Omega,\beta) = V(\Omega - \frac{1}{2}\Theta P_\beta, \beta - \frac{1}{2}\Theta P_\Omega)\psi(\Omega,\beta). \quad (24)$$

Thus, noncommutative Wheeler–DeWitt equation turns out to be

$$[-\frac{\partial^2}{\partial\Omega^2} + \frac{\partial^2}{\partial\beta^2} + 48\exp(-2\sqrt{3}\Omega + \sqrt{3}\Theta P_\beta)]\psi(\Omega,\beta) = 0. \quad (25)$$

Assuming a separation of variables with the following ansatz $\psi(\Omega,\beta) = \exp(\sqrt{3}\nu\beta)\chi(\Omega)$, where $\chi(\Omega)$ satisfy the equation

$$\left[-\frac{d^2}{d\Omega^2} + 48\exp(-3i\nu\Theta)\exp(-2\sqrt{3}\Omega) + 3\nu^2\right]\chi(\Omega) = 0. \quad (26)$$

Thus, the solution of equation (25) is given by the wave function [16]

$$\psi_\nu^\pm(\Omega,\beta) = e^{\pm i\sqrt{3}\nu\beta}K_{i\nu}\left\{4\exp\left[-\sqrt{3}\left(\Omega \mp \frac{\sqrt{3}}{2}\nu\Theta\right)\right]\right\}. \quad (27)$$

Note that noncommutativity induces a difference of the arguments of the Bessel functions in these solutions. Moreover, from its form, we can expect that the noncommutativity effects are enhanced for ψ^+. Thus, for the particular model we have chosen, the solution (27) allows an exact analysis, without the need of a Θ expansion. In order to see the consequences of noncommutativity, let us consider a wave packet weighted by a Gaussian,

$$\Psi(\Omega, \beta) = \mathcal{N} \int_{-\infty}^{\infty} e^{-a(\nu-b)^2} \psi_\nu^+ (\Omega, \beta) \, d\nu. \tag{28}$$

This integral is performed numerically. We are interested to see which is the influence of the Θ parameter, but we are as well interested on consequences for the values of the Ω and β variables, which in particular can provide information about the anisotropy.

As mentioned, a minimal size should exist for quanta of space and time configuration, and we are considering the very early time of the universe, where the influence of noncommutativity could have played a role in its quantum behavior. We are interested to see which is the influence of the Θ parameter, but we are as well interested in the β variable, which reflects the anisotropy of the solution. For illustrative purposes we show the graphics of $|\Psi|^2$ against Ω for fixed values $a = 8$, $b = 1.3$ and: figure 1, $\beta = 1$ and $\Theta = 0$, figure 2, $\beta = 1$, $\Theta = 4$. These figures show the dramatic changes that the universe could have had if noncommutativity was present. Figure 1 corresponds to $\Theta = 0$, the standard commutative case, and is presented for Ω in the range [0,6]. In figure 2, $\Theta = 4$ and things have drastically changed, more peaks appear that seem *to compete to be the most probable state* of the universe. In figure 3 it can be appreciate the same behavior of $|\Psi|^2$ for $\beta = 8$ and Ω varying in the range $\{0, 50\}$ and Θ varying in the range $\{0, 20\}$. The same behavior as in figure 2 is observed as the value of Θ increases.

Noncommutativity in minisuperpace creates then, new possible states of the universe. So the universe we live today could have evolved not only from the state for the commutative case (figure 1), but from any of the other states of the noncommutative model, like those of figure 2 or 3. It could have jumped, by means of a tunnelling process from one universe (one state) to other universe (other state).

5. Concluding remarks

As already mentioned, there are in the literature proposals for a noncommutative theory of gravity [10, 11, 12, 13] in four dimensions and [14, 15] in the three–dimensional case. In particular, in reference [12] the Seiberg–Witten map was used to construct a deformed Einstein gravity.

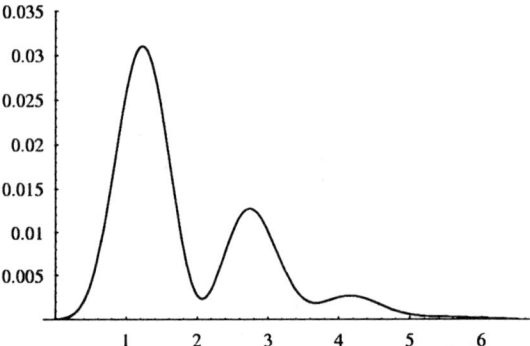

Figure 1. Variation of $|\Psi|^2$ with respect to Ω and $\beta = 1$, at the value $\Theta = 0$. This corresponds with the commutative ordinary quantum cosmology.

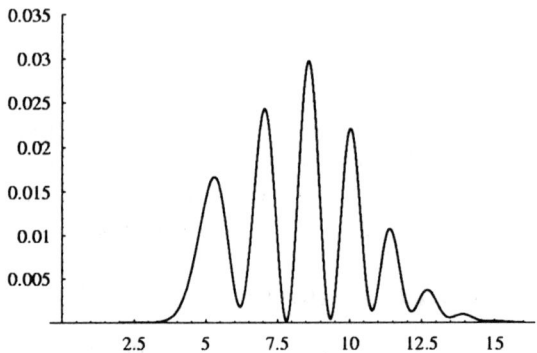

Figure 2. Variation of $|\Psi|^2$ with respect to Ω and $\beta = 1$, at the value $\Theta = 4$. New many peaks corresponding to different universe.

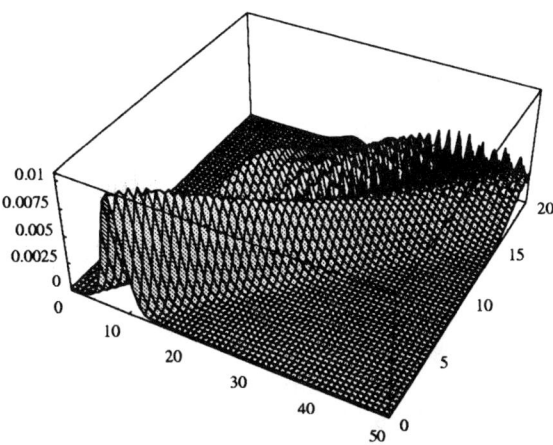

Figure 3. Variation of $|\Psi|^2$ for $\beta = 8$, with Ω varying in the range $\{0, 50\}$ and Θ in the range $\{0, 20\}$.

By means of this result, one could try to find the corresponding noncommutative Wheeler–DeWitt equation for specific cosmological models. In that case the Θ terms corresponding to spacetime noncommutativity could be considered in order to search for another way to define a noncommutative cosmological model, even though cosmology depends only on time. The computation of this Wheeler–DeWitt equation is a very complicated task, because at each Θ order higher derivative terms will appear. If such a noncommutative Hamiltonian model could be defined, its quantum cosmological solutions and the corresponding states could be compared with those obtained by means of our noncommutative minisuperspace proposal. In this context the Θ parameter we have proposed, could be a kind of effective noncommutative parameter. Such an approach is currently under exploration.

Although quantum cosmology, as discussed above, is only a limited model in an attempt to describe some of the features of the quantum theory of the universe, the consideration of noncommutativity seems to be one way to take into account the presence of constant Neveu–Schwarz background B–fields in M(atrix) [6] and string theory, at early times in the universe.

Further work in more realistic cosmological models, including matter, will be needed to search for constraints on the range of values of the Θ parameter in the early stages of evolution of the universe. In our proposal this is in principle possible, because it means to enlarge the minisuperspace configuration space and then consider an appropriate commutativity among its coordinates. Following these lines, we will consider also the supersymmetric extension [33] to these models.

In previous works, the Wheeler–DeWitt equation of the Kantowski-Sachs model has been related to quantum black holes [29, 30] and wormholes [31, 32]. It could be interesting to search for an extension of our results to possible noncommutative versions of these quantum gravitational systems.

Our simple proposal provides a picture of the dramatic influence that noncommutativity could have played at early stages of the universe. We have been able to obtain and work with an exact quantum solution for the Kantowski–Sachs metric, without the need to expand on the Θ parameter. The corresponding wave packet exhibits new stable states of the universe with similar probabilities, showing peaks around β values different from zero. A tunnelling process could happen between these states. By these means there are different possible universes (states) from which our present universe could have evolved and also could have tunnelled in the past, from one universe (state) to other one. Further work is needed to analyze other physical consequences of this and other more realistic quantum cosmological models, taking into account the influence of matter. It will be also necessary to extend these ideas to other, more general gravitational models. In particular, it would be very interesting to reinterpret the results of reference [34], concerning the influence of a primordial magnetic field in classical cosmology. The corresponding quantum model should be constructed and compared with our proposal in terms of the noncommutativity of the minisuperspace. Results in these directions and those mentioned above will be presented elsewhere.

Acknowledgments

This work was supported in part by CONACyT Mexico Grant Nos. 28454E and 33951E.

References

[1] H. Snyder, *Phys. Rev.* **71** (1947) 38.

[2] A. Connes, *Noncommutative Geometry,* (Academic Press, New York, USA, 1994).

[3] J.C. Varilly, physics/9709045.

[4] M.R. Douglas and N.A. Nekrasov, "Noncommutative Field Theory", hep-th/0106048

REFERENCES

[5] R.J. Szabo, "Quantum Field Theory on Noncommutative Spaces", hep-th/0109162.

[6] A. Connes, M.R. Douglas and A. Schwarz, *JHEP* **02** (1998) 003.

[7] V. Schomerus, *JHEP* **06** (1999) 030.

[8] N. Seiberg and E. Witten, *JHEP* **09** (1999) 032.

[9] N. Nekrasov and A. Schwarz, *Commun. Math. Phys.* **198** (1998) 689.

[10] J.W. Moffat, *Phys. Lett.* **B491** (2000) 345.

[11] J.W. Moffat, *Phys. Lett.* **B493** (2000) 142.

[12] A.H. Chamseddine, *Phys. Lett.* **B504** (2001) 33.

[13] V.P. Nair, "Gravitational Fields on a Noncommutative Space, hep-th/0112114.

[14] M. Bañados, O. Chandia, N. Grandi, F.A. Schaposnik and G.A. Silva, *Phys. Rev.* **D64** (2001) 084012.

[15] S. Cacciatori, D. Klemm, L. Martucci and D. Zanon, "Noncommutative Einstein-AdS Gravity in Three Dimensions", hep-th/0201103.

[16] H. García-Compeán, O. Obregón and C. Ramírez, "Noncommutative Quantum Cosmology", hep-th/0107250.

[17] L. Bonora, M. Schabl, M.M. Sheikh-Jabbari and A. Tomasiello, *Nucl. Phys.* **B589** (2000) 461.

[18] M. Ryan, Hamiltonian Cosmology (Springer-Verlag, Berlin, Germany, 1972).

[19] J. Hartle and S. W. Hawking, *Phys. Rev.* **D28** (1983) 2960.

[20] S. Sinha and B.L. Hu, *Phys. Rev.* **D44** (1991) 1028; B.L. Hu and S. Sinha, in: *Directions in General Relativity*, ed. B.L. Hu, M.P. Ryan and C.V. Vishveshwara (Cambridge University Press, Cambridge, UK, 1993).

[21] L. Susskind and J. Uglum, Talk presented at the PASCOS meeting in Syracuse, New York, May 1994, hep-th/9410074.

[22] J.J. Halliwell, in: Proc. 13th. Int. Conf. on General Relativity, ed R.J. Gleisser, C.N. Kozameh and O.M. Moreschi, Bristol (IOP Publishing, 1993).

[23] C. Grojean, F. Quevedo, G. Tasinato and I. Zavala, *JHEP* **08** (2001) 005.

[24] J. Gamboa, M. Loewe and J.C. Rojas, *Phys. Rev.* **D64** (2001) 067901; M. Chaichian, M.M. Sheikh-Jabbari and A. Tureanu, *Phys. Rev. Lett.* **86** (2001) 2716.

[25] A. Mazumdar and M.M. Sheikh-Jabbari, *Phys. Rev. Lett.* **87** (2001) 011301.

[26] Chong-Sun Chu, B.R. Greene and G. Shiu, *Mod. Phys. Lett.* **A16** (2001) 2231.

[27] F. Lizzi, G. Mangano, G. Miele and G. Sparano, *Int. J. Mod. Phys.* **A11** (1996) 2907.

[28] C. Misner, in: *Magic without magic*, ed. John Archibald Wheeler (Freeman, USA, 1972).

[29] O. Obregón and M.P. Ryan, *Mod. Phys. Lett.* **A13** (1998) 3251.

[30] M. Cavaglia, V. De Alfaro and A.T. Filippov, *Int. J. Mod. Phys.* **D4** (1995) 661.

[31] M. Cavaglia, *Mod. Phys. Lett.* **A9** (1994) 1897.

[32] L.M. Campbell and L. Garay, *Phys. Lett.* **B254** (1991) 49.

[33] A. Macías, O. Obregón and M.P. Ryan, *Class. Quantum Grav.* **4** (1987) 1477; O. Obregón and C. Ramírez, *Phys. Rev.* **D57** (1998) 1015; P.D. D'Eath, S.W.

Hawking and O. Obregón, *Phys. Lett.* **B300** (1993) 44; R. Graham, *Phys. Rev. Lett.* **67** (1991) 1381; P.D. D'Eath, *Supersymmetric Quantum Cosmology*, (Cambridge University Press, Cambridge, UK, 1996); P.V. Moniz, *Int. J. Mod. Phys.* **A11** (1996) 4321.

[34] C.G. Tsagas and Roy Maartens, *Class. Quantum Grav.* **17** (2000) 2215; D. R. Matravers and C.G. Tsagas, *Phys. Rev.* **D62** (2000) 103519; C.G. Tsagas, *Phys. Rev. Lett.* **86** (2001) 5421.

CONSISTENT DISCRETIZATIONS IN CLASSICAL AND QUANTUM GENERAL RELATIVITY

Rodolfo Gambini
Instituto de Física, Facultad de Ciencias, Universidad de la República
Iguá 4225, CP 11400 Montevideo, Uruguay
rgambini@fisica.edu.uy

Jorge Pullin
Department of Physics and Astronomy, Louisiana State University,
202 Nicholson Hall, Baton Rouge, LA 70803-4001
pullin@phys.lsu.edu

Abstract   Discretizations are used in various contexts in general relativity. In the classical theory, they are used to evolve the Einstein equations numerically. In quantum gravity, they are used to regularize expressions both at the canonical and path integral levels. Somewhat surprisingly, in most cases the resulting discrete theories are inconsistent. We discuss the nature and implications of these inconsistencies and propose methods to generate consistent discretizations. This is a short write up of a talk given at the First Mexican Meeting on Mathematical and Experimental Physics. Details will be published elsewhere.

Keywords: Discretization, general relativity, quantum gravity.

1.   Introduction

There are many situations in which continuum expressions are replaced by discrete expressions that approximate them in a limit in which the discretization is refined. In general relativity this is done when numerically integrating the Einstein equations. It is also used when one regularizes expressions in quantum gravity via the introduction of lattices. This has implications both in canonical quantum gravity, as when one regularizes the Hamiltonian constraint, and in the path integral quantization when the integral is discretized, as in spin foam approaches.

We wish to point out that if one examines carefully the resulting discrete theories, generically they are inconsistent. That is, they do not make sense as theories. In numerical relativity this can, to a certain extent, be ignored. In the quantum cases we believe these problems need to be fixed. In the next few sections we will outline the general ideas in all the relevant cases.

2. Numerical relativity

It is perhaps pedagogically clearer to start with the example of numerical relativity, where the inconsistency we refer to is well known. In numerical relativity one splits the Einstein equations in $3+1$ form. One is then left with two sets of equations, one of them evolves the variables and the other constrains the variables at each instant of time. The continuum equations have the feature that if one starts with initial data that satisfies the constraint equations, if one evolves the data with the evolution equations some time into the future, the resulting quantities will automatically solve the constraints evaluated at such instant. If this were not the case, then one could not find a simultaneous solution to all equations and the theory would not make sense. It is well known that if one discretizes the equations, the situation changes. If one starts with initial data that solves the discrete constraint equations, generically the evolution produced by the discrete evolution equations will fail to solve the discretized constraints in the future.

In numerical relativity, this inconsistency is not viewed as a fundamental problem. The point of view is that one is not really interested in the discrete theory, which is only viewed as a tool to generate an approximation to the continuum theory. One already is introducing an error (with respect to the continuum solution) in using the discrete evolution equations. Therefore one can relativize the importance of, in addition to that, failing to solve the discrete constraints. In fact, it can be proved that the errors in the evolution and in the constraints are commensurate [1], and therefore a plausible point of view is to forget about the inconsistency altogether.

It might still prove useful to construct discretizations such that the discrete constraints are preserved in evolution. The errors in numerical relativity have a tendency to blow up in time. This has been, for instance, the motivation for various rewritings of the evolution equations (in the continuum) such that they "return to the constraint surface" [2]. Ensuring that at least the discrete constraints are exactly preserved could help control the instabilities that appear in most numerical codes.

How does one construct a consistent discretization? One possible way is to use the freedom in lapse and shift. One starts from a given set of initial data and evolves them with the discrete evolution equations with unspecified lapse and shift. One then takes the evolved data and solves the discrete constraint equations by specifying the lapse and shift (notice that in the continuum such a construction would not work since the constraints are automatically preserved; the construction is unique to the discrete theory). One has four constraint equations and four quantities to solve for, so this is in principle feasible. One can repeat this construction in all evolution steps. At the end one will have an evolved spacetime that solves the discrete constraints in all evolution steps. The lapse and the shift will be fixed. This appears quite surprising, since it seems to imply that coordinate freedom was lost. A lengthier discussion will be published elsewhere, but basically what has happened is that the coordinates were fixed by the choice of grid. It is not reasonable to expect that one can evolve with arbitrary lapses and shifts on a grid, since generically evolution will map points on the grid to points that are not on it if one allows arbitrary lapse and shift. Coordinate freedom is still present in the choice one makes when spreading the grid points.

It is not at present known if using consistent discretizations will prove advantageous in numerical relativity. This can only be tested empirically. Although it is true that the consistent evolutions remain (to machine accuracy) on the (discrete) constraint surface in phase space, it is not clear that such surface always remains close to the continuum constraint surface. Worse, even if both constraint surfaces remain close to each other, it could be that the discretized trajectory, while remaining on the discrete constraint surface, veers away from the continuum trajectory. Moreover, one generically has to give initial and boundary data for the lapse and shift for the scheme to work, and the specification of such conditions is unclear. We are currently testing the scheme in spherical symmetry to understand this and other issues. Maurice Van Putten [3] has explored a related idea in the context of the four dimensional formulation and claims success in a model example of a Gowdy spacetime.

3. Canonical quantization

A few years ago Thiemann [4] presented the first finite, well defined, non-trivial, anomaly free quantization of the Hamiltonian constraint of general relativity using the mathematical techniques of Ashtekar and Lewandowski [5] and the spin network states of Rovelli and Smolin [6]. To describe Thiemann's achievement in a few lines we can say that he started by proposing a novel expression for the Hamiltonian constraint in the continuum theory. He then discretized space via the introduction of a simplicial decomposition and wrote an expression in terms of holonomies that approximates the Hamiltonian constraint in the limit in which the decomposition is refined. He then proceeded to promote the resulting expression to an operator. Since the quantities involved in writing the expression (the volume operator and holonomies) are all well defined expressions in the space of spin network states, the construction of a well defined finite operator for the Hamiltonian is almost immediate. Thiemann actually produces an operator that acts on the space of diffeomorphism invariant spin network states. So in reality he does not just quantize the classical discrete expression he started with, but actually achieves that while at the same time taking a non-trivial limit.

This last point is important, since the classical theory he started with is a discrete theory that is inconsistent, in a sense related to the one we discussed in the previous section. In particular, if one naively tries to compute the constraint algebra for the discrete classical Hamiltonian constraint that is quantized, one will find that the algebra does not close. The final quantum Hamiltonian constraint is consistent in the sense that it commutes with itself, as one expects for the Hamiltonian constraint in the space of diffeomorphism invariant spin network states. It appears that the limiting process has had the remarkable property of producing a consistent quantum theory out of an apparently inconsistent classical theory.

Let us suppose we wanted to quantize Thiemann's discrete Hamiltonian without taking the limit. This raises the question of how does one quantize canonically a discrete theory? Our proposal is to adhere strictly to Dirac's quantization procedure. We have worked out this in detail for lattice QCD as a first step before tackling the more complex gravitational case.

Although the phrase "Hamiltonian lattice QCD" has been used since the 70's, in reality no one ever performed a proper Hamiltonian treatment of the theory in the lattice. Instead, the "transfer matrix" [7] method was used to read–off from the discrete action a Hamiltonian for the theory that was promoted to a quantum operator. This requires a certain degree of cleverness, and the hindsight of the vast experience available with QCD. It is unlikely that a similar shortcut will be available in quantum gravity.

It turns out that one can indeed construct a proper Hamiltonian formulation of lattice QCD. One starts by writing the Wilson action and performing a $3+1$ split. Canonical variables can be defined and a Legendre transform yields a Hamiltonian. The theory however, has a quite involved canonical structure. There are constraints that need to be dealt with, including second class constraints. Strict application of the Dirac procedure, including the introduction of Dirac brackets and the strong imposition of the second class constraints finally yields the usual Hamiltonian that was produced by the "transfer matrix" method. This example is instructive in that if one had naively attempted to quantize the initial Hamiltonian that was produced by the Legendre transform, it is very unlikely that one would have ended up with a correct theory of quantum QCD. It could well be the case that one is incurring in the same behavior when handling the Hamiltonian constraint of quantum gravity without analyzing in detail the canonical structure of the discrete theory, which appears to be quite involved.

We are taking further steps in this direction by analyzing BF theory and introducing irregular lattices. The latter point has additional complications, particularly the definition of canonical momenta. It is therefore desirable to control these issues before attempting a quantization of discretized general relativity. One possibility is to start by attempting a quantization of general relativity on square lattices while at the same time developing irregular lattice formulations for BF theories and QCD. We are currently studying all these issues.

4. The path integral

Similar mathematical tools to the ones that led to the construction of a finite quantum Hamiltonian constraint have also been used to try to compute the path integral of general relativity, in an approach known as "spin foams". Roughly speaking, what is done is to write the action in terms of connection-type variables and then construct an approximation to it via the introduction of a simplicial decomposition of space and constructing holonomies along its links. The functional integral has then been reduced by the discretization to a finite sum of integrals along the group, which can be computed. The final result is the sum of terms that depend on the "colors" of the links of the spin network, summed for all colors.

There are several difficulties that arise in these constructions. To begin with, even for a finite lattice, the resulting sum over colors can diverge. There are several proposals to fix this and our construction does not add anything new in this point so we will not discuss it further. A more serious problem is that the result depends on the simplicial decomposition chosen. In BF theories, since they describe flat connections, the dependence on the discretization disappeared. This accounts for the significant progress that this approach has made in the context of BF theories. In gravity it is unlikely that one could perform a decomposition-independent construction, since the theory has local degrees of freedom. One can envision two possible solutions to this problem. The most immediate one would be to take the limit in which the decomposition is refined. This is what is done in usual lattice QCD. Except that in that case one has the fact that the theory is renormalizable and this manifests itself in that there is a phase transition that makes the details of the discretization irrelevant. This is unlikely to happen for gravity. A second possibility would be to sum over all discretizations. The result would therefore be independent of any particular discretization. But it is difficult to see how such sum could be made convergent. First of all, it is obvious that one is summing over many "gauge equivalent" configurations and this is not correct when computing a path integral. Disentangling the gauge independent modes could be hard for gravity.

An additional problem has to do with the issues we have been discussing in this paper. If one takes an action and discretizes it, generically it will yield inconsistent equations of motion. This can be particularly tricky in the context of the path integral, since one can remain completely oblivious to this fact and compute the integral anyway. It is unlikely however, that the integral will have any physical meaning, since there is no consistent classical theory to recover in from the quantum theory constructed. That is, for a given discretization, there is no sense in which one can use the computed path integral to represent a physical system corresponding to the discrete action, since the latter is inconsistent. As a consequence, it appears to us as very unlikely that this could be fixed by summing over all possible discretizations.

We therefore see that it appears desirable to construct discretizations of the action that are consistent. In such a case, for any given discretization, one could calculate a path integral of which one could study its semi-classical limit and one should recover the solutions of the discrete theory. Discrete theories have many more solutions than continuum ones, but in particular include solutions that approximate those of the continuum theory. Since such solutions will be present in all discretizations, when one sums over all discretizations, the sum will be dominated by the solutions that approximate the continuum theory (since they appear in all terms in the sum) and one could "divide by N" and that would suppress the spurious solutions. This construction is, at the moment, only wishful thinking. It involves all the usual problems of infinite dimensional sums. But it provides a starting point for discussing the "summing over all discretizations" idea that is simply unthinkable if one is using inconsistent discretizations.

But even if the above idea works, one is still left with the problem of summing over gauge related configurations. In other field theories, this problem is best handled by understanding the canonical path integral. We therefore think that it is of great importance to combine the consistent canonical approaches discussed in the previous section with consistent discretizations of the path integral. That could allow to properly compute a path integral in the sense of fixing the gauge a la Fadeev–Popov. We have only started in this route, but it appears as encouraging that the discretization of the Plebanski action proposed by Reisenberger [8] appears to be consistent in the sense discussed in this paper.

5. Conclusions

We have noted that most discretizations used in general relativity — both in classical and in quantum contexts— yield inconsistent theories. The inconsistencies can be cured. In numerical relativity it is still to be tested if consistency makes codes more stable. In quantum gravity it appears that consistency is mandatory in order to build sensible theories. We are in the process of applying these ideas both at the canonical and path integral levels.

Acknowledgments

This work was supported in part by grants NSF-PHY0090091, NSF-PHY-9800973, NSF-INT-9811610, by funds of the Horace C. Hearne Jr. Institute for Theoretical Physics, and the Uruguay Fulbright commission.

References

[1] M. Choptuik, Phys. Rev. **D44** (1991) 3124.

[2] O. Brodbeck, S. Frittelli, P. Hubner and O. A. Reula, J. Math. Phys. **40** (1999) 909.

[3] M. Van Putten, "Past and future gauge in numerical relativity" gr-qc/0108056.

[4] T. Thiemann, Class. Quantum Grav. **15** (1998) 839; 875; 1207; 1249; 1281; 1463.

[5] A. Ashtekar and J. Lewandowski, J. Geom. Phys. **17** (1995) 191.

[6] C. Rovelli and L. Smolin, Phys. Rev. **D52** (1995) 5743.

[7] See for instance H. Rothe, "Lattice gauge theories, an introduction", (World Scientific, Singapore, 1997).

[8] M. P. Reisenberger, Class. Quantum Grav. **14** (1997) 1753.

THE FIELD–TO–PARTICLE TRANSITION PROBLEM

Jerónimo Cortez
Instituto de Ciencias Nucleares, Universidad Nacional Autónoma de México
A. P. 70-543 México D.F. 04510
cortez@nuclecu.unam.mx

Leonardo Patiño
Instituto de Física, Universidad Nacional Autónoma de México
A. P. 20-364, México D. F. 01000
leonardo@ft.ifisicacu.unam.mx

Hernando Quevedo
Instituto de Ciencias Nucleares, Universidad Nacional Autónoma de México
A. P. 70-543 México D.F. 04510
quevedo@nuclecu.unam.mx

Abstract
We formulate in an intuitive manner several conceptual aspects of the field–to–particle transition problem which intends to extract physical properties of elementary particles from specific field configurations. We discuss the possibility of using the conceptual basis of the holographic principle and the mathematical fundamentals of nonlinear sigma models for the field–to–particle transition. It is shown that certain classical gravitational configurations in vacuum may contain physical parameters with discrete values, and that they behave under rotations as particle-like objects.

Keywords: Field–to–particle transition problem, holographic principle, sigma models, geon.

1. Introduction

Probably, the oldest dream of many generations of physicists has been to find a consistent and simple way for describing the phenomena we observe in nature. Simplicity is in this context an important feature. In all our research proposals we intent to extract the simplest aspects of a phenomenon and to put them as the fundamentals for constructing a consistent theory. It is in this connection that physicists today argue that all phenomena in nature should be described by only four types of interactions. One important goal of today's research is to show that in fact we are dealing with probably only one interaction which manifests itself differently at different levels. If this turns out to be true, we will be left with one fundamental interaction which will be responsible for all the physical processes.

General relativity and quantum mechanics can be considered as the most important conceptual cornerstones in the development of physics during the last century. General relativity gave us the possibility to understand the physical phenomena related with the gravitational interaction at the classical level, whereas quantum mechanics gave rise to the development of quantum field theories which describe the fundamentals of the weak and strong interactions. The standard model which includes also the electromagnetic interaction constitutes the basis of modern elementary particle physics.

However, in our opinion, in the process of development of the modern physical theories we have been mixing different concepts. On the one hand, we know from experience that two ingredients are necessary in order to construct a realistic model of a physical interaction, namely, fields and matter. In gravity, for example, when we want to describe the gravitational field of a matter distribution, we take an action with a term which contains the field itself and a second term which represents matter. Whereas we know (more or less) from physical and mathematical arguments how to select the correct term that plays the role of field, the term corresponding to matter has to be imposed *ad hoc*. Matter is an external entity that has to be postulated in accordance with our understanding of its specific properties. The question arises, do we really need to postulate matter in such a rough manner? Of course, we cannot argue that this approach is not correct. In fact, it is one of the best we know, the standard model of elementary particles being an example of its great success. But we can argue that this method is not simple. If our goal is to simplify the way we do physics, we are going in the wrong direction. It is difficult to imagine that at the end of the job, we will have one fundamental interaction with all possible types of matter and all possible types of different properties. Instead, it would be more convenient to end with one fundamental interaction which gives rise (through a variational principle) to many special field configurations, each of them corresponding to a different particle or constituent of matter. This dream is known as the field–to–particle transition problem.

It is the aim of this work to present several intuitive ideas about the methods one could apply in order to attack the field–to–particle transition problem. We believe that it is necessary to start with the analysis of old and new ideas in which the tendency to simplicity is inherent. In Section II we will briefly review the main aspects of the recently formulated holographic principle. We will see that the intrinsic aim of the holographic principle is to replace complex physical systems by other simpler systems without loosing information. Section III is devoted to a review of nonlinear sigma models from which we expect to obtain some of the mathematical tools necessary in order to compare different theories in different spaces. A concrete example of classical gravitational field configurations from which particle–like properties can be derived is presented in Section III. We end with a discussion in Section IV.

2. The holographic principle

The conceptual origin of the holographic principle was settled in the 70's by Bekenstein [1] and Hawking [2], who formulated the second law of thermodynamics for black holes and the process of particle creation by black holes, respectively. These ideas were implemented in the context of quantum field theories, specially in quantum gravity and cosmology by t'Hooft [3] and Susskind [4].

The holographic principle is a statement about the counting of the quantum states of a physical system. Let us begin with some intuitive ideas that should clarify the meaning of this statement. Consider a region B of space. No condition is imposed on the topology or geometry of the region B, but for the sake of simplicity one can begin by identifying B with a sphere. Then, let V be the volume of that sphere. Suppose that we introduce in the interior of B an arbitrary physical system so that the region exterior to B is empty. Now consider the space $\mathcal{H}$ of states that describe the physical system inside B. One can then ask the question about the dimensionality of the state space $\mathcal{H}$. Obviously, the answer will depend on the physical characteristics of the system. Suppose that it consists of a lattice of spins with spacing d. If the lattice fills entirely the region B, then the maximum number of spins contained in it is V/d^3. Since each spin can be in two different states, the total number of orthogonal states is $2^{V/d^3}$ and this is also the dimensionality of the state space $\mathcal{H}$. This number also determines the maximum entropy S_{max} of the system which is defined as the logarithm of the total number of states, $S_{max} = \ln N_{states} = \ln \dim(\mathcal{H})$. So, in this example we have that $S_{max} = (V \ln 2)/d^3$. This counting process and its relation to the maximum entropy of a system is what holography intents to do in a general setting. In our example, no further information can be extracted because we are dealing with a system with a finite number of degrees of freedom and, consequently, the dimensionality of $\mathcal{H}$ (Hilbert space) is finite.

The interesting cases are those in which the number of degrees of freedom is infinite as in field theory. To handle this case one has to determine the entropy density s as a function of the energy density ρ of the field. Then $S_{max} = s(\rho_{max})V$ and the total number of states is $N_{states} = \exp(S_{max})$. Now suppose that in the region B we have a set of fields sources including gravity. Let A be the area of the boundary ∂B. The maximum mass of the system contained within ∂B cannot exceed the mass of black hole of horizon area A. This is a crucial point. According to our theoretical understanding of field theories, we do not know of any physical system that, being localized within a certain region, could possess a mass greater than that of black hole whose horizon area coincides with the area of the boundary of that region. In other words, black holes are the most massive objects in nature. Now, we know from the second law of black hole thermodynamics that the maximum entropy of a black hole of area A is $S_{max} = A/4$ (we use Planck units with $c = G = \hbar = 1$). Consequently, the total number of states has the bound $N_{states} = \exp(A/4)$. This is the statement of the holographic principle in this particular case. This simple relationship has deep implications. It relates a quantity in the bulk (N_{states}) with a different quantity on the boundary (A). On the other hand, it predicts an upper bound for the dimensionality of the state space $\mathcal{H}$. Notice that we did not impose any conditions on the fields and matter distribution which fills B. That is, the fields can be classical or quantum. Both cases have been analyzed in the literature.

In the above description we did not specify the region B and we have freely "inserted" fields into it. However, it is well known that especially relativistic fields affect the topology and geometry of the space where they live. Therefore, one of the main challenges when trying to develop specific examples is to define the region B in accordance with the existing fields. In the case of cosmological models, this has been done for the Friedmann–Robertson–Walker (FRW) spacetime [5] and some classical generalizations [6]. It has been found that the dynamics of the FRW spacetime "in the bulk" is governed by an entropy relationship "on the boundary". This result has been generalized to include different types of quantum corrections [7]. For the case where the fields in B are quantum, the most studied example is that of string theory in $B = AdS_5 \otimes S_5$ which turned out to be completely equivalent to a Super–Yang–Mills theory on the boundary of the AdS space $\partial B = \partial AdS$ [8].

We now turn back to the field–to–particle transition problem. First of all, let us mention that this problem is within the conceptual idea of the holographic principle. The internal properties of the field (in the bulk) should become represented by external properties of the particle (on the boundary). Let φ_0 be a specific field configuration, which is solution of a classical field theory, that describes the spin of a particle. The total number of states is $N_{states} = 2$. If we now fill the region where the particle lives with a black hole, then, according with the above discussion, the "area" of the particle is $A = 4\ln 2$. Up to here, nothing especial seems to happen. However, when we try to analyze the dynamics of this system, we find the problem of the zero–modes [9]. The fluctuations of the field $\varphi_0 + \delta\varphi_0$ lead to unstable configurations for the elementary particle. This contradicts our daily experience since elementary particles are stable with respect to infinitesimal perturbations. This problem is due to the fact that fluctuations occur in the field, which has an infinite number of degrees of freedom, and affect the particle, a system with a finite number of degrees of freedom. To handle this problem properly and in accordance to the conceptual idea of the holographic principle, one has to follow the following steps: (i) Select the field φ, together with its underlying theory, and a specific field configuration φ_0 that describes the spin of the particle; (ii) Define the region B in accordance with the geometrical properties of φ_0; (iii) Find the theory on ∂B and the configuration $\tilde{\varphi}_0$ which are the counterparts of the theory in B and the field φ_0, respectively; (iv) Perform a perturbation around the new specific configuration $\tilde{\varphi}_0 + \delta\tilde{\varphi}_0$. If the entire procedure works correctly, it could be expected that the configuration "on the boundary", $\tilde{\varphi}_0$, is stable. At the moment, all this is just a speculation. However, we will see in Section III that in the framework of Einstein's gravity one can find a large class of gravitational configurations that show a spin–like behavior.

We think that one of the main obstacles in performing the above procedure is that no exact mathematical tools are known for "going from the bulk to the boundary". In the next Section, we will review a mathematical construction that could help to a better understanding of the difficulties.

3. Nonlinear sigma models

Nonlinear sigma models (NLSM) are an important theoretical laboratory in the framework of field theory, particularly those defined on Riemannian symmetric spaces, which are the integrability condition of the classical theory [11] (they admit an infinite number of conservation laws and are examples of completely integrable field theories). Nonlinear sigma models are, in several ways, closely related to Yang–Mills theories and they have points of resemblance with QCD [12] (in $1+1$ spacetime, some of these models have asymptotic freedom [13] and instantons [14]). The Einstein equations for some gravitational fields, for instance axisymmetric fields, Einstein–Rosen gravitational waves, T^3 and $S^1 \times S^2$ Gowdy cosmological models, etc., are related to the equations for nonlinear models in $1+1$ spacetime [15]. NLSM also appear in string theory and the similarities between the gravitational and sigma fields [16] are enough to make them of interest as toy models for quantum gravity.

Roughly speaking, a NLSM is a field theory of maps between manifolds with the following properties: (a) the fields are subject to nonlinear constraints and (b) the Lagrangian density and the constraints are invariant under the action of a global symmetry (Lie) group G. More precisely [10, 11], the classical field configurations in such a model are maps $\phi: B \to M$, where B is a given base space and M is a given target space. The description "nonlinear" is reserved to those models where the physical fields for all points $p \in B$ take values in a (Riemannian) manifold M which is not a linear space. This restriction is to guarantee positivity of the energy in the corresponding NLSM. Since any Riemannian manifold M can be isometrically embedded into a Euclidean vector space E, the Lagrangian density of the model (rewritten in terms of E-valued fields) has to be supplemented by the constraints expressing the fact that the E-valued fields must be restricted to lie on the embedded submanifold M.

In most of these models the global invariance group G acts transitively on M; here we shall assume that this is indeed the case (thus, M is a Riemannian homogeneous space for G). If H is the stability group of a point $m \in M$, then M can be identified with the space of left cosets G/H (i.e., $M = \{gH\}$, $g \in G$).

General methods exist for constructing Lagrangians for these theories, the idea is simply to represent the field configurations of the model not by maps ϕ from B to M but by maps g from B to G, with $\phi(x) = g(x)H$. Now, the Lagrangian density $\mathcal{L}$ in any nonlinear model is a function of g and $\partial_\mu g$, $\mathcal{L} = \mathcal{L}(g, \partial_\mu g)$, that is invariant under the gauge transformation

$$g(x) \mapsto g(x)h(x), \quad h(x) \in H, \tag{1}$$

in such a way that the gauge invariant fields have values in G/H and the Lagrangian density can be regarded as a function of fields with values in G/H.

The construction of $\mathcal{L}$ proceeds as follows. Let G be a faithful representation of a compact semisimple Lie group G (global symmetry group) and let $\{b(\rho)\}$ be a basis for the Lie algebra $\mathcal{G}$ of G with the following property: $\text{Tr}(b(\rho)b(\sigma)) = \delta_{\rho,\sigma}$, where $\rho, \sigma = \{1, ..., [G] := \dim G\}$. For $\alpha \leq [H] := \dim H$, the generators $b(\alpha)$ are taken to span the Lie algebra $\mathcal{H}$ of H and we denote them by $t(\alpha)$. The remaining generators are called $s(i)$ with $[H] + 1 \leq i \leq [G]$. Thus the commutation relations are

$$[t(\alpha), t(\beta)] = i f_{\alpha\beta\gamma} t(\gamma) , \tag{2}$$

$$[t(\alpha), s(i)] = i \bar{f}_{\alpha ij} s(j) , \tag{3}$$

$$[s(i), s(j)] = i(\bar{f}_{\alpha ij} t(\alpha) + f_{kij} s(k)) . \tag{4}$$

Let ω be the one–form defined on G with components $\omega_\mu(g) = g^{-1}\partial_\mu g$ which under a gauge transformation of the form (1) transforms as

$$\omega_\mu(gh) = h^{-1}\omega_\mu(g)h + h^{-1}\partial_\mu h . \tag{5}$$

It is not difficult to see that one can write ω_μ as a sum of two terms (actually, projections of ω_μ into the Lie algebra $\mathcal{H}$ and its orthogonal complement):

$$\omega_\mu = A_\mu + B_\mu , \tag{6}$$

where $A_\mu(g) = t(\alpha)\text{Tr}(t(\alpha)\omega_\mu(g))$ and $B_\mu(g) = s(i)\text{Tr}(s(i)\omega_\mu(g))$. Accordingly, under a gauge transformation these components behave as

$$A_\mu(gh) = h^{-1}A_\mu(g)h + h^{-1}\partial_\mu h , \quad B_\mu(gh) = h^{-1}B_\mu(g)h . \tag{7}$$

It is worth noticing that we have the structure of a principal fibre bundle $(\mathbf{E}, \mathbf{B}, \mathbf{F}, \mathbf{G}, \Pi)$ with total space $\mathbf{E} = G$, base space $\mathbf{B} = G/H$, fiber $\mathbf{F} \simeq H$, structure group $\mathbf{G} = H$ and projector $\Pi : G \to G/H$, $g \mapsto [g]$. By construction all fields lie in G and the physical fields lie in the base space G/H. Due to this structure we have then that the one–form A with components A_μ transforms like a gauge potential (for the gauge group H) and therefore it acts as a connection one–form. Thus, from A we can construct the curvature two-form F which under (1) transforms as $F' = h^{-1}Fh$.

The structure of the one-form ω allows us to construct different quantities satisfying the above given requirements to be Lagrangian densities for a NLSM. In particular,

$$\mathcal{L} = \sqrt{-g}g^{\mu\nu}\text{Tr}(B_\mu B_\nu) , \tag{8}$$

and
$$\mathcal{L} = \sqrt{-g}g^{\mu\tau}g^{\nu\rho}\text{Tr}(F_{\mu\nu}F_{\tau\rho}) \,, \tag{9}$$

where $g^{\mu\nu}$ are the components of the given metric tensor on (the spacetime, for example) B, in (local) coordinates x^μ, and g its determinant.

If Φ is a gauge invariant field, then we can express it as follows

$$\Phi = g\left(\sum_\alpha t(\alpha)\right)g^{-1} \,. \tag{10}$$

Since the r.h.s of (10) belongs to $\mathcal{G}$, then

$$g\left(\sum_\alpha t(\alpha)\right)g^{-1} = \sum_\rho \phi_\rho b(\rho) \,, \tag{11}$$

where the fields ϕ_ρ are also physical fields. From Eq.(11) and the normalization property for the generators of $\mathcal{G}$, one can show that the following relationship must hold

$$\sum_\alpha \text{Tr}(t(\alpha)t(\alpha)) = \sum_\rho \phi_\rho^2 \text{Tr}(b(\rho)b(\rho)) \,. \tag{12}$$

It follows from Eq.(12) that the fields ϕ_ρ are subject to a nonlinear constraint and there are $[G]-1$ independent physical fields. The manifold M is defined by Eq.(12) and, due to the nonlinear character of this constraint, is not a vector space.

An interesting case of a NLSM is the so-called harmonic map. Considering x^μ as the local coordinates in B, the fields ϕ_b define the original map $\phi : B \to M$ which is called harmonic if the corresponding Lagrangian (8) or (9) satisfies a minimum action principle. This condition leads to a set of (partial) differential equations in M that, in general, can be made to be related to the field equations of the fields in B. In particular, when the field $g_{\mu\nu}$ in B is taken to satisfy Einstein's vacuum equations with two Killing vector fields, the fields ϕ_b in M turn out to satisfy a geodesic equation. Thus, a gravitational field (with the appropriate symmetry) in B is represented by a geodesic in M. This also corresponds to a reduction of the number of degrees of freedom, at least at the level of differential equations, and, accordingly, to a "projection" of a theory in the bulk B to an equivalent theory on the boundary M.

In the context of holography we can proceed as follows. Suppose we have a field living in B and we want to "project" it into its boundary ∂B. If we intend to apply a harmonic map for this projection, we need to look at the case in which the target space M is the boundary of the base space B, i.e., $\phi : B \to \partial B$. This is allowed in the construction explained above. The next step is the identification of the target space ∂B with the space of left cosets G/H and the construction of the corresponding Lagrangian density. The details of this construction will vary, depending on the original field and on B itself. In general, one could expect some arbitrariness in the choice of the space of left cosets since for a given boundary space ∂B the choice of the group G that acts transitively on it is not unique. Each selection would lead, in general, to a different Lagrangian density and hence to a different theory on the boundary. It is not clear under which criteria the selection of the "right" theory should be done. One possibility could be to impose some relationships between the field equations in B and the corresponding equations in ∂B.

It is interesting to note that in general a harmonic map does not impose any connection between the base and target spaces. This allows us to analyze specific examples of the holographic principle in which the boundary is not really a boundary in the topological sense. For instance, in the case of string theory in $AdS_5 \otimes S_5$ one might think that its boundary is a (8+1)–dimensional space. However, its boundary turns out to be a (3+1) space that coincides with the boundary of the AdS spacetime. All these kinds of possibilities are allowed in the context of harmonic maps.

The intuitive ideas described above about the implementation of the holographic principle and nonlinear sigma models in the context of the field–to–particle transition problem are all based upon the assumption that there are field configurations that emulate physical properties of elementary particles. In the next Section we will show that in fact such configurations exist at the level of classical fields.

4. The field–to–particle transition problem in gravity

An interesting possibility from the point of view of gravity is to think of an elementary particle as a specific gravitational configuration. The conceptual idea was first proposed by Wheeler [17], who introduced the term "geon" as an abbreviation for "gravitational–electromagnetic entity", an electromagnetic field configuration which would keep together by its own gravitational interaction having the approximate properties of a particle. In a general sense, a geon is considered today as a special gravitational configuration with a nontrivial topology.

A topological geon as a gravitational field configuration is easy to construct; difficulties arise when we try to obtain a topological geon which reproduces the behavior of a particle. To this end, we could use the great richness of the field of exact solutions to Einstein-Maxwell equations; however, the problem is that there is not a recipe to construct such a geon and, of course, it is not appropriate to apply the method of trial and error with all the possible spacetimes. The real task is to find generic conditions which should be satisfied by spacetimes which are supposed to represent a particle, and then to restrict the analysis only to the subset of solutions to Einstein–Maxwell equations which satisfy those conditions.

One of the first differences that comes to mind between a particle and a gravitational configuration is that the intrinsic properties of a particle, such as mass, charge, and spin are quantized, wile the parameters involved in a gravitational configuration can take values in a continuum. In view of this fact, the consequent question is, how can a configuration with continuous parameters account for an entity with discrete parameters? This is precisely the question we have addressed in a previous work [18], and we will briefly review our arguments here.

There are different approaches to answer this question. Here, our proposal is to extend Dirac's argument about the quantization of the electric charge [19] to the case of gravitational configurations. Remember that the quantum phase acquired by a charged particle describing a closed path in the presence of a magnetic field can be computed by means of the expression

$$\Phi = e^{iq \int_s B}, \qquad (13)$$

where B is the magnetic field, q the charge of the particle travelling along a closed loop, and s a surface having the path of the particle as its boundary. It is possible to use different surfaces in Eq.(13) for the same path, and as long as the surfaces are homotopic, the resulting phases will be identically the same. Dirac used the fact that the form of the field of a magnetic monopole allows the construction of non homotopic surfaces with the same boundary, and then, by requiring that the phases computed using two such surfaces be equal, he got to quantize the electric charge.

The key point to notice is that the integral in Eq.(13) is in fact the integral of the electromagnetic tensor $F_{\alpha\beta}$ over a spatial surface. Furthermore, from the geometrical point of view of field theory, the electromagnetic tensor is just the curvature associated to the electromagnetic connection. This gives us a clue about how to extend the argument to gravitational configurations. Consider a pseudo–Riemannian manifold $(M, g_{\mu\nu})$, where $g_{\mu\nu}$ is the underlying metric. Then we can introduce a phase–like object

$$\Phi = e^{\int R}, \qquad (14)$$

where R is the curvature associated with the metric, namely the Riemann tensor, which is an endomorphism valued two–form (see, for instance, [20]). Now we will use this object to repeat Dirac's argument in the case of gravity.

The field-to-particle transition problem 65

The first problem we have is that the components of the Riemann tensor, when understood as an endomorphism valued two–form, are endomorphisms, so they live in the tangent and cotangent spaces to the manifold at the specific point of evaluation. If we perform the integral in (14) just as it appears, we will be adding objects which live in different spaces and this summation will not be justified. What we need is to perform a parallel translation of the endomorphism from the point of evaluation to a specific point where we will say that the integral is based. So instead of Eq.(14) we will use

$$\Phi = e^{\int H^{-1}RH}, \tag{15}$$

where H is the holonomy resulting from the parallel transport that depends explicitly on the point of evaluation.

Of course there are infinite different paths along which the parallel transport can be done. Therefore, the specific form of H will have to be fixed by using physical arguments [21]. For the moment the important point is that there are some conclusions that can be stated independently of the explicit form of H; the only thing we will require is that it preserves some fundamental symmetries of the curvature. Since H is directly related with the metric and its symmetries, this behavior is not unnatural to be expected.

To repeat Dirac's argument we need to use spacetimes which allow the existence of non homotopic surfaces, for instance, spacetimes with curvature singularities. Moreover, we will restrict our study to vacuum solutions to Einstein equations. For the sake of generality, we consider the Petrov classification [22] of the curvature tensor, and analyze each Petrov type separately. Since the components of the Riemann tensor are endomorphisms which map a four dimensional space into itself, they must have four eigenvalues λ_i ($i = 1, 2, 3, 4$), and it can be shown [18] that for all Petrov types these eigenvalues satisfy the relationships

$$\lambda_1 = -\lambda_2 \quad \text{and} \quad \lambda_3 = -\lambda_4. \tag{16}$$

Furthermore, there are ways to guarantee [18, 21] that the property (16) is preserved by the eingenvalues of the endomorphism obtained by the integral in Eq.(15).

In terms of the eigenvalues of the integral in (15) the phase obtained can be expressed as

$$\Phi = T \begin{pmatrix} e^{\lambda_1} & 0 & 0 & 0 \\ 0 & e^{-\lambda_1} & 0 & 0 \\ 0 & 0 & e^{\lambda_3} & 0 \\ 0 & 0 & 0 & e^{-\lambda_3} \end{pmatrix} T^{-1}, \tag{17}$$

where T is a real matrix that describes a coordinate transformation.

Let us now consider spacetimes with metrics that are invariant with respect to the change $\phi \to \phi + \pi$ (in spherical coordinates). This condition is satisfied by a large class of gravitational configurations, for instance, by all the axially symmetric solutions. To perform the calculation of the phase–like object let us consider two non homotopic surfaces with common boundary (this is the case, for example, when a curvature singularity exists between the two surfaces). Using Dirac's argument about the equality of the phases calculated along the two surfaces, we obtain the following conditions for the λ's

$$\lambda_1 = i n_1 \pi \quad \text{and} \quad \lambda_3 = i n_2 \pi, \qquad (18)$$

where n_1 and n_2 are arbitrary integers. Since the λ's are functions of the parameters of the metric, Eqs.(18) can be considered as relationships similar to those obtained by Dirac. The importance of this result is that we have reached a discretization in the continuum of the parameters that determine the gravitational configuration. The explicit form of these conditions depend on the exact expression for the holonomy H. Nevertheless, the result obtained in Eqs.(18) is quite general because it does not depend on any explicit value of the holonomy.

Another interesting result arises when we insert the discrete values (18) in Eq.(17). It can be shown that the only possible phases are either the identity $\Phi = 1_{4 \times 4}$, or an expression different from the identity but whose square, however, becomes the identity $\Phi^2 = 1_{4 \times 4}$. If we understand the travelling of an observer around an object as an active diffeomorphism, this is equivalent to a rotation of the object by 2π. Therefore, saying that the phase acquired by such an observer is restricted as we have mentioned above, is equivalent to saying that some of the particles we are modelling are invariant under 2π rotations and others under 4π rotations. This is a surprising result, because the first case corresponds to the behavior of a boson, and the second one to a fermion, being the only options allowed. At the beginning we were not looking for this prediction, but it arose in a natural way when extending Dirac's argument to gravity. It clearly provides another important known characteristic of elementary particles. If we were to consider strictly the bosonic or fermionic nature of these models, we would need to analyze other important conditions such as the stability of this particle–like behavior under fluctuations of the underlying field, that is, the problem of zero–modes mentioned above. A further crucial condition is that imposed by the spin-statistics theorem, that is, the behavior of the model under the interchange of identical particles according to this perspective.

This has been already studied in the context of geons by using different approaches (see, for instance [23]).

5. Conclusions

In this work we have reviewed some of the main conceptual aspects of the field–to–particle transition problem. The main goal is to find a different approach to the study of physical systems in which fields and matter are involved. In this approach, matter should not be an external entity that enters the theory in an *ad hoc* manner. Instead, matter should be an additional field component that arises as a specific field configuration. The first step in this approach is to explore the possibility of reproducing the physical properties of elementary particles from a field configuration.

We have described some introductory aspects of the holographic principle, and we have shown that it can conceptually be used to understand the intrinsic problems of the field-to-particle transition. In particular, the problem of zero–modes could be investigated by specifying an equivalent theory in a different space such that the fluctuations of the field become described by fluctuations of an equivalent entity with a finite number of degrees of freedom. This would help to handle the divergences that appear in the zero–modes.

As the mathematical tool to formulate correctly the field–to–particle transition and the inherent problem of zero–modes, we propose to use nonlinear sigma models. We have described how harmonic maps allow us to project a theory from a space to a different theory in a different space. Although this procedure is quite arbitrary in general, we see this as an advantage for the formulation of apparently different theories which can then be analyzed under additional restrictions in order to find out their physical equivalence.

As an explicit example for a field–to–particle transition, we have analyzed certain gravitational field configurations by using a phase–like object. It was proven that the physical parameters entering these configurations become discretizated when we demand that the phase–like object be equal on two non homotopic surfaces with a common boundary. Additionally, we saw that these configurations behave under rotations either as bosons or as fermions. No other options are allowed! It is interesting that classical field configurations show the fermionic behavior, a property which is usually associated with quantum systems.

The proposals presented in this work are all very rough and have no deep physical explanations. A more detailed investigation will be necessary in order to formulate them in a more consistent manner from the physical and mathematical points of view. They should be interpreted more as a first attempt to formulate questions which bother the authors. However, we consider that these questions have to situated in the conceptual kernel of most modern field theories.

Acknowledgments

This work has been supported by DGAPA–UNAM, grant No. 112401, and CONACYT–Mexico, grant No. 36581. J. C. was supported by a CONACYT–UNAM (DGEP) Graduate Fellowship. L. P. was supported by a UNAM–DGEP Graduate Fellowship.

References

[1] J. D. Bekenstein, *Lett. Nuovo Cim.* **4** (1972) 737.

[2] S. Hawking, *Commun. Math. Phys.* **43** (1975) 199.

[3] G. t' Hooft, *Dimensional reduction in quantum gravity*, gr–qc/9310026.

[4] L. Susskind, *J.Math.Phys.* **36** (1995) 6377.

[5] W. Fischler and L. Susskind, *Holography and cosmology*, hep–th/9806039.

[6] R. Bousso, JHEP 9906 (1999) 028.

[7] O. Nojiri, S. Odintsov, O. Obregon, H. Quevedo and M. Ryan, *Mod.Phys.Lett.* **A16** (2001) 1181.

[8] J. Maldacena, *Int. J. Theor. Phys.* **38** (1999) 1113.

[9] P. Rajaraman, *Solitons and instantons* (Noth–Holland Press, Amsterdam, 1988).

[10] A.P. Balachandran, A. Stern and G. Trahern, *Phys.Rev.* **D19** (1978) 2416; A.P.Balachandran, G. Marmo, B.S. Skagerstam and A. Stern, *Classical Topology and Quantum States* (World Scientific, 1991).

[11] E. Abdalla, M.C.B. Abdalla and K.D. Rothe, *Non–perturbative methods in 2–dimensional quantum field theory* (World Scientific, Singapore, 1991).

[12] W. Marciano and H. Pagels, *Phys.Rep.* **36C** (1978) 137; F.J. Ynduráin, *Quantum Chromodynamics* (Springer–Verlag, Berlin, Germany, 1983).

[13] A.M. Polyakov, *Phys.Lett.* **59B** (1975) 79.

[14] A.A. Belavin and A.M. Polyakov, *JETP Lett.* **22** (1975) 245; A. D'Adda, M. Lüsher and P. Di Vecchia, *Nucl.Phys.* **B146** (1978) 63.

[15] C. Misner, *Phys.Rev.* **D18** (1978) 4510; M. Hirayama, H. Chia Tze, J. Ishida and T. Kawabe, *Phys.Lett.* **A66** (1978) 352; A.Ashtekar and V.Husain, *Int. J. Mod. Phys* **D7** (1998) 549; J. Cortez, D. Núñez and H. Quevedo, *Int. J. Theo. Phys.* **40** (2001) 251.

[16] B. DeWitt, in: *Geometrical and Algebraic Aspects of Nonlinear Field Theory*, ed. S. De Filippo, M. Marinaro, G. Marmo and G. Vilasi, (Elsevier Science Publishers B.V. North–Holland, Netherlands, 1989).

[17] J. A. Wheeler, *Phys. Rev.* **97** (1955) 511; *Geometrodynamics* (Academic, New York)

[18] L. Patiño and H. Quevedo, *submitted*.

REFERENCES

[19] P.A.M. Dirac, *Proc. Roy. Soc.* **A133** (London, 1931) 60.

[20] J. Baez and J. Muniain, *Gauge Fields, Knots and Gravity*, (World Scientific, Singapore, 1994).

[21] L. Patiño and H. Quevedo, *in preparation*.

[22] D. Kramer, H. Stephani, M. MacCallum and E. Herlt, *Exact Solutions of Einstein's Field Equations*, (Cambridge University Press, Cambridge, UK, 1980).

[23] J. Friedman and R. Sorkin, *Phys. Rev. Lett.* **44** (1980) 1100; *Gen. Rel. Grav.* **14** (1982) 615.

TOWARDS NON–COMMUTATIVE TOPOLOGICAL GAUGE THEORY OF GRAVITY

H. Garcia–Compeán
Departamento de Física, Centro de Investigación y de Estudios Avanzados del IPN
P.O. Box 14-740, 07000, México D.F., México
compean@fis.cinvestav.mx

O. Obregón
Instituto de Física de la Universidad de Guanajuato
P.O. Box E-143, 37150, León Gto., México
octavio@ifug3.ugto.mx

C. Ramirez
Facultad de Ciencias Físico Matemáticas, Universidad Autónoma de Puebla
P.O. Box 1364, 72000, Puebla, México
cramirez@fcfm.buap.mx

M. Sabido
Instituto de Física de la Universidad de Guanajuato
P.O. Box E-143, 37150, León Gto., México
msabido@ifug3.ugto.mx

Abstract The possibility of non–commutative gravity arising in the same manner as Yang–Mills theory is conjectured. Using the Seiberg–Witten map we obtain a non–commutative version of topological gravity, also a self–dual version is constructed, in both cases to order θ^2.

Keywords: non–commutative gravity, general relativity.

1. Introduction

The idea of quantized spacetime, or the non–commutative nature of space–time coordinates, is quite old [1], it has been extensively studied by many authors from a mathematical [2]and physical point of view.

Recently non–commutative gauge theory has attracted much attention, specially in connection with string theory [3]; non–commutative theories arise naturally from string theory describing the low energy excitations of open strings on D–branes on a constant Neveau–Schwarz two–form B field [4].

Basically, Seiberg and Witten have observed the following: ordinary non–commutative gauge fields can be induced by the same $2D$ σ–model, regularized in different ways. They obtain a system of first order differential equations, that relates non–commutative gauge fields deformed by the star Moyal product, with those in commutative space (*Seiberg–Witten map*). It is argued that besides the $U(N)$ gauge group, there seems to be no way of obtaining other gauge groups.

Considering that string theory is sought to be the fundamental theory of everything, and the correct quantum theory of gravity, and taking into account that it includes non commutative Yang Mills theory, we can ask ourselves if non–commutative gravity would arise from it. This is a difficult question, but we can go the other way around, we can start by looking to non commutative extensions of different theories of gravity, and see if any direct connection to string theory exists. To start on fair ground, we should take a road as close as possible to the Yang–Mills construction. Thus, the basic ingredients should be, a Yang Mills type theory of gravity, a suitable gauge group, and the Seiberg–Witten map. In [5], the group $U(N)$ has been considered. However, as shown in [6], gauge theories based on $SO(N)$ can be constructed, by defining subgroups of orthogonal and symplectic subalgebras on non–commutative unitary gauge transformations, or by direct use of the Seiberg–Witten map [4, 7].

The object of this paper is to present a first step to defining non–commutative gravity, in a way inspired on the construction of non–commutative Yang Mills theories based on the Seiberg–Witten map. We will consider a non dynamical sector of gravity, given by topological gravity.

The paper is organized as follows. In section 2 a quick review of the Seiberg–Witten map is given. In section 3 the main features of non–dynamical or topological gravity are given, as well as for self dual topological gravity, for the $SO(4,1)$ gauge group. In section 4 we present non–commutative topological gravity up to order θ^2. Finally, section 5 contains conclusions.

2. Non commutative gauge symmetry and the Seiberg–Witten map

We start this section with a few conventions and properties of non–commutative spaces.

For non–commutative spaces we can generalize the usual quantum mechanical commutation relations, to include non–commutativity of the coordinates

$$[x^i, x^j] = i\theta^{ij}, \quad (1)$$

where x^i are coordinates of $\mathbf{R}^n$ and θ^{ij} are real. This spoils Lorentz invariance, but considering that this perturbation arises at very short distances, we should recover Lorentz invariant theories at the limit $\theta \to 0$. Given this algebra, we can try to relate the functions or fields in $\mathbf{R}^n$ with those in non commutative space. In order to consider fields depending on these variables, one redefines the function multiplication law by the Moyal product,

$$f(x) \star g(x) \equiv \left[e^{\frac{i}{2}\theta_{ij}\partial_{\varepsilon i}\partial_{\eta j}} f(x+\varepsilon)g(x+\eta) \right]_{\varepsilon=\eta=0}. \quad (2)$$

It is associative

$$[(f \star g) \star h] = [f \star (g \star h)], \quad (3)$$

and the complex conjugation is given by

$$(f \star g)^* = g^* \star f^*, \quad (4)$$

Due to the fact that we will be working with non–Abelian groups, we must include also matrix multiplication, so $*$ will be used as the external product of matrix multiplication with $\star$ product. In this case hermitian conjugation is given by

$$(f * g)^\dagger = g^\dagger * f^\dagger, \quad (5)$$

Inside integrals it has the cyclical property (for more extensive review on Moyal product properties see [8])

$$Tr \int (f_1 * f_2 * f_3 * \cdots * f_n) = Tr \int (f_n * f_1 * f_2 * f_3 * \cdots * f_{n-1}), \quad (6)$$

For ordinary Yang–Mills theory we write the gauge transformation of a gauge field A_i and its field strength $F_{ij} = \partial_i A_j - \partial_j A_i - i[A_i, A_j]$ as

$$\begin{aligned} \delta_\lambda A_i &= \partial_i \lambda + i[\lambda, A_i] \\ \delta_\lambda F_{ij} &= i[\lambda, F_{ij}]. \end{aligned} \quad (7)$$

For non commutative Yang–Mills, the same formulas can be used, but we replace matrix multiplication by star product $*$. Thus, the gauge transformations for the non commutative theory can be written as

$$\widehat{\delta_\lambda \widehat{A}_i} = \partial_i \widehat{\lambda} + i\left(\widehat{\lambda} * \widehat{A}_i - \widehat{A}_i * \widehat{\lambda}\right)$$
$$\widehat{\delta_\lambda \widehat{F}_{ij}} = i\left(\widehat{\lambda} * \widehat{F}_{ij} - \widehat{F}_{ij} * \widehat{\lambda}\right), \quad (8)$$

where $\widehat{F}_{ij} = \partial_i \widehat{A}_j - \partial_j \widehat{A}i - i\left(\widehat{A}_i * \widehat{A}_j - \widehat{A}_j * \widehat{A}_i\right)$

From which the following relation, the Seiberg–Witten map, between commutative gauge fields A_i and non-commutative ones $\widehat{A}_i$,

$$\widehat{A}(A) + \widehat{\delta_{\widehat{\lambda}}} \widehat{A}(A) = \widehat{A}(A + \delta_\lambda A) \quad (9)$$

which, expanded in powers of θ and using (2), has the solution

$$\widehat{A}_i(A) = A_i + A'_i(A) = A_i - \frac{1}{4}\theta^{kl}\{A_k, \partial_l A_i + F_{li}\} + \mathcal{O}(\theta^2)$$
$$\widehat{\lambda}(\lambda, A) = \lambda + \lambda'(\lambda, A) = \lambda + \frac{1}{4}\theta^{ij}\{\partial_i\lambda, A_j\} + \mathcal{O}(\theta^2)$$
$$\widehat{F}_{ij} = F_{ij} + \frac{1}{4}\theta^{kl}\left(2\{F_{ik}, F_{jl}\} - \{A_k, D_l F_{ij} + \partial_l F_{ij}\}\right)$$
$$+ \mathcal{O}(\theta^2). \quad (10)$$

We finally write the differential equations that generate the map to all finite orders in θ.

$$\delta \widehat{A}_i(\theta) = -\frac{1}{4}\delta\theta^{kl}\left\{\widehat{A}_k, \partial_l \widehat{A}_i + \widehat{F}_{li}\right\}$$
$$\delta\widehat{\lambda}(\lambda, A) = \frac{1}{4}\theta^{kl}\left\{\partial_k\widehat{\lambda}, \widehat{A}_l\right\}$$
$$\delta \widehat{F}_{ij} = \frac{1}{4}\theta^{kl}\left(2\left\{\widehat{F}_{ik}, \widehat{F}_{jl}\right\} - \left\{\widehat{A}_k, \widehat{D}_l\widehat{F}_{ij} + \partial_l\widehat{F}_{ij}\right\}\right), \quad (11)$$

where the bracket represents $\{A, B\} = A * B + B * A$, which as already explained, includes matrix multiplication and Moyal product. To obtain the solution to different orders in θ, first the ordinary fields are inserted, the obtained result will be for order θ this result is then inserted and the θ^2 order turns out. The $U(1)$ case can be explicitly solved as shown in [2, 9].

3. Topological gravity

In this section we shortly review topological gravity, with an action of the general form

$$I_{TOP} = \frac{\Theta_G^E}{2\pi} Tr \int_X R \wedge \widetilde{R} + \frac{\Theta_G^P}{2\pi} Tr \int_X R \wedge R, \quad (12)$$

where X is a four dimensional closed Lorentzian manifold. Here, the coefficients are the gravitational analogues of the Θ vacuum in QCD[10, 11]. Other actions including gravitational Θ-terms have been analyzed [12]. The first term in (12) is the Euler term, and the second is the Pontrjagin one. Another motivation for using this action in connection to gravity is that it arises naturally from MacDowell–Mansouri (MM) type action based on a self dual spin connection and the gauge group $SO(3,2)$ [13]. Similarly, for 2+1 Chern–Simons gravity, this type of construction permits to define a Θ-term [14]. Keeping this philosophy in mind, (12) can be rewritten in terms of the self–dual and anti–self–dual parts of the Riemann tensor as follows:

$$I_{TOP} = Tr \int_X \left[\tau^+ \left(R^+ \wedge R^+ \right) - \tau^- \left(R^- \wedge R^- \right) \right], \quad (13)$$

with $\tau^{\pm} = \left(\frac{1}{2\pi} \right) \left(\Theta_G^E \mp \Theta_G^P \right)$. In local coordinates on X, this action is written as

$$I_{TOP} = \int_X dx^4 \epsilon^{\mu\nu\rho\sigma} \left[\tau^+ \left(R_{\mu\nu}^{+\ ab} R_{\rho\sigma ab}^+ \right) - \tau^- \left(R_{\mu\nu}^{-\ ab} R_{\rho\sigma ab}^- \right) \right], \quad (14)$$

where $R_{\mu\nu}^{\pm\ ab} = \frac{1}{2} \left(R_{\mu\nu}^{\ ab} \mp \frac{i}{2} \epsilon^{ab}_{\ cd} R_{\mu\nu}^{\ cd} \right)$ and satisfies

$$\epsilon^{ab}_{\ cd} R_{\mu\nu}^{\pm\ cd} = \pm 2i R_{\mu\nu}^{\pm\ ab}, \quad (15)$$

The self–dual and anti–self–dual Riemann tensors can be written as well in terms of the self dual (anti–self–dual) components of the spin connection $\omega_{\mu}^{\pm\ ab} = \frac{1}{2} \left(\omega_{\mu}^{ab} + \frac{i}{2} \epsilon^{ab}_{\ cd} \omega_{\mu}^{cd} \right)$ as

$$R_{\mu\nu}^{\pm\ cd} = \partial_\mu \omega_\nu^{\pm\ ab} - \partial_\nu \omega_\mu^{\pm\ ab} + \frac{1}{2} f^{[ab]}_{[cd][ef]} \omega_\mu^{\pm\ cd} \omega_\nu^{\pm\ ef}, \quad (16)$$

with

$$f^{[ab]}_{[cd][ef]} = \frac{1}{2} \left[\eta_{ce} \delta_d^a \delta_b^b - \eta_{cf} \delta_d^a \delta_e^b + \eta_{df} \delta_c^a \delta_e^b - \eta_{de} \delta_c^a \delta_f^b \right] - (a \longleftrightarrow b). \quad (17)$$

4. Non–commutative topological gravity

In this section we present a non–commutative version of topological gravity. To start with, we expand the gauge fields in powers of θ

$$\widehat{\omega}_\mu^{AB} = \omega_\mu^{AB} - i\theta_{\mu\nu\rho}^{AB} + \cdots. \quad (18)$$

The basic assumption is that the deformed fields are related to the undeformed ones by the Seiberg–Witten map (10). Thus, we only need to insert (18) in (10), to obtain the corrections to the gauge fields and field strength to any order in θ. To determine the non commutative action of topological gravity, we first write down

$$\widehat{R}^{AB}_{\mu\nu} = R^{AB}_{\mu\nu} + i\theta^{op} R^{AB}_{\mu\nu op} + \theta^{op}\theta^{lm} R^{AB}_{\mu\nu oplm}, \tag{19}$$

where

$$R^{AB}_{\mu\nu op} = \partial_\mu \omega^{AB}_{\nu op} + \omega^{AC}_\mu \omega_{\nu op C}{}^B + \omega^{AC}_{\mu op}\omega_{\nu C}{}^B - \frac{1}{2}\partial_o \omega^{AC}_\mu \partial_p \omega_{\nu C}{}^B$$
$$- (\mu \leftrightarrow \nu) \tag{20}$$

$$R^{AB}_{\mu\nu oplm} = \partial_\mu \omega^{AB}_{\nu oplm} + \omega^{AC}_\mu \omega_{\nu oplm C}{}^B + \omega^{AC}_{\mu oplm}\omega_{\nu C}{}^B - \omega^{AC}_{\mu op}\omega_{\nu lm C}{}^B$$
$$- \frac{1}{4}\partial_o\partial_l \omega^{AC}_\mu \partial_p\partial_m \omega_{\nu C}{}^B - (\mu \leftrightarrow \nu), \tag{21}$$

from this we can obtain the non commutative version of the Pontrjagin term

$$\frac{\Theta^P_G}{2\pi} Tr \int_{M_4} \widehat{R} \wedge *\widehat{R} = \frac{\Theta^P_G}{2\pi} \int_{M_4} d^4x \epsilon^{\mu\nu\rho\sigma} \widehat{R}^{AB}_{\mu\nu} * \widehat{R}_{\rho\sigma AB}, \tag{22}$$

using (19), and the properties of the Moyal product we get

$$\frac{\Theta^P_G}{2\pi} \int_{M_4} d^4x \epsilon^{\mu\nu\rho\sigma} \widehat{R}^{AB}_{\mu\nu} * \widehat{R}_{\rho\sigma AB} = \frac{\Theta^P_G}{2\pi} \int_{M_4} d^4x \epsilon^{\mu\nu\rho\sigma} \left[R^{AB}_{\mu\nu} R_{\rho\sigma AB} \right.$$
$$\left. + \theta^{op}\theta^{lm} \left(2 R^{AB}_{\mu\nu} R_{\mu\nu oplm AB} - R^{AB}_{\mu\nu op} R_{\rho\sigma lm AB} \right) \right], \tag{23}$$

which as can be see corresponds to the Pontrjagin term plus a correction to order θ^2. To construct the full topological theory we use the arguments presented in section III. We start by defining a self–dual (anti–self–dual) spin connection as

$$\widehat{\omega}^{+AB}_\mu = \widehat{\omega}^{AB}_\mu \left(\omega^{+AB}_\mu\right), \qquad \widehat{\omega}^{-AB}_\mu = \widehat{\omega}^{AB}_\mu \left(\omega^{-AB}_\mu\right), \tag{24}$$

where ω^{+AB}_μ and ω^{-AB}_μ are the commutative (anti)self dual spin connections. Thus we can define $R^{\pm AB}_{\mu\nu}$

$$\widehat{R}^{\pm AB}_{\mu\nu} = \partial_\mu \widehat{\omega}^{\pm AB}_\nu - \partial_\nu \widehat{\omega}^{\pm AB}_\mu + \widehat{\omega}^{\pm AC}_\mu \widehat{\omega}^{\pm B}_{\nu C} - \widehat{\omega}^{\pm AC}_\nu \widehat{\omega}^{\pm B}_{\mu C}, \tag{25}$$

And the non commutative action is given by,

$$I_{TOP} = Tr \int_X \left[\tau^+ \left(\widehat{R}^+ \wedge *\widehat{R}^+\right) - \tau^- \left(\widehat{R}^- \wedge *\widehat{R}^-\right) \right]$$
$$= \int_X dx^4 \epsilon^{\mu\nu\rho\sigma} \left[\tau^+ \left(\widehat{R}^{+AB}_{\mu\nu} * \widehat{R}^+_{\rho\sigma ab}\right) - \tau^- \left(\widehat{R}^{-AB}_{\mu\nu} * \widehat{R}^-_{\rho\sigma ab}\right) \right]. \tag{26}$$

Taking into account the Seiberg–Witten map and the properties of the Moyal product, we get

$$\int_{M_4} d^4x \epsilon^{\mu\nu\rho\sigma} \tau^+ \left[R^{+AB}_{\mu\nu} R^{+}_{\rho\sigma AB} + \theta^{op}\theta^{lm} \left(2R^{+AB}_{\mu\nu} R^{+}_{\mu\nu oplmAB} \right. \right.$$
$$\left. - R^{+AB}_{\mu\nu op} R^{+}_{\rho\sigma lmAB} \right) \right] - \tau^- \left[R^{-AB}_{\mu\nu} R^{-}_{\rho\sigma AB} + \theta^{op}\theta^{lm} \right.$$
$$\times \left. \left(2R^{-AB}_{\mu\nu} R^{-}_{\mu\nu oplmAB} - R^{-AB}_{\mu\nu op} R^{-}_{\rho\sigma lmAB} \right) \right]. \tag{27}$$

The zeroth order terms in this action correspond to standard topological gravity, and the extra terms give the corrections to order theta θ^2.

5. Conclusions

In this work, by means of the Seiberg–Witten map applied to the Yang–Mills like formulation of topological self dual gravity, a formulation of non–commutative (topological) gravity is given. It is shown that the first order term in the θ–expansion vanishes, and the second order term is computed. It would be interesting to have a similar formulation for dynamical gravity.

Acknowledgments

This work was supported in part by CONACyT Mexico Grant Nos. 28454E and 33951E.

References

[1] H. Snyder, *Phys. Rev.* **71** (1947) 38.

[2] A. Connes, *Non commutative geometry*, Academic Press (1994).

[3] A. Connes, M. R. Douglas, and A. Schwarz, *JHEP* (1998) 9802:003.

[4] N. Seiberg and E. Witten, *JHEP* (1999) 9909:032.

[5] Ali H. Chamseddine, *Commun. Math. Phys.* **218** (2001) 283, *Phys. Lett.* **B504** (2001) 33.

[6] L. Bonora, M. Schnabl, M. Sheikh–Jabbari and A. Tomasiello, *Nucl. Phys.* **B589** (2000) 461.

[7] J. Wess, *Commun. Math. Phys.* **219** (2001) 247. B. Jurco, S. Schraml, P. Schupp, J. Wess, *Eur. Phys. J.* **C17** (2000) 521.

[8] A. Micu and M. Sheikh–Jabbari, *JHEP* (2001) 0101:025.

[9] M. Kreuzer and J. Zhou. *JHEP* (2000) 0001:011.

[10] S. Deser, M. J. Duff, and C. J. Isham, *Phys. Lett.* **B93** (1980) 419.

[11] A. Ashtekar, A. P. Balachandran, and So Jo, *Int. J. Mod. Phys.* **A4** (1989) 1493.

[12] L. Smolin, *J. Math. Phys.* **36** (1995) 6417.
[13] J. A. Nieto, O. Obregón, and J. Socorro, *Phys. Rev.* **D50** (1994) R3583.
[14] H. García-Compean, O. Obregón, C. Ramírez, and M. Sabido, *Phys. Rev.* **D61** (2000) 085022.

II

COSMOLOGY AND BLACK HOLES

IMPROVING THE "NO–HAIR" THEOREM FOR THE PROCA FIELD

Eloy Ayón–Beato
Departamento de Física, Centro de Investigación y Estudios Avanzados del IPN
Apdo. Postal 14–740, 07000 México D.F., MEXICO
ayon@fis.cinvestav.mx

Abstract This paper reconsider the problem of a Proca field in the exterior of a static black hole. The original Bekenstein's demonstration on the vanishing of this field, based on an integral identity, is improved by using more natural arguments at the event horizon. In particular, the use of the so–called *standard* integration measure in the horizon is fully justified. Accordingly, the horizon contribution to the Bekenstein integral identity is more involved and its vanishing can be only established using the related Einstein equations. With the new reasoning the "no–hair" theorem for the Proca field now rest on better founded grounds.

Keywords: Black holes, "no–hair" theorems, horizon measure, Proca field.

1. Introduction

Massive fields are forbidden in the exterior of any stationary black hole. Such statement rest in the fact that the strongest version of the "no–hair" conjecture establishes that a stationary black hole is uniquely determined by global charges (conserved Gauss–like surface integrals at spatial infinity) [1], but massive fields exponentially fall–off at infinity and made no contributions to the corresponding surface integrals; there are no global charges associated with them. This kind of unobserved–from–infinity configuration is what is called "hair" in the literature. The first results explicitly showing the nonexistence of massive "hair" were due to Bekenstein who studied massive scalar fields, Proca–massive spin–1 fields, and massive spin–2 fields [2]. This paper intents to improve the original demonstration of Bekenstein for Proca fields in the presence of static black holes. Such demonstration depends on an integral identity only built from matter field equations; no Einstein equation is used. The fundamental changes introduced in the proof are related to the arguments concerning the event horizon, which are usually the more involved. On the one hand, an essentially appropriate integration measure is introduced on the event horizon, which is a degenerate hypersurface. As a consequence the horizon contribution to the cited integral identity is more elaborate. On the other hand, the Bekenstein proofs are usually announced as independent of the particular metric theory of gravity (see [3]); due to they involve only matter field equations. However, we shall show that in order to vanishing properly the horizon contributions to the Bekenstein's identity, it is imperative also the use of Einstein's equations. We would like to point out that the appropriate justification of the "no–hair" conjecture for massive vector fields has been a useful tool for excluding the existence of new black hole configurations from very complicated theories as metric–affine gravity, where a relevant sector of this theory reduces to an effective Einstein–Proca system [4]. Secondly, but not least important, the methods developed here has also been a start point in the exclusion of "hair" for more complex system where the mass terms appears dynamically through spontaneous symmetry breaking [5].

In the following Sec. 2 the Einstein–Proca system is introduced and the consequences of considering this system in the exterior of a static black hole are highlighted. In Sec. 3 the "no–hair" theorem for the Proca field is established using the Bekenstein argument but with a different integration measure in the horizon, and helping us of the Einstein equations in the reasoning. Section 4 is devoted to the relevant conclusions. The final Appendix is dedicated to properly justify the use of the standard integration measure in the horizon mentioned above.

2. Proca fields on static spacetimes

In this section we introduce some fundamental properties of a Proca field lying in the domain of outer communications $\mathcal{I}$ of a static black hole. The Einstein–Proca action describing such interaction is given by

$$S = \int \left(\frac{1}{2\kappa} R - \frac{1}{16\pi} H_{\mu\nu} H^{\mu\nu} - \frac{m^2}{8\pi} B_\mu B^\mu \right) dv, \qquad (1)$$

where R stands for scalar curvature, and $H_{\mu\nu} \equiv 2\nabla_{[\mu} B_{\nu]}$ is the field strength of the Proca field B_μ. From (1) the Einstein and Proca equations are established

$$\frac{4\pi}{\kappa} R_{\mu\nu} = H_\mu{}^\alpha H_{\nu\alpha} + m^2 B_\mu B_\nu - \frac{1}{4} g_{\mu\nu} H_{\alpha\beta} H^{\alpha\beta}, \qquad (2)$$

$$\nabla_\beta H^{\beta\alpha} = m^2 B^\alpha. \qquad (3)$$

In a static black hole the Killing field **k** coincides with the null generator of the event horizon $\mathcal{H}^+$. At the same time this field is timelike and hypersurface orthogonal in all the domain of outer communications $\mathcal{I}$. These properties of the Killing field together with the simply connectedness of $\mathcal{I}$ [6] allow us to choose a global coordinate system (t, x^i), $i = 1, 2, 3$, in all $\mathcal{I}$ [7] such that $\mathbf{k} = \partial/\partial \mathbf{t}$ and

$$\mathbf{g} = -V\mathbf{dt}^2 + \gamma_{ij}\mathbf{dx}^i\mathbf{dx}^j, \tag{4}$$

where V and γ are t-independent, γ is positive definite in all $\mathcal{I}$, and the function V is positive in all $\mathcal{I}$ and vanishes in $\mathcal{H}^+$. From (4) it can be note that staticity is equivalent to the existence of a time–reversal isometry $t \mapsto -t$ in all $\mathcal{I}$.

We shall assume that the Proca field shares the same symmetries of the metric; firstly, that it is stationary $\mathcal{L}_k B = 0$. Secondly, that the staticity of the metric is also extended to the Proca field B^α and its field equations (3) in the sense of requiring they are all invariant under time–reversal transformations (electromagnetic staticity). The condition of time–reversal invariance for Proca equations (3) written in the coordinates of Eq. (4) demands that the components B^t and H^{ti} remain unchanged while B^i and H^{ij} change sign, or the opposite scheme, i.e., B^t and H^{ti} change sign as long as B^i and H^{ij} remain unchanged under time reversal [2]. Therefore, for a time–reversal invariant Proca field the components B^i and H^{ij} must vanish in the first case mentioned above, and the components B^t and H^{ti} vanish in the second one. Hence, time–reversal invariance implies the existence of two separated cases: a purely electric case (I) and a purely magnetic case (II).

3. The "no–Proca–hair" theorem

Now we are ready to proof the "no–hair" theorem for the Proca field and we start by obtaining the corresponding integral identity mentioned in the introduction. Let $\mathcal{V} \subset \mathcal{I}$ be the open region bounded by the spacelike hypersurface Σ, the spacelike hypersurface Σ', and the pertinent portions of the horizon $\mathcal{H}^+$, and the spatial infinity i^o. The spacelike hypersurface Σ' is obtained by shifting each point of Σ a unit parametric value along the integral curves of the Killing field **k**. Multiplying the Proca equations (3) by B_α and integrating by parts over $\mathcal{V}$ using the Gauss law, one obtains

$$\left[\int_{\Sigma'} - \int_{\Sigma} + \int_{\mathcal{H}^+ \cap \overline{\mathcal{V}}} + \int_{i^o \cap \overline{\mathcal{V}}} \right] B_\alpha H^{\beta\alpha} d\Sigma_\beta$$
$$= \int_\mathcal{V} \left(\frac{1}{2} H_{\alpha\beta} H^{\alpha\beta} + m^2 B_\alpha B^\alpha \right) dv. \tag{5}$$

The boundary integral over Σ' cancels out the corresponding one over Σ, since Σ' and Σ are isometric hypersurfaces taken with reversed normals in the Gauss law. The boundary integral over the infinity $i^o \cap \overline{V}$ vanishes by the usual Yukawa fall–off of massive fields asymptotically.

We will show that the integrand of the remaining boundary integral at the portion of the horizon $\mathcal{H}^+ \cap \overline{V}$ also vanishes. To achieve this goal we use the standard measure at the horizon [8],

$$d\Sigma_\beta = 2n_{[\beta} l_{\mu]} l^\mu d\sigma, \qquad (6)$$

where **l** is the null generator of the horizon, **n** is the other future–directed null vector ($n_\mu l^\mu = -1$), orthogonal to the spacelike cross sections of the horizon, and $d\sigma$ is the surface element. We shall justify the use of the standard measure on the horizon in the final Appendix, see Eq. (A.4). By using the quoted measure the horizon integrand can be written as

$$B_\alpha H^{\beta\alpha} d\Sigma_\beta = \left(B_\alpha H^{\beta\alpha} l_\beta + B_\alpha H^{\beta\alpha} n_\beta l_\mu l^\mu \right) d\sigma. \qquad (7)$$

In order to show that the last integrand is vanishing it is sufficient to prove that the quantities inside the parenthesis at the right–hand side of Eq. (7) satisfy the following conditions: $B_\alpha H^{\beta\alpha} l_\beta$ vanishes and $B_\alpha H^{\beta\alpha} n_\beta$ remains bounded at the horizon. The behavior of these quantities at the horizon can be established by studying some invariants constructed from the curvature. Using Einstein equations (2), we obtain, $4\pi R/\kappa = m^2 B_\mu B^\mu$ and

$$\frac{16\pi^2}{\kappa^2} R_{\mu\nu} R^{\mu\nu} = 3H^2 + 4I^2 + \left(H - m^2 B_\mu B^\mu \right)^2 + 2m^2 H_\mu{}^\alpha B^\mu H_{\nu\alpha} B^\nu, \qquad (8)$$

where $H \equiv H_{\alpha\beta} H^{\alpha\beta}/4$, $I \equiv {}^*H_{\alpha\beta} H^{\alpha\beta}/4$, and ${}^*H_{\alpha\beta} = \eta_{\mu\nu\alpha\beta} H^{\mu\nu}/2$ is the usual Hodge dual. Since the horizon is a smooth surface curvature invariants are bounded there, from which it follows first that $B_\mu B^\mu$ is bounded at the horizon. The last term in Eq. (8) is nonnegative in both cases (I) and (II), the remaining terms are also nonnegative, and consequently each one is bounded at the horizon, in particular the invariants H and I. Other invariants can be built from the Ricci curvature (2) by means of **l** and **n**, which are well–defined smooth vector fields on the horizon. The first invariant reads

$$\frac{4\pi}{\kappa} R_{\mu\nu} n^\mu n^\nu = J_\mu J^\mu + m^2 (B_\mu n^\mu)^2 - n_\mu n^\mu H, \qquad (9)$$

where $J^\mu \equiv H^{\mu\nu} n_\nu$. The last term above vanishes because the bounded behavior of the invariant H. Since **J** is orthogonal to the null vector **n** it must be spacelike or null ($J_\mu J^\mu \geq 0$), therefore each one of the remaining terms in the right–hand side of Eq. (9) must be bounded. The next invariant to be considered, which vanishes at the horizon by applying the Raychaudhuri equation to the null generator [9], reads

$$0 = \frac{4\pi}{\kappa} R_{\mu\nu} l^\mu l^\nu = D_\mu D^\mu + m^2 (B_\mu l^\mu)^2 - l_\mu l^\mu H, \quad (10)$$

where $D^\mu \equiv H^{\mu\nu} l_\nu$ is the electric field at the horizon. Once again the bounded behavior of the invariant H can be used to vanishing the last term of relations (10). The vector $\mathbf{D}$ is orthogonal to the null generator $\mathbf{l}$ hence must be spacelike or null ($D_\mu D^\mu \geq 0$). Consequently each term on the right–hand side of Eq. (10) vanishes independently, which implies that $B_\mu l^\mu = 0$ and that $\mathbf{D}$ is proportional to the null generator $\mathbf{l}$ at the horizon, i.e., $\mathbf{D} = -(D_\alpha n^\alpha)\mathbf{l}$. The following relation arise from the last invariant to be studied

$$\frac{4\pi}{\kappa} R_{\mu\nu} l^\mu n^\nu - H = (D_\mu n^\mu)^2 + m^2 (B_\mu n^\mu)(B_\nu l^\nu), \quad (11)$$

where it has been used that $\mathbf{D} = -(D_\alpha n^\alpha)\mathbf{l}$. Since $B_\mu l^\mu = 0$ and $B_\mu n^\mu$ is bounded at the horizon, it follows that the second term on the right–hand side of Eq. (11) vanishes. Therefore, $D_\mu n^\mu$ is bounded at the horizon as consequence of the bounded behavior of the related left–hand side in Eq. (11).

Summarizing, the study of the horizon behavior of all the above invariants leads to the following conclusions: the quantities $D_\mu n^\mu$, $B_\mu n^\mu$, $B_\mu B^\mu$, and $J_\mu J^\mu$ are bounded at the horizon, and the relations $B_\mu l^\mu = 0$, and $\mathbf{D} = -(D_\alpha n^\alpha)\mathbf{l}$ are satisfied in the same region.

Now we are in position to show the fulfillment of the sufficient conditions for the vanishing of the integrand (7) over the horizon, i.e., that $B_\alpha H^{\beta\alpha} l_\beta$ vanishes and $B_\alpha H^{\beta\alpha} n_\beta$ remains bounded at the horizon. Using the definition $D^\mu \equiv H^{\mu\nu} l_\nu$ and that $\mathbf{D} = -(D_\alpha n^\alpha)\mathbf{l}$, we obtain for the first quantity at the horizon

$$B_\alpha H^{\beta\alpha} l_\beta = (D_\mu n^\mu)(B_\nu l^\nu) = 0, \quad (12)$$

where the vanishing follows from the fact that, as we just establish, $D_\mu n^\mu$ is bounded and $B_\nu l^\nu$ vanishes at the horizon. For the second quantity we note that $\mathbf{B}$ and $\mathbf{J}$ are orthogonal to the null vectors $\mathbf{l}$ and $\mathbf{n}$, respectively. Therefore, $\mathbf{B}$ must be spacelike or proportional to $\mathbf{l}$, and $\mathbf{J}$ must be spacelike or proportional to $\mathbf{n}$. Using a null tetrad basis constructed with $\mathbf{l}$, $\mathbf{n}$, and a pair of linearly independent spacelike vectors, spanning the spacelike cross sections of the horizon, the $\mathbf{B}$ and $\mathbf{J}$ vectors can be written as

$$\mathbf{B} = -(B_\alpha n^\alpha)\mathbf{l} + \mathbf{B}^\perp, \quad \mathbf{J} = -(J_\alpha l^\alpha)\mathbf{n} + \mathbf{J}^\perp, \quad (13)$$

where $\mathbf{B}^\perp$ and $\mathbf{J}^\perp$ are the projections, orthogonal to l and n, on the spacelike cross sections of the horizon. Using expressions (13) it is clear that $B_\mu B^\mu = B_\mu^\perp B^{\perp \mu}$ and $J_\mu J^\mu = J_\mu^\perp J^{\perp \mu}$, i.e., the contribution to these bounded magnitudes comes only from the spacelike sector orthogonal to l and n. With the help of Eqs. (13) the other quantity appearing in the integrand (7) can be written as

$$B_\alpha H^{\beta \alpha} n_\beta = -B_\alpha J^\alpha = -(B_\alpha n^\alpha)(D_\beta n^\beta) - B_\alpha^\perp J^{\perp \alpha}, \qquad (14)$$

where the identity $J_\alpha l^\alpha = -D_\alpha n^\alpha$ has been used. The first term in (14) is bounded because $B_\alpha n^\alpha$ and $D_\beta n^\beta$ are bounded. For the second term we can apply the Schwarz inequality since $\mathbf{B}^\perp$ and $\mathbf{J}^\perp$ belong to a spacelike subspace. Thus, $(B_\alpha^\perp J^{\perp \alpha})^2 \leq (B_\mu^\perp B^{\perp \mu})(J_\nu^\perp J^{\perp \nu}) = (B_\mu B^\mu)(J_\nu J^\nu)$ and since $B_\mu B^\mu$ and $J_\nu J^\nu$ are bounded at the horizon we conclude that the second term of Eq. (14) is also bounded.

Finally, the vanishing of the term (12) and the bounded behavior of the other term (14), together with the null character of l at the horizon lead to the vanishing of the integrand (7) over the event horizon. With no contribution from boundary integrals in the identity (5) we shall write the volume integral, using the coordinates from Eq. (4), for each one of the different cases discussed at the beginning of this section.

For the purely electric case (I) we have

$$\int_\mathcal{V} -V \left(\frac{1}{2} \gamma_{ij} H^{ti} H^{tj} + m^2 (B^t)^2 \right) dv = 0. \qquad (15)$$

The non–positiveness of the above integrand, which is minus the sum of squared terms, implies that the integral is vanishing only if H^{ti} and B^t vanish everywhere in $\mathcal{V}$, and hence in all $\mathcal{I}$. For the purely magnetic case (II) the volume integral reads as

$$\int_\mathcal{V} \left(\frac{1}{2} \gamma_{ik} \gamma_{jl} H^{kl} H^{ij} + m^2 B_i B^i \right) dv = 0, \qquad (16)$$

in this case the non–negativeness of the above integrand is responsible for the vanishing of H^{ij} and B^i in all $\mathcal{I}$.

4. Conclusions

Concluding, we have proved that the Proca field **B** is trivial in the presence of a static black hole. It must be pointed that we improve the original proof of Bekenstein on the subject by using an appropriate integration measure on the event horizon of the black hole, and also making explicit use of the gravitational field equations. The vanishing of **B** implies that the action (1) reduces to the Einstein–Hilbert one, for which the only static black hole is the Schwarzschild solution (see [10, 11] for references on improvements to the original proofs). The existence of static soliton (particle–like) configurations can be also excluded using similar arguments, since the only change in the proof is that in this case the boundary of the volume $\mathcal{V}$ only consists of the isometric surfaces Σ and Σ', and a portion of the spatial infinity i^o, i.e., there is no interior boundary corresponding to the event horizon.

Acknowledgments

The author thanks Alberto Garcia, Alfredo Macias, Hernando Queve-do, and Thomas Zannias for useful discussions and hints. This research was partially supported by the CONACyT Grant 38495E. The author also thanks Isabel Negrete for typing the manuscript.

Appendix: On the suitable integration measure of the horizon

In this appendix we justify the use of the standard integration measure (6) in the boundary integrals on the event horizon. In the derivation of the basic identity (5) we make use of Gauss's law, which is a well–known particular form of Stokes's theorem

$$\int_\mathcal{V} d\alpha = \int_{\partial\mathcal{V}} \alpha \quad \Longrightarrow \quad \int_\mathcal{V} \nabla_\beta v^\beta dv = \int_{\partial\mathcal{V}} v^\beta d\Sigma_\beta, \qquad (A.1)$$

for some volume $\mathcal{V}$ with boundary $\partial\mathcal{V}$. The relation between both theorems rest on that the three–form α is the Hodge dual of the vector field **v** [9], we shall explore such relation in order to find the horizon integration measure. In the Stokes version we can write the boundary integrand as $\alpha = h\,\eta_3$ using that the three–form α must be proportional to the volume three–form η_3 of the boundary $\partial\mathcal{V}$. For example, in the case of a boundary consisting of non–null surfaces the induced metric there is nondegenerate, and we can choose as volume three–form on $\partial\mathcal{V}$ the one associated with the induced metric. It is given by $\eta_{3\,\alpha\beta\gamma} = \pm{}^*\tilde{n}_{\alpha\beta\gamma} \equiv \pm\eta_{\mu\alpha\beta\gamma}\tilde{n}^\mu$, where η is the four–dimensional volume form, $\tilde{n}$ is the unit normal to $\partial\mathcal{V}$, and we use the plus sign if $\tilde{n}$ is spacelike and the minus one if is timelike; in both cases the normal is chosen to be "outward pointing" in the volume $\mathcal{V}$ in order to keep the orientation needed in Stokes's theorem [9]. For this election of the boundary volume three–form we have the relation $\alpha = {}^{**}\alpha = \pm h\,{}^*\tilde{n}$, from which it follows that $h = {}^*\alpha^\beta \tilde{n}_\beta$, and we recover the Gauss form of the boundary integral

$$\int_{\partial\mathcal{V}} \alpha = \int_{\partial\mathcal{V}} {}^*\alpha^\beta \tilde{n}_\beta\, \eta_3 = \int_{\partial\mathcal{V}} v^\beta d\Sigma_\beta, \qquad (A.2)$$

here $v^\beta = {}^*\alpha^\beta = \eta^{\mu\nu\gamma\beta}\alpha_{\mu\nu\gamma}/3!$, and the integration measure at the boundary is the traditional one for non–null surfaces $d\Sigma_\beta = \tilde{n}_\beta\,\eta_3 = \tilde{n}_\beta d\sigma$, where $d\sigma$ stands for the volume element ($d\sigma = \eta_3$) following the usual notation. The above situation does not apply to null surfaces, which is our case of interest when we try to integrate on the horizon. In this case the induced metric is degenerate, and a priori there is no natural choice for the volume form. However, in the case of the event horizon we can use other geometrical objects naturally defined on it to specify a volume three-form. Let l be the null generator of the horizon, and n be the other linearly independent and future-directed null vector orthogonal to the spacelike cross sections of the horizon, and normalized in such a way that $n_\mu l^\mu = -1$. For smooth event horizon they are well-defined smooth vector fields along it. In this case the volume three-form must not be orthogonal to l since such vector is tangent to the horizon, in fact, their interior product must coincide with the two-form expanding the area of spacelike cross sections of horizon which is obviously given by $*(l \wedge n)$. Hence, the volume three-form η_3 must satisfy the relation $\eta_{3\mu\alpha\beta} l^\mu = *(l \wedge n)_{\alpha\beta}$, using now the identity $\eta_{3\mu\alpha\beta} l^\mu = -*(l \wedge *\eta_3)_{\alpha\beta}$ [10] we conclude that a natural election for the volume three-form at the horizon is $\eta_3 = -*n$. Now we can find the function h inside the boundary integrand; multiplying the relation $\alpha_{\alpha\beta\gamma} = -h\eta_{\rho\alpha\beta\gamma} n^\rho$ by the three-form $[*(l \wedge n) \wedge l]^{\alpha\beta\gamma} = 3\eta^{\mu\nu[\alpha\beta} l_\mu n_\nu l^{\gamma]}$ we obtain

$$3\alpha_{\alpha\beta\gamma} \eta^{\mu\nu\alpha\beta} l_\mu n_\nu l^\gamma = -3 h \eta_{\rho\alpha\beta\gamma} n^\rho \eta^{\mu\nu\alpha\beta} l_\mu n_\nu l^\gamma = 3! h,$$

and expanding the left-hand side above using that $\alpha = **\alpha$ we have finally

$$h = \frac{1}{2} \eta_{\rho\alpha\beta\gamma} {}^*\alpha^\rho \eta^{\mu\nu\alpha\beta} l_\mu n_\nu l^\gamma = 2^* \alpha^\rho n_{[\rho} l_{\gamma]} l^\gamma. \quad (A.3)$$

Hence the boundary integral can be expressed in the Gauss form as

$$\int_{\partial V} \alpha = \int_{\partial V} {}^*\alpha^\beta 2 n_{[\beta} l_{\mu]} l^\mu \, \eta_3 = \int_{\partial V} v^\beta d\Sigma_\beta, \quad (A.4)$$

where again $v^\beta = *\alpha^\beta$, but this time the boundary integration measure is expressed as $d\Sigma_\beta = 2n_{[\beta} l_{\mu]} l^\mu d\sigma$, and we use the standard notation for the volume element $d\sigma = \eta_3$. This is the boundary integration measure introduced in Eq. (6) for the boundary integral at the event horizon and also used in previous references [8, 4, 5].

References

[1] P. Bizoń, *Acta Phys. Polon.* **B25** (1994) 877.
[2] J.D. Bekenstein, *Phys. Rev. Lett.* **28** (1972) 452; *Phys. Rev.* **D5** (1972) 1239; **D5** (1972) 2403.
[3] J.D. Bekenstein, in: *Proceedings of the 9th Brazilian School of Cosmology and Gravitation*, Rio de Janeiro, Brazil (1998) gr-qc/9808028.
[4] E. Ayón-Beato, A. García, A. Macías and H. Quevedo, *Phys. Rev.* **D61** (2000) 084017; **D64** (2001) 024026.
[5] E. Ayón-Beato, *Phys. Rev.* **D62** (2000) 104004.
[6] P.T. Chruściel and R.M. Wald, *Class. Quant. Grav.* **11** (1994) L147.
[7] B. Carter, in: *Gravitation in Astrophysics (Cargèse Summer School 1986)*, eds. B. Carter, J.B. Hartle (Plenum, New York 1987).
[8] T. Zannias, *J. Math. Phys.* **36** (1995) 6970; **39** (1998) 6651.
[9] R.M. Wald, *General Relativity* (Univ. of Chicago Press, Chicago 1984).
[10] M. Heusler, *Black Hole Uniqueness Theorems* (Cambridge Univ. Press, Cambridge 1996); *Living Rev. Rel.* **1**, 1998-6,
http://www.livingreviews.org/Articles/Volume1/1998-6heusler.
[11] P.T. Chruściel, "Black Holes," gr-qc/0201053.

NEW MODEL CALCULATIONS OF PROTOSTELLAR COLLAPSE AND FRAGMENTATION

Jaime Klapp
Instituto Nacional de Investigaciones Nucleares, ININ,
Km. 36.5 Carretera México-Toluca, Ocoyoacac, 52045 Estado de México, México
klapp@nuclear.inin.mx

Leonardo Di G. Sigalotti
Centro de Física, Instituto Venezolano de Investigaciones Científicas,
IVIC, Apartado 21827, Caracas 1020A, Venezuela
lsigalot@cassini.ivic.ve

Abstract   A new generation of hydrodynamical collapse calculations employing a spatial resolution higher than the local Jeans length have started to appear. Compared to previous low-resolution models, the new calculations are effectively making a superior job in evaluating the likelihood of fragmentation in molecular cloud cores. The results of these new models show that binary fragmentation depends sensitively on both the numerical resolution and the detailed thermodynamical treatment. Clarifying the issue of fragmentation is of fundamental importance to explain the observed duplicity of young stellar objects and couple the processes of binary and star formation.

Keywords: hydrodynamics - methods: numerical - stars: formation - binaries: general

1. Introduction

The way in which molecular cloud cores — the dense clumps that form at scales less than a parsec within large molecular clouds [1, 2] — condense and fragment into binary and multiple protostars, and the subsequent evolution of these systems to the observed distribution and duplicity of pre–main–sequence (PMS) stars, are central problems in modern astronomy. Over the past few years, our understanding of the processes of binary and star formation has greatly improved both observationally and theoretically. In particular, surveys of PMS stars show that binary systems are at least as common among young visible stars as among main–sequence (MS) stars [3] – [7]. Thus an important issue for current star formation models is to account for the relatively high frequency of binaries and predict their physical properties. While fragmentation of cloud cores during gravitational collapse has been invoked as the likely mechanism for forming binary stars [8, 9], the only direct observational support to this hypothesis is at present provided by a small number of observations aimed at identifying both protobinaries [10] – [13] and small protoclusters [14]. Among these observations only a few close binary systems with separations < 100 AU have been detected [12, 13]. This low frequency may be the result of inadequate angular resolution available in earlier observations rather than an intrinsic sparsity of binaries. Clarifying the likelihood of fragmentation is of fundamental importance because it represents a natural way of coupling the processes of binary and star formation.

Much of what we know about fragmentation has come primarily from the results of three–dimensional hydrodynamical collapse calculations. A huge number of such models exist which use different methods and a variety of initial conditions. Earlier work on protostellar collapse and fragmentation was largely based on low–resolution calculations. However, the bulk of these models suffered from an inherent numerical viscosity which in most cases caused artificial fragmentation to occur. It was only recently, with the work of Truelove et al. [15], that high–resolution calculations have started to appear. They found that working at a resolution higher than the local Jeans length, the effects of the numerical viscosity are minimized avoiding artificial fragmentation. In particular, for a Gaussian cloud model they obtained the formation of a singular filament during the isothermal collapse in clear contrast with previous calculations which predicted fragmentation into a binary or quadruple system [16] – [19]. Further calculations [20, 21] have shown that satisfying the Jeans length resolution constraint, not only changes the nature of the solution from a binary to a singular filament for the same isothermal Gaussian cloud model of Truelove et al. [15], but also guarantees convergence of the results regardless of the numerical methods employed.

In this paper we outline some recent observations that constrain current star formation models, review the results of recent high–resolution (adaptive), finite–difference (FD) calculations and discuss their implications on binary formation. Future work in this field and applications of the Smoothed Particle Hydrodynamics (SPH) method to fragmentation models are also commented.

2. Observational constraints

Various surveys of low–mass binaries have been performed in recent years. All these surveys have recently been reexamined by Duchêne [6] who has clarified the issue of the possible binary excess in star–forming regions. The binary fraction in loose associations such as Taurus exceeds the MS value by a factor of ~ 1.7, implying that almost 95% of stars in Taurus are multiple systems. Other star–forming regions such as Ophiuchus, Chamaleon and possibly Lupus show similar excesses. The binary fraction in these associations seems to be established after ~ 1 Myr, i.e., very soon in the history of star formation. On the other hand, dense young clusters such as Trapezium and NGC in Orion have binary fractions similar to the MS. The same conclusion applies to the Pleiades and Hyades. The precise nature of the difference in the overall binary fraction between the various regions is still unclear.

An analysis of a sample of 14 spatially resolved PMS binaries in low-mass star–forming regions, have shown that for all pairs with separations from 90 AU to 250 AU, the individual components appear to be coeval [22]. This finding is similar to that of Hartigan et al. [23] who detected that 2/3 of the 26 binaries sampled (with separations ≥ 400 AU) are coeval. Thus a significant fraction of these PMS binaries very likely formed through fragmentation shortly before or during the collapse phase of a molecular cloud. In a separate survey, Brandner & Köhler [24] concluded that the distribution of binary separations is not a universal quantity since both the peak and the width of the distribution might vary from one region to the other.

The search for precollapse cores has in recent years finally been crowned with success, thanks to the emerging high–resolution observations in the millimeter and submillimeter [25] – [27]. The properties of molecular cloud cores are of great importance for realistic fragmentation models since they constrain the initial conditions for the protostellar collapse phase. One important result from these observations is that starless cores are centrally condensed with flat inner density profiles, similar to the profiles predicted by calculations of magnetized cores in the ambipolar diffusion stage [28]. Flat inner density profiles are also characteristic of precollapse cores in cluster–forming regions [29, 30]. Another property concerns the three–dimensional shape of cloud cores. As was first deduced by Myers et al. [31], recent analyses continue to indicate that such cores are preferentially elongated with inferred axial ratios of $\sim$ 2:1 [32, 33]. Evidence for rotation has also been reported by Barranco & Goodman [34], who measured angular velocities in the range 10^{-14}–10^{-13} s^{-1} in dense cloud cores, corresponding to ratios of the rotational kinetic energy to the absolute value of the gravitational energy in the range $0.001 \leq \beta \leq 0.03$. Rotation becomes dynamically important once a core collapses to form a single star or binary system with associated disks.

Most hydrodynamical collapse calculations, including the newest ones, have employed idealized initial conditions by assuming either an isothermal or polytropic equation of state. By contrast, McLaughlin & Pudritz [35] have adopted a logotropic equation of state of the form

$$\frac{p}{p_c} = 1 + A \ln\left(\frac{\rho}{\rho_c}\right) , \qquad (1)$$

to model the internal structure of dense cores, where p_c and ρ_c are the central pressure and density in a cloud, and A is a free parameter. This equation of state provides a unified treatment of low– and high–mass cores and entire giant molecular clouds. It can account quantitatively for the global size line width and mass–radius relations between cloud complexes, the typical mass and density contrasts between large molecular clouds and the clumps within them, and the observed dependence of line width on radius inside dense cores of any mass. Relation (1) is meant to account for all contributions to the total pressure, including the effects of disordered magnetic fields (MHD turbulence). The internal velocity–dispersion profiles of real clumps are seen to be consistent with Eq. (1) for $A \approx 0.2$ [36, 37].

3. New protostellar collapse calculations

Recent collapse models consist of high–resolution calculations that solve in three–space dimensions the familiar equations of hydrodynamics for a self–gravitating, rotating gas cloud with either an isothermal or polytropic equation of state. These equations are: the continuity equation

$$\frac{\partial \rho}{\partial t} + \nabla \cdot (\rho \mathbf{v}) = 0 , \qquad (2)$$

the momentum equation

$$\frac{\partial \mathbf{v}}{\partial t} + (\mathbf{v} \cdot \nabla) \mathbf{v} = -\frac{1}{\rho} \nabla p - \nabla \Phi , \qquad (3)$$

and the Poisson equation

$$\nabla^2 \Phi = 4\pi G \rho , \qquad (4)$$

where ρ, p, $\mathbf{v}$ and Φ denote the density, pressure, velocity and gravitational potential, respectively. Assuming an ideal isothermal gas, Eqs.

(2)–(4) are closed by the simple relation

$$p = c_s^2 \rho , \qquad (5)$$

where c_s is the isothermal sound speed. A logotropic gas could be modelled by using Eq. (1) instead of relation (5).

3.1. Adaptive calculations: Jeans resolution constraint

A new generation of protostellar collapse and fragmentation calculations has begun to appear as a consequence of the need of appropriate resolution requirements. In a seminal paper Truelove et al. [15] demonstrated, with a new FD adaptive mesh refinement (AMR) Cartesian code [38], that perturbations arising from discretization of Eqs. (2)–(4) can grow and induce artificial fragmentation in multiple–grid simulations. In this case, the discretization errors act as an inherent numerical viscosity that artificially retards the collapse of the densest regions allowing the growth of non–axisymmetric perturbations. The effects of this numerical viscosity are minimized only if the cell size always remains smaller than one–fourth of the local Jeans length

$$\lambda_J = \left(\frac{\pi c_s^2}{\rho G}\right)^{1/2} . \qquad (6)$$

While these results specialize to uniform Cartesian grids, Boss [39] generalized the above constraint for a non–uniform spherical–coordinate grid by demanding that

$$\Delta x = \left(r^2 \sin\theta \Delta r \Delta\theta \Delta\phi\right)^{1/3} < \frac{\lambda_J}{4} , \qquad (7)$$

coupled with the three Jeans length requirements

$$\Delta r < \frac{\lambda_J}{4} \quad , r\Delta\theta < \frac{\lambda_J}{4} \quad , r\sin\theta\Delta\phi < \frac{\lambda_J}{4} . \qquad (8)$$

Truelove et al. [38] performed AMR calculations of the standard (uniform density) isothermal test case for a 50% and a 10% amplitude $m = 2$ density perturbation. In the former case, a binary system formed as in previous low–resolution models. As they continued the evolution to higher densities by keeping the gas isothermal, each fragment condensed to a singular filamentary state without subfragmenting in accordance with the findings of Inutsuka & Miyama [40], who predicted by linear analysis that as long as the mass per unit length of an infinitely long cylinder is greater than the equilibrium value $(M/L)_e = 2c_s^2/G$, fragmentation of the cylinder will not occur as long as the collapse will remain isothermal. In the second case, the cloud collapsed to form a binary but this time linked by a thin bar which also never subfragmented. Using a spherical–coordinate based code, Boss et al. [20] presented high–resolution calculations for the 10% standard test case and obtained convergence to the results of Truelove et al. [38]. Similar SPH calculations by Bate & Burkert [41] adhering to the Jeans constraints also predicted a long connecting bar for the 10% standard model. This time, however, the bar fragmented into four small clumps as a result of allowing heating of the central cloud regions to guarantee preservation of the Jeans mass resolution beyond densities of $\approx 10^{-13}$ g cm^{-3} in their calculations. Further AMR calculations of the 10% standard test were reported by Klein et al. [42] but this time using a barotropic equation of state which makes the transition from the isothermal to the adiabatic regime in a smooth fashion. Again, they found a forming long bar which, after becoming optically thick, fragmented into a binary core. A first implication of these results is that gradual heating of the densest regions increases the pressure with a consequent retardation of the collapse. In contrast with purely isothermal calculations, nonisothermal heating allows the growth of non–axisymmetric perturbations causing fragmentation of the bar.

Calculations starting from centrally condensed, Gaussian density variations are of greater interest for binary formation. For a particular spherical Gaussian cloud model, perturbed with a 10% amplitude $m = 2$ density variation, Truelove et al. [38] first found that this model collapses isothermally to form a singular filament rather than a binary or quadruple system as predicted by all previous low–resolution models. This result questioned the validity of earlier fragmentation calculations and showed the importance of working with a spatial resolution higher than the local Jeans length. Convergence to this result was also achieved independently by Boss et al. [20] with a refined spherical–coordinate code. More recently, Sigalotti & Klapp [21] employed an adaptive, spherical–coordinate code based on the "zooming" coordinates to investigate the isothermal collapse of both spherical and prolate Gaussian core models. In particular, the filament solution of Truelove et al. [38] and Boss et al. [20] for the spherical Gaussian model was reproduced to very high central density contrasts, as shown in Fig. 1. Most importantly, the prolate clouds all collapsed self–similarly to produce singular filaments which did not fragment. This result also questions the reliability of previous prolate collapse calculations which predicted binary fragmentation for a wide range of the initial conditions.

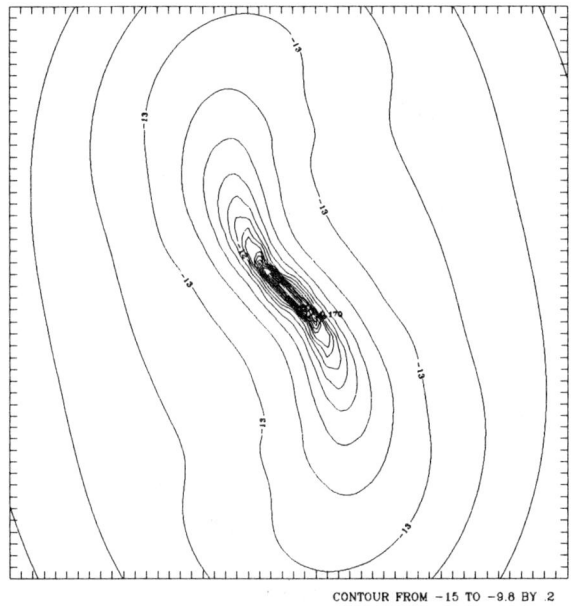

Figure 1. Density contours in the equatorial plane during the isothermal collapse of a 1 $M_\odot$ Gaussian cloud model at 1.361 t_{ff}, where $t_{ff} = (3\pi/32G\rho_0)^{1/2}$. Initially, the cloud has a radius $R = 0.016$ pc, a central density $\rho_0 = 1.7 \times 10^{-17}$ g cm^{-3} and amounts of the thermal and rotational energies given by $\alpha = 0.265$ and $\beta = 0.16$, respectively. The maximum density shown is $1.01 \times 10^7 \rho_0$ and the box radius is 5.0×10^{14} cm.

3.2. Binary vs filament formation

Evidently, the new adaptive calculations are making a superior job in evaluating the likelihood of fragmentation during the collapse of Gaussian condensations. While this type of initial conditions are chosen to resemble the structure of real cloud cores, it is surprising to learn that fragmentation seems unlikely during the isothermal collapse phase. Some physical mechanism capable of retarding the collapse appears to be necessary to induce fragmentation at roughly the same maximum densities where the isothermal calculations predict the formation of a singular filament. Since the filaments take form during the transition from the isothermal to the nonisothermal phase of collapse, the increasing pressure forces as the filament region heats up may provide an effective retardation mechanism which may ultimately favors the fragmentation process. This possibility was more realistically considered by Boss et al. [20], who used nonisothermal thermodynamics with Eddington approximation radiative transfer and detailed equations of state. They found that the gas heating retards the in fall and allows fragmentation in the spherical Gaussian cloud model. Unfortunately, their results are inconclusive because the calculation was not followed deep enough into the nonisothermal regime to show the stability of the forming binary.

Except for rotation, all the above models have started from static initial conditions. Recent high-resolution SPH calculations of the spherical Gaussian cloud model seem to indicate that binary fragmentation may still occur during the isothermal collapse phase as a result of nonlinear density fluctuations present in the SPH initial conditions [43]. We may expect that similar fluctuations in the velocity field (subsonic and supersonic turbulence) at the beginning of collapse may therefore have implications on the outcome of fragmentation. A further exciting possibility is to consider the collapse of pressure–truncated, logotropic spheres with $A \approx 0.2$. Such models would be consistent with the line width measurements in both low- and high-mass cores from a variety of molecular clouds [35]. A pure logotrope is the simplest barotropic model that can fit on average the structure of real cloud cores.

4. Concluding remarks

A more precise understanding of the process of binary formation must require at least a clarification of the above issues. Future detailed observations, coupled with the increasing sophistication of numerical calculations, should point toward this direction. In particular, the high spatial resolution afforded by the new generation of collapse calculations allows for a better evaluation of the likelihood of binary fragmentation during protostellar collapse. These models have shown that the transition from the isothermal phase to the adiabatic phase is relevant for the outcome of fragmentation. The effects of magnetic fields should also be included. Although magnetic fields may not influence the fragmentation process, they are crucial to determine the appropriate conditions at the onset of collapse.

High-resolution SPH calculations are also required to increase the reliability of present FD results. Further intercomparisons between independent calculations based on different computational methods should also help discovering new limitations of presently available hydrocodes.

References

[1] R. Chini, B. Reipurth, D. Ward–Thompson, J. Bally, L.-A. Nyman, A. Sievers and Y. Billawala, *Astrophys. J.* **474** (1997) L135.

[2] T. L. Wilson, R. Mauesberger, P. D. Gensheimer, D. Muders and J. H. Bieging, *Astrophys. J.* **525** (1999) 343.

[3] A. M. Ghez, D. W. McCarthy, J. L. Patience and T. L. Beck, *Astrophys. J.* **481** (1997) 378.

[4] A. M. Ghez, R. J. White and M. Simon, *Astrophys. J.* **490** (1997) 353.

REFERENCES

[5] M. G. Petr, V. C. du Foresto, S. V. W. Beckwith, A. Richichi and M. J. McCaughrean, *Astrophys. J.* **500** (1998) 825.

[6] G. Duchêne, *Astron. Astrophys.* **341** (1999) 547.

[7] G. Duchêne, J. Bouvier and T. Simon, *Astron. Astrophys.* **343** (1999) 831.

[8] P. Bodenheimer, A. Burkert, R. I. Klein and A. P. Boss, in: *Protostars and Planets IV*, ed. V. G. Mannings, A. P. Boss and S. S. Russell, (Tucson: Univ. Arizona Press, 2000), pp. 675.

[9] L. Di G. Sigalotti and J. Klapp, *IJMP D* **10(2)** (2001) 115.

[10] O. P. Lay, J. E. Carlstrom and R. E. Hills, *Astrophys. J.* **452** (1995) L73.

[11] G. A. Fuller, E. F. Ladd and K. W. Hodapp, *Astrophys. J.* **463** (1996) L97.

[12] L. W. Looney, L. G. Mundy and W. J. Welch, *Astrophys. J.* **484** (1997) L157.

[13] S. Terebey, D. van Buren, D. L. Padgett, T. Hancock and M. Brundage, *Astrophys. J.* **507** (1998) L71.

[14] M. Tafalla, P. C. Myers, D. Mardones and R. Bachiller, *Astron. Astrophys.* **348** (1999) 479.

[15] J. K. Truelove, R. I. Klein, C. F. McKee, J. H. Holliman, L. H. Howell and J. A. Greenough, *Astrophys. J.* **489** (1997) L179.

[16] A. P. Boss, *Nature* **351** (1991) 298.

[17] A. P. Boss, *Astrophys. J.* **410** (1993) 157.

[18] J. Klapp, L. Di G. Sigalotti and F. de Felice, *Astron. Astrophys.* **273** (1993) 175.

[19] A. Burkert and P. Bodenheimer, *MNRAS* **280** (1996) 1190.

[20] A. P. Boss, R. T. Fisher, R. I. Klein and C. F. McKee, *Astrophys. J.* **528** (2000) 325.

[21] L. Di G. Sigalotti and J. Klapp, *Astron. Astrophys.* **378** (2001) 165.

[22] W. Brandner and H. Zinnecker, *Astron. Astrophys.* **321** (1997) 220.

[23] P. Hartigan, K. M. Strom and S. E. Strom, *Astrophys. J.* **427** (1994) 961.

[24] W. Brandner and R. Köhler, *Astrophys. J.* **499** (1998) L79.

[25] D. Ward-Thompson, D. Scott, P. F. Hills and P. André, *MNRAS* **268** (1994) 276.

[26] P. André, D. Ward-Thompson and F. Motte, *Astron. Astrophys.* **314** (1996) 625.

[27] P. André, A. Bacmann, F. Motte and D. Ward-Thompson, in: *The Physics and Chemistry of the Interstellar Medium*, ed. V. Ossenkopf, J. Stutzki and G. Winnerwisser, (Zermatt: GCA-Verlag, 1998), pp. 241.

[28] S. Basu and T. Ch. Mouschovias, *Astrophys. J.* **432** (1994) 720.

[29] F. Motte, P. André and R. Neri, *Astron. Astrophys.* **336** (1998) 150.

[30] F. Motte and P. André, in: *The Physics and Chemistry of the Interstellar Medium*, ed. V. Ossenkopf, J. Stutzki and G. Winnerwisser, (Zermatt: GCA-Verlag, 1998), pp. 249.

[31] P. C. Myers, G. A. Fuller, A. A. Goodman and P. J. Benson, *Astrophys. J.* **376** (1991) 561.

[32] B. S. Ryden, *Astrophys. J.* **471** (1996) 822.

[33] J. Jijina, P. C. Myers and F. C. Adams, *Astrophys. J. Suppl. Series* **125** (1999) 161.

[34] J. A. Barranco and A. A. Goodman, *Astrophys. J.* **504** (1998) 207.

[35] D. E. McLaughlin and R. E. Pudritz, *Astrophys. J.* **469** (1996) 194.

[36] P. C. Myers and A. A. Goodman, *Astrophys. J.* **329** (1988) 392.

[37] P. Caselli and P. C. Myers, *Astrophys. J.* **446** (1995) 665.

[38] J. K. Truelove, R. I. Klein, C. F. McKee, J. H. Holliman, L. H. Howell, J. A. Greenough and D. T. Woods, *Astrophys. J.* **495** (1998) 821.

[39] A. P. Boss, *Astrophys. J.* **501** (1998) L77.

[40] S.-I. Inutsuka and S. M. Miyama, *Astrophys. J.* **388** (1992) 392.

[41] M. R. Bate and A. Burkert, *MNRAS* **288** (1997) 1060.

[42] R. I. Klein, R. T. Fisher, C. F. McKee and J. K. Truelove, in: *Numerical Astrophysics*, ed. K. Tomisaka and S. Miyama, (Kluwer: Dordrecht, 1998), pp. 131.

[43] A. Burkert, 2001, private communication.

INFLATIONARY COSMOLOGY AND THE BRANEWORLD SCENARIO

James E. Lidsey
Astronomy Unit, School of Mathematical Sciences,
Queen Mary, University of London, Mile End Road, LONDON E1 4NS, UK
J.E.Lidsey@qmul.ac.uk

Abstract   An overview of cosmology in the second Randall–Sundrum scenario is presented, focusing on the derivation of the Friedmann equation and the consistency equation between the density and gravitational wave perturbations generated during inflation.

Keywords: Inflatiom, Cosmology, Branes, Braneworld.

The fundamental aim of particle cosmology is the development of a unified picture of the very early universe that is consistent both with known particle physics theory and cosmological observations. A synthesis of these two disciplines provides a unique window onto physics at energies inaccessible to any terrestrial form of experiment.

Given the recent rapid advances in the availability of high precision data from a variety of cosmic microwave background (CMB) and large-scale structure observations, the constraints on models of the early universe have become ever more stringent [1, 2, 3, 4]. This trend is certain to continue, given the expected avalanche of data from surveys planned for the near future. In particular, recent measurements by the Boomerang and Maxima balloon–borne experiments indicate that the universe is very nearly spatially flat, with its total density being very close to the critical value [1, 2, 3]. On the other hand, there is now ample evidence from observations – including the spectrum of CMB anisotropies, galaxy clustering statistics, peculiar velocities and the baryon mass fraction in clusters of galaxies – that the density of the clumped baryonic and dark matter in the universe is substantially lower, being of order 0.2–0.3 of the critical value. Added to this is new evidence from spectral and photometric observations of Type Ia supernovae that the expansion of the universe may be accelerating at the present epoch [5, 6].

The most common way of explaining this diverse set of observations has been to postulate that the universe underwent an epoch of inflationary expansion in its most distant past at energy scales at or above the electroweak phase transition, and that at the present epoch a substantial proportion of the energy density of the universe is in the form of a dark component, which is smooth on cosmological scales and which possesses a negative pressure. The inflationary scenario has become the cornerstone of modern, early universe cosmology. (For a recent review, see, e.g., Refs. [7]). It is now widely believed that the observed structures in the universe evolved via gravitational instability from tiny quantum fluctuations that were generated during this period [8]. Moreover, inflation results in an effectively flat universe, so the current data supports but also provides strong constraints on such models. A favored candidate for the dark energy is a scalar field, termed 'quintessence' which slowly evolves down a potential [9]. Interestingly, it has also been shown that some quintessence models have the further appealing feature that they exhibit 'tracking', i.e, they become independent of their initial conditions as they evolve, thus resolving the fine tuning problem present in models based purely on a cosmological constant [10].

What these two paradigms share is the need for new degrees of freedom, typically in the form of self–interacting scalar fields. Despite the success of many specific potential–driven inflationary and tracking models, however, there is as yet no canonical theory for either of these two scenarios. This provides strong motivation for establishing that inflation and tracking can arise generically and robustly within the context of unified field theories of the fundamental interactions [11]. Superstring theory has emerged as the primary candidate for such a theory. The subject has undergone a major revolution in recent years, following the discovery that the five perturbative string theories are related non–perturbatively by duality symmetries, with each theory now viewed as a different limit of a more fundamental theory, known as 'M–theory' [12, 13]. The low–energy limit of M–theory is eleven–dimensional supergravity, implying that it is more than simply a theory of superstrings. This reestablishes the importance of eleven–dimensional supergravity in cosmology.

Given this fundamental change of perspective, there is a pressing need to study the cosmological implications of string/M–theory [14]. The crucial point about these new developments is that such theories naturally give rise to multiple moduli fields, i.e., scalar fields which describe flat directions in the compactified vacuum manifold of the theory. Moreover, recent studies indicate that multiple combinations of scalar fields typically interact in cosmological environments in such a way that all of the fields can remain dynamically important [15]. This effect – known as assisted dynamics – implies that the previously considered classes of single–field models may not be sufficiently generic.

String/M–theory requires the universe to be higher–dimensional. An exciting theoretical development has been the realization that the standard model interactions may be confined to a four–dimensional membrane or domain wall (indeed corresponding to our observable universe), whereas gravitational interactions may also propagate through the extra dimensions (the bulk) [16, 17]. This change in viewpoint is motivated in part by the discovery that the quantum dynamics of the D–branes can be described in terms of open strings whose ends are fixed on the brane [18].

There has been intense activity in this rapidly developing field and it is therefore crucial to derive cosmological models in this context. One of the main obstacles to realizing a realistic inflationary model in standard cosmology has been the severe fine–tuning necessary to ensure that the inflaton potential is sufficiently flat. For realistic parameters, the potentials are generally too steep. A striking feature of the braneworld scenario, however, is that the Friedmann equations receive high–energy corrections due to brane–bulk interactions [19, 20, 21, 22]. These result in enhanced friction on the scalar fields [23] and may allow inflation in situations where it was previously thought impossible [24, 25]. Moreover, due to the steepness of the potential, inflation ends naturally as standard Einstein gravity is recovered at low energies.

Within the context of M–theory, there are numerous candidates for a scalar field with such a steep potential, including the dilaton originating in compactified heterotic M–theory [26], the dilaton that arises in self–tuning mechanisms for cancelling the cosmological constant [27], as well as combinations of dilaton and moduli fields [28]. The last possibility is particularly interesting as it may allow inflation to be realized before the moduli fields become stabilized. However, much work remains to be done in this scenario.

Of particular interest in braneworld scenarios is the production of scalar and gravitational wave spectra. The brane–bulk corrections modify the scale–dependence and amplitudes of the perturbations, leading to potentially observable signatures [23, 29]. Indeed, the amplitude of tensor perturbations is larger than in the standard scenario. In models of steep inflation, these perturbations may be detectable by the Planck satellite [24]. The potential importance of this cannot be overemphasized – the gravitational waves are uniquely determined by the inflaton's potential and an observation of the former would lead to unique information on the nature of the latter. There is also the related question of the nature of the consistency equations that relate the amplitudes and tilts of the spectra in a model–independent way [25, 30].

To date, the vast majority of work has centered on models consisting of a single brane or of two or more parallel branes. However, supergravity configurations that represent the intersection of p–branes have played an important role in advances in string theory [31, 32] and the development of intersecting braneworlds is therefore well motivated. One approach is to consider the compactification of the antisymmetric form fields that arise in the string/M–theory effective actions. When the components of such fields depend on the compactifying manifold, the field has a non–trivial flux on that space that can support a p–brane. Depending on the degree of the form field and the nature of the internal manifold, the corresponding models represent brane intersections. Configurations representing curved, intersecting domain walls were recently found [33].

In this talk we focus on the cosmology of the second Randall–Sundrum model (RSII). Randall and Sundrum (RS) embedded a codimension one brane in five–dimensional Anti–de Sitter space with a Z_2 reflection symmetry imposed [17]. The action is

$$S = \int_{\mathcal{M}} d^5x \sqrt{G}\left[2M^{3(5)}R - \Lambda\right] + \int_{\partial\mathcal{M}} d^4x \sqrt{h}\left(\lambda + L_{\text{matter}}\right), \quad (1)$$

where $^{(5)}R$ is the Ricci curvature scalar of the bulk spacetime, $\mathcal{M}$, with metric G_{AB}, Λ is the five–dimensional (negative) cosmological constant, $G \equiv \det G_{AB}$, h is the determinant of the metric induced on the boundary of $\mathcal{M}$, λ is the tension of the brane, L_{matter} is the Lagrangian density of the matter on the brane and M is the five–dimensional Planck mass.

The bulk solution is given by

$$ds^2 = e^{[-2k|y|]}\left(-dt^2 + dx_3^2\right) + dy^2, \quad (2)$$

where k is a constant. The exponential 'warp factor' implies that the geometry of the bulk is non–factorizable and the extent of the fifth dimension (y) can be infinite. A fine–tuning between the bulk cosmological constant and the brane tension (vacuum energy) is required for Poincaré invariance in four dimensions [17]:

$$\Lambda = -24M^3k^2, \quad \lambda = 24M^3k. \quad (3)$$

RS derived and solved the graviton equation of motion for this background. Remarkably, they found that the solution has a zero energy ground state that is localized around the domain wall [17]. This is interpreted as the four–dimensional, massless spin–2 graviton. There is also a continuum of massive states that result in corrections to the Newtonian potential of the form

$$V(r) \approx \frac{G_N m_1 m_2}{r}\left(1 + \frac{1}{k^2 r^2}\right). \tag{4}$$

However, these massive states are suppressed near the brane and are therefore not dangerous if the constant k is sufficiently high, i.e., if the bulk geometry is sufficiently warped. The key point is that *the curvature of the five-dimensional world effectively determines the four-dimensional physics*. This is in stark contrast to the standard Kaluza–Klein compactification scheme based on the 'cylinder' condition, where the extra dimensions are Ricci flat.

The radical proposal then is that *our universe is a brane embedded in a higher-dimensional space*. This has profound implications for cosmology and braney effects are expected to become important in the early universe sometime before the electroweak phase transition, $t < 10^{-15}$sec. This has led to an active programme of research. The key objectives from the astrophysical and cosmological points of view are:

- To derive families of cosmological solutions and determine their early- and late-time behavior.
- To establish whether inflation occurs and, if so, whether it is generic.
- To investigate the production of scalar (density), vector (electromagnetic) and tensor (gravitational wave) perturbations during inflation.
- To determine whether these perturbations are compatible with cosmological observations, in particular, the CMB power spectrum and large-scale structure

Cosmic dynamics in the braneworld scenario arises when the brane follows a timelike trajectory in a *static* bulk spacetime [34, 35, 36]. An observer confined to the surface of the domain wall interprets this motion in terms of cosmic expansion (or contraction). The equations of motion describing this expansion are derived by employing the standard thin wall formalism of General Relativity extended to five dimensions. For simplicity, we choose the bulk to have the symmetries of the spatially flat FRW universe: this corresponds to AdS$_5$. The line element is given in static coordinates by

$$ds^2 = G_{AB}dx^A dx^B = -\frac{r^2}{L^2}dt^2 + \frac{L^2}{r^2}dr^2 + r^2 dE_3^2, \tag{5}$$

where the constant L is related to the bulk cosmological constant. The induced metric on the wall then has the desired form:

$$h_{AB} = G_{AB} + n_A n_B$$
$$ds_4^2 = -d\tau^2 + a^2(\tau)dE_3^2, \qquad (6)$$

where n^A is the unit normal vector to the brane and cosmic time on the brane is parametrized by τ, defined such that

$$d\tau^2 = \frac{r^2}{L^2}dt^2 - \frac{L^2}{r^2}dr^2. \qquad (7)$$

The scale factor, $a(\tau)$, is represented by the 'radial coordinate':

$$r = r\left[a(\tau)\right]. \qquad (8)$$

The Israel junction conditions relate the extrinsic curvature, K_{AB}, of the brane to the energy–momentum tensor, T_{AB}, of the matter confined to it [37]:

$$K_{AB} \equiv h^C{}_{(A} h_{B)}{}^D \nabla_C n_D \qquad (9)$$

$$K_{AB} = -4\pi G_5 \left(T_{AB} - \frac{1}{3}Th_{AB}\right), \qquad (10)$$

where G_5 is the five–dimensional Newton's constant, $T \equiv T_A^A$ and we have taken the Z_2 symmetry into account. A further constraint – the Codazzi equation – must also be imposed on the dynamics. This equation is given by

$$\nabla_B K_A^B - \nabla_A K = {}^{(5)}R_{BC} G_A^B n^C \qquad (11)$$

and, for the AdS_5 bulk geometry that we are considering here, the right–hand side of Eq. (11) is identically zero. Thus, substituting the Israel junction conditions (9) and (10) into Eq. (11) implies conservation of energy–momentum on the brane:

$$^{(4)}\nabla_\mu T^{\mu\nu} = 0. \qquad (12)$$

For a prefect fluid matter source on the brane, $T_\nu^\mu = \mathrm{diag}\left[-\rho, p, p, p\right]$, this results in the familiar fluid equation:

$$\dot{\rho} + 3\frac{\dot{a}}{a}(\rho + p) = 0, \qquad (13)$$

where a dot denotes $d/d\tau$.

The effective Friedmann equation on the brane is derived from the spatial components of the Israel junction conditions (10). These are given by

$$K_{ij} = -\sqrt{\frac{1}{L^2} + \frac{\dot{a}^2}{a^2}}\delta_{ij} \qquad (14)$$

and imply that

$$\frac{\dot{a}^2}{a^2} = \left(\frac{4\pi G_5 \rho}{3}\right)^2 - \frac{1}{L^2}. \qquad (15)$$

It can be shown that the time–time components of Eq. (10) are automatically satisfied by Eq. (15).

The main feature of Eq. (15) is that the right–hand side depends *quadratically* on the energy density, rather than linearly as in conventional cosmology. There is also an effective negative cosmological constant term arising from the bulk cosmological constant. At first, this appears to be highly problematic for early universe cosmology, particularly in satisfying the stringent bounds from primordial nucleosynthesis. However, the wall has a tension, λ, associated with it and we can therefore separate the energy density, ρ, into two components, one associated with the matter, ρ_B, and the other with the tension, i.e., $\rho = \rho_B + \lambda$. Substituting into Eq. (15) then implies that

$$H^2 = \frac{8\pi G_4}{3}\rho_B\left(1 + \frac{\rho_B}{2\lambda}\right) + \left(\frac{4\pi G_5 \lambda}{3}\right)^2 - \frac{1}{L^2}, \qquad (16)$$

where $4\pi\lambda G_5^2 = 3G_4$. Crucially, the cosmological terms cancel each other if the Randall–Sundrum fine–tuning condition (3) is imposed. We therefore arrive at the famous Friedmann brane equation for the RSII brane cosmology [19, 20, 21, 34, 36]:

$$H^2 = \frac{8\pi}{3m_4^2}\rho\left[1 + \frac{\rho}{2\lambda}\right], \qquad (17)$$

where we drop the subscript B for notational simplicity and define the four–dimensional Planck mass, $m_4 \equiv G_4^{-1/2}$.

Eqs. (13) and (17) uniquely determine the cosmic dynamics on the brane. In the case where the brane matter is dominated by a scalar field with a self–interaction potential, $V(\phi)$, Eq. (13) takes the Klein–Gordon form

$$\ddot{\phi} + 3H\dot{\phi} + V' = 0, \qquad (18)$$

where a prime denotes $d/d\phi$.

We now study some aspects of inflationary cosmology within the context of this new paradigm. Further insight may be gained by defining new variables [38]

$$x \equiv \frac{\dot{\phi}}{\sqrt{2\rho}}, \qquad y \equiv \sqrt{\frac{V}{\rho}}, \qquad (19)$$

where, by definition, the variables satisfy the constraint $x^2+y^2 = 1$. This allows us to write Eqs. (17) and (18) in the form of a plane autonomous system:

$$\frac{dx}{dN} = -3x + 3x^3 - \sqrt{\frac{3}{2}}\tilde{\alpha}y^2$$

$$\frac{dy}{dN} = \sqrt{\frac{3}{2}}\tilde{\alpha}xy + 3yx^2, \qquad (20)$$

where $N \equiv \ln a$ and

$$\tilde{\alpha}^2 \equiv \frac{2\lambda}{\rho+2\lambda}\alpha^2, \qquad \alpha^2 = \frac{m_4^2}{8\pi}\frac{V'^2}{V^2}. \qquad (21)$$

This parametrization is helpful because the field equations are formally equivalent to those of the standard inflationary cosmology, where the 'slow–roll' parameter, $\tilde{\alpha}^2$, replaces the corresponding parameter, α^2, of the latter scenario. The parameter, $\tilde{\alpha}^2$, has received a correction due to the brane effects. Specifically, for a given value of α^2, the effective value of $\tilde{\alpha}^2$ is reduced by the factor $[1 + (\rho/2\lambda)]^{-1}$. Thus, in the region of parameter space where the quadratic correction to the Friedmann equation (17) is dominant, inflation may proceed even if $\alpha^2 \gg 2$, i.e., if the potential is too steep to drive conventional inflation. This is the basis of the recently proposed *steep inflationary scenario* [24, 25]. In effect, the brane corrections enhance the value of the Hubble parameter, thereby introducing extra friction on the scalar field and resisting its motion down the potential. Inflation ends naturally once the linear terms begin to dominate.

We refer the reader to Refs. [24, 25] for the details of this new scenario and focus here on the density and gravitational wave perturbations that are generated quantum mechanically during braneworld inflation. We adopt the normalization conventions of Ref. [30]. To lowest–order in the slow–roll approximation, the amplitudes of the scalar and tensor fluctuations are given respectively by [23, 29]

$$A_S^2 = \frac{512\pi}{75m_4^6}\frac{V^3}{V'^2}\left(1 + \frac{V}{2\lambda}\right)^3 \qquad (22)$$

$$A_T^2 = \frac{4}{25\pi m_4^2}H^2 F^2(H/\mu), \qquad (23)$$

where
$$F(x) \equiv \left[\sqrt{1+x^2} - x^2 \sinh^{-1}\left(\frac{1}{x}\right)\right]^{-1/2} \tag{24}$$
and
$$x \equiv \left(\frac{3}{4\pi\lambda}\right)^{1/2} Hm_4. \tag{25}$$

The standard expressions are recovered in the low–energy limit, $\rho \ll \lambda$. The right–hand sides of these expressions are evaluated when a given mode, k, goes beyond the Hubble radius, $k = aH$. Thus,

$$k(\phi) = a_e H(\phi) \exp[-N(\phi)], \tag{26}$$

where a subscript 'e' denotes values at the end of inflation and the number of e–foldings between a scalar field value, ϕ, and ϕ_e is given by

$$N \approx -\frac{8\pi}{m_4^2} \int_\phi^{\phi_e} \frac{V}{V'}\left(1 + \frac{V}{2\lambda}\right) d\phi. \tag{27}$$

Finally, the spectral indices of the two spectra are defined by $n_S \equiv 1 + d\ln A_S^2/d\ln k$ and $n_T \equiv d\ln A_T^2/d\ln k$. They are determined in the high–energy limit by the inflaton potential and its first and second derivatives:

$$n_S - 1 \approx -\frac{m_4^2 \lambda}{2\pi V}\left[3\frac{V'^2}{V^2} - \frac{V''}{V}\right] \tag{28}$$

$$n_T \approx -\frac{3m_4^2}{4\pi}\frac{\lambda V'^2}{V^3}. \tag{29}$$

In the standard cosmology, there is a consistency equation [30]:

$$\frac{A_T^2}{A_S^2} = -\frac{1}{2}n_T \tag{30}$$

that relates the ratio of the tensor and scalar amplitudes with the tensorial spectral index. Eq. (30) is independent of the inflaton potential. It therefore relates potentially observable quantities in a model–independent way, in the sense that it is a generic prediction of single field inflationary cosmology, where structure arises through adiabatic density perturbations. This equation allows a degeneracy to be lifted when determining best–fit cosmological parameters to the CMB power spectrum, because it removes a freedom in choosing n_T once the other three fundamental parameters $\{A_S, A_T, n_T\}$ have been specified.

We now derive the corresponding consistency equation in the RSII braneworld. Throughout, we invoke the slow–roll approximation. In the braneworld scenario, this corresponds to the constraint $\tilde{\alpha}^2 \ll 1$ on the modified parameter (21). Given that the spectra (22) and (23) are modified from that of the standard scenario, one would anticipate a different form for the consistency equation, thus leading in principle to a method for distinguishing between the different scenarios. Differentiation of Eq. (23) with respect to k and substitution of Eqs. (24) and (26) implies that

$$n_T = -2 \frac{1}{N'} \frac{x'}{x} \frac{F^2}{\sqrt{1+x^2}}. \tag{31}$$

Substituting the Friedmann equation (17) and Eqs. (23) and (27) into Eq. (31) then implies that the tensor and scalar perturbations are related to the tensorial spectral index [25]:

$$\frac{A_T^2}{A_S^2} = -\frac{1}{2} n_T. \tag{32}$$

Remarkably, Eq. (32) has the same form as that of the consistency equation (30) for the standard inflationary scenario. This is surprising given that the spectra themselves are modified by the brane effects. Different consistency equations arise in other scenarios where the spectra are altered, such as in the warm inflationary scenario [40] and models of inflation driven by higher–order terms in the curvature invariants [43]. Corrections also appear when cross–correlations between adiabatic and isocurvature modes are accounted for [42].

From a mathematical point of view, the degeneracy between Eqs. (30) and (32) arises because the normalization mode function (24) is constrained to obey a specific differential equation [39]. This implies that the final expression must be independent of the brane tension, λ. The immediate question that should be addressed, therefore, is whether this degeneracy between the consistency equations is a coincidence, or whether it is indicative of some deeper physics. One possible way to lift the degeneracy is to consider the next–to–leading order terms in the slow–roll approximation. In this case, the scale dependence of the scalar spectral index becomes important [30, 44] and an open question is whether terms of the same form arise in the braneworld scenario. Finally, it has recently been argued that short–distance physics can result in corrections to the consistency relation in the standard scenario [41] and it is clearly of much interest to consider these effects further within the context of the braneworld scenario.

Acknowledgments

JEL is supported by the Royal Society. He thanks A. Garcia and the local organizing committee for the invitation to speak at the First Mexican Meeting on Mathematical and Experimental Physics.

References

[1] C. B. Netterfield et al., astro-ph/0104460; A. T. Lee et al., astro-ph/0104459; R. Stompor et al., astro-ph/0105062; N. W. Halverson et al., astro-ph/0104489; C. Pryke et al., astro-ph/0104490.

[2] M. Tegmark, Astrophys. J. **514** (1999) L69; M. Tegmark, M. Zaldarriaga, and A. J. Hamilton, Phys. Rev. **D63** (2001) 043007; X. Wang, M. Tegmark, and M. Zaldarriaga, astro-ph/0105091.

[3] G. Efstathiou and J. R. Bond, Mon. Not. Roy. Astron. Soc. **304** (1999) 75; G. Efstathiou, astro-ph/0109151; G. Efstathiou et al., astro-ph/0109152.

[4] W. Hu and S. Dodelson, astro-ph/0110414.

[5] B. P. Schmidt et al., Astrophys. J. **507** (1998) 46 ; A. G. Riess et al. [Supernova Search Team Collaboration], Astron. J. **116** (1998) 1009.

[6] S. Perlmutter et al. [Supernova Cosmology Project Collaboration], Astrophys. J. **517** (1999) 565; S. Jha et al., astro-ph/0101521; A. H. Jaffe et al., Phys. Rev. Lett. **86** (2001) 3475; R. Bean and A. Melchiorri, astro-ph/0110472.

[7] D. H. Lyth and A. Riotto, Phys. Rep. **314** (1999) 1; M. Kamionkowski and A. Kosowsky, Ann. Rev. Nucl. Part. Sci. **49** (1999) 77; A. R. Liddle and D. H. Lyth, Cosmological Inflation and Large-Scale Structure (Cambridge University Press, Cambridge, 2000).

[8] V. Mukhanov and G. Chibisov, Pis'ma Zh. Eksp. Teor. Fiz. **33** (1981) 549; [JETP Lett. **33** (1981) 532]; A. H. Guth and S. Y. Pi, Phys. Rev. Lett. **49** (1982) 1110 ; S. Hawking, Phys. Lett. **115B** (1982) 295; A. A. Starobinsky, Phys. Lett. **117B** (1982) 175; A. D. Linde, Phys. Lett. **116B** (1982) 335; J. M. Bardeen, P. J. Steinhardt, and M. S. Turner, Phys. Rev. **D28** (1983) 679.

[9] R. R. Caldwell, R. Dave, and P. J. Steinhardt, Phys. Rev. Lett. **80** (1998) 1582; L. Wang, R. R. Caldwell, J. P. Ostriker, and P. J. Steinhardt, Astrophys. J. **530** (2000) 17.

[10] I. Zlatev, L. Wang, and P. J. Steinhardt, Phys. Rev. Lett. **82** (1999) 896; P. J. Steinhardt, L. Wang, and I. Zlatev, Phys. Rev. **D59** (1999) 123504; I. Zlatev and P. J. Steinhardt, Phys. Lett. **B459** (1999) 570.

[11] J. Polchinski, String Theory (Cambridge University Press, Cambridge, 1998).

[12] E. Witten, Nucl. Phys. **B443**(1995) 85.

[13] C. M. Hull and P. K. Townsend, Nucl. Phys. **B438** (1995) 109.

[14] J. E. Lidsey, D. Wands, and E. J. Copeland, Phys. Rep. **337** (2000) 343.

[15] A. R. Liddle, A. Mazumdar and F. E. Schunck, Phys. Rev. **D58** (1998) 061301; K. Malik and D. Wands, Phys. Rev. **D59** (1999) 123501; E. J. Copeland, A. Mazumdar, and N. J. Nunes, Phys. Rev. **D60** (1999) 083506; A. M. Green and J. E. Lidsey, Phys. Rev. **D61** (2000) 167301.

[16] K. Akama, hep-th/0001113; V. A. Rubakov and M. E. Shaposhnikov, *Phys. Lett.* **B159** (1985) 22; N. Arkani-Hamed, S. Dimopoulos, and G. Dvali, *Phys. Lett.* **B429** (1998) 263; I. Antoniadis, N. Arkani-Hamed, S. Dimopoulos, and G. Dvali, *Phys. Lett.* **B436** (1998) 257; M. Gogberashvili, *Europhys. Lett.* **49** (2000) 396; L. Randall and R. Sundrum, *Phys. Rev. Lett.* **83** (1999) 3370.

[17] L. Randall and R. Sundrum, *Phys. Rev. Lett.* **83** (1999) 4690.

[18] J. Polchinski, *Phys. Rev. Lett.* **75** (1995) 4724.

[19] P. Binétruy, C. Deffayet, and D. Langlois, *Nucl. Phys.* **B565** (2000) 269; P. Binétruy, C. Deffayet, U. Ellwanger, and D. Langlois, *Phys. Lett.* **B477** (2000) 285.

[20] E. E. Flanagan, S. -H. Tye, and I. Wasserman, *Phys. Rev.* **D62** (2000) 044039.

[21] J. M. Cline, C. Grojean, and G. Servant, *Phys. Rev. Lett.* **83** (1999) 4245; C. Csáki, M. Graesser, C. Kolda, and J. Terning, *Phys. Lett.* **B462** (1999) 34.

[22] T. Shiromizu, K. Maeda, and M. Sasaki, *Phys. Rev.* **D62** (2000) 024012.

[23] R. Maartens, D. Wands, B. Bassett, and I. Heard, *Phys. Rev.* **D62** (2000) 041301.

[24] E. J. Copeland, A. R. Liddle, and J. E. Lidsey, *Phys. Rev.* **D64** (2001) 023509.

[25] G. Huey and J. E. Lidsey, *Phys. Lett.* **B514** (2001) 217.

[26] A. Lukas, B. A. Ovrut, K. S. Stelle and D. Waldram, *Phys. Rev.* **D59** (1999) 086001; *Nucl. Phys.* **B552** (1999) 246.

[27] G. T. Horowitz, I. Low and A. Zee, *Phys. Rev.* **D62** (2000) 086005.

[28] T. Barreiro, B. de Carlos, and N.J. Nunes, *Phys. Lett.* **B497** (2001) 136.

[29] D. Langlois, R. Maartens, and D. Wands, *Phys. Lett.* **B489** (2000) 259.

[30] J. E. Lidsey, A. R. Liddle, E. W. Kolb, E. J. Copeland, T. Barreiro, and M. Abney, *Rev. Mod. Phys.* **69** (1997) 373.

[31] R. Argurio, hep-th/9807171.

[32] J. P. Gauntlett, hep-th/9705011; and references therein.

[33] J. E. Lidsey, *Phys. Rev.* **D64** (2001) 063507.

[34] P. Kraus, *J. High En. Phys.* **9912** (1999) 11; D. Ida, *J. High En. Phys.* **0009** (2000) 14.

[35] H. A. Chamblin and H. S. Reall, *Nucl. Phys.* **B562** (1999) 133; H. A. Chamblin, M. J. Perry, and H. S. Reall, *J. High En. Phys.* **9909** (1999) 14.

[36] C. Barcelo and M. Visser, *Phys. Lett.* **B482** (2000) 183.

[37] W. Israel, *Nuovo Cim.* **44B** (1966) 1.

[38] G. Huey and R. Tavakol, *Phys. Rev.* **D65** (2002) 043504.

[39] A. N. Taylor and A. R. Liddle, *Phys. Rev.* **D65** (2002) 041301.

[40] A. N. Taylor and A. Berera, *Phys. Rev.* **D62** (2000) 083517.

[41] L. Hui and W. H. Kinney, astro-ph/0109107.

[42] N. Bartola, S. Matarrese, and A. Riotto, *Phys. Rev.* **D64** (2001) 123504.

[43] H. Noh and J. Hwang, *Phys. Lett.* **B515** (2001) 231.

[44] E. Stewart and D. H. Lyth, *Phys. Lett.* **B302** (1993) 171.

GALAXIES FORMATION FROM THE SCALAR FIELD DARK MATTER MODEL

Tonatiuh Matos
Departamento de Física,
Centro de Investigación y de Estudios Avanzados del IPN,
AP 14-740, 07000 México D.F., México.
tmatos@fis.cinvestav.mx

Abstract In recent works, we presented a model for the dark matter in cosmos supposing that dark matter is a scalar field endowed with a cosh scalar potential. We obtained that the model fit well the observed mass and angular power spectrums, is consistent with the observed acceleration of the expansion of the Universe and contains and natural cutoff for small structure at $\sim 0.1\ Mpc^{-1}$. In other works we presented a model for the dark matter in galaxies supposing that dark matter is a scalar field endowed with an exponential scalar potential. We found that the effective energy density goes like $1/(r^2 + b^2)$ and the resulting circular velocity profile of tests particles in it is in good agreement with the observed one in spiral galaxies. In this work we give a connection between both models, supposing that the cosmological scalar field fluctuations are the halos of the galaxies and discuss some of the physical consequences of the model.

Keywords: Cosmology, Galaxy Formation, Dark Matter

1. Introduction

Doubtless we are living very exiting times in cosmology. At this end of Millennium many new crucial observations have been carry out giving the human beings a new vision of the Universe. Among others, observations of the luminosity–redshift relation of Ia Supernovae suggest that distant galaxies are moving slower than predicted by Hubble's law, implying an accelerated expansion of the Universe[1, 2]. These observations open the possibility to the existence of an energy component in the Universe with a negative equation of state, $\omega < 0$, being $p = \omega\rho$, called dark energy. Current observations of CMBR anisotropy by BOOMERANG [3] and MAXIMA[4], could implies a flat, homogenous and isotropic Universe. The counting of the distribution of cluster and supercluster of galaxies gives us a new vision of the large scale structure of the universe [5]. The most successful cosmological model until now seems to be the Λ Cold Dark Matter (ΛCDM). This model consists of cold dark matter $\sim 25\%$, whose nature is unknown, and $\sim 75\%$ of cosmological constant Λ.

Nevertheless, in addition to the old problems of the cosmological constant, the Λ CDM model over predicts subgalactic structure and singular cores for the halos of galaxies[6]. Some problems of the cosmological constant paradigm can be ameliorated by Quintessence: a fluctuating, inhomogeneous scalar field (Q) rolling down a scalar potential $V(Q)$. However, it has not been agreement about which scalar potential $V(Q)$ is the correct one. It is assumed that flat models with $\Omega_M = 0.33 \pm 0.05$ and $\omega_Q = -0.65 \pm 0.07$ are the most consistent with all observations[7].

The matter component Ω_M of the Universe decomposes itself in baryons, neutrinos, etc. and dark matter. Observations indicate that stars and dust (baryons) represent something like 0.3% of the whole matter of the Universe. The new measurements of the neutrino mass indicate that neutrinos contribute with a same quantity like matter. On other words, say $\Omega_M = \Omega_m + \Omega_{DM} = \Omega_b + \Omega_\nu + \cdots + \Omega_{DM} \sim 0.05 + \Omega_{DM}$, where Ω_{DM} represents the dark matter part of the matter contributions which has a value of $\Omega_{DM} \sim 0.25$. This value of the amount of baryonic matter is in concordance with the limits imposed by nucleosynthesis (see for example [8]). But we do not know the nature of the dark matter component Ω_{DM}.

The flat profile of the rotational curves is maybe the main feature observed in many galaxies. There are some particles with nice features in super-symmetric theories which could be candidates to be the dark matter, they are called WIMP'S (Weak Interacting Massive Particles). However, since these candidates behave just like standard CDM, they can not explain the observed scarcity of dwarf galaxies and the smoothness of the galactic-core matter densities [11]. This is the reason why we need to look for alternative candidates that can explain both the structure formation at cosmological level, the observed amount of dwarf galaxies, and the dark matter density profile in the core of galaxies.

2. Scalar field as cosmological dark matter

We have proposed a scalar field model for dark matter in galaxies [12]. This model has created a great expectation for solving the problem of the nature of dark matter [13, 14]. The scalar field not only gives the correct energy density for the required matter in galaxies to predict the rotation curves of stars, but it is obtained the correct distribution of dark matter in galaxies as well. It also has the advantage to be a particle predicted by fundamental theories like superstrings, Kaluza–Klein, etc. At the galactic level, attention has been put on the quadratic potential Φ^2, because of the well known fact that it behaves as pressureless matter due to its oscillations around the minimum of the potential[15], implying that $\omega_\Phi \simeq 0$, for $<p_\Phi>=\omega_\Phi<\rho_\Phi>$. Following an analogous procedure to the one used in particle physics, we may write a phenomenological Lagrangian with all the terms we need in order to reproduce the observed Universe. In particular, for modelling the dark matter of the Universe we use a minimally coupled real scalar field Φ with a self-interaction potential of the form ($\kappa_0 = 8\pi G$, we use natural units such that $\hbar = c = 1$)[16, 17]

$$V(\Phi) = V_o[\cosh(\lambda\sqrt{\kappa_o}\Phi) - 1] \qquad (1)$$
$$= \begin{cases} \frac{1}{2}m_\Phi^2\Phi^2 + \frac{1}{24}\lambda^2 m_\Phi^2 \kappa_o \Phi^4 & |\lambda\sqrt{\kappa_o}\Phi| \ll 1 \\ (V_o/2)\exp(\lambda\sqrt{\kappa_o}\Phi) & |\lambda\sqrt{\kappa_o}\Phi| \gg 1 \end{cases}.$$

The mass of the scalar field Φ is defined as $m_\Phi^2 = V''|_{\Phi=0} = \lambda^2 \kappa_o V_o$. In this case, we have a massive scalar field. The components of the Universe are baryons, radiation, three species of light neutrinos, etc., and two minimally coupled and homogenous scalar fields Φ and Ψ, which represent the dark matter and the dark energy, respectively.

The scalar field has been proposed as a viable candidate, since it mimics standard CDM above galactic scales very well, reproducing most of the features of the standard (ΛCDM) model [17, 18, 13]. However, at galactic scales, the scalar field model presents some advantages over the standard ΛCDM model. For example, it can explain the observed scarcity of dwarf galaxies since it produces a sharp cut-off in the Mass Power Spectrum. Also, its self-interaction can, in principle, explain the smoothness of the energy density profile in the core of galaxies [17, 19].

At the cosmological scale, it is found that the mass of the boson is not the only parameter that determines the power spectrum. The self-interaction of the scalar field is also important. The free parameters of the scalar potential, V_0 and the scalar field mass $m_\Phi = \lambda\sqrt{V_0 \kappa_0}$, can be fitted by cosmological observations. Doing this one finds that [17]

$$\lambda \simeq 20.28, \qquad (2)$$
$$V_0 \simeq (3 \times 10^{-27} m_{Pl})^4, \qquad (3)$$
$$m_\Phi \simeq 9.1 \times 10^{-52} m_{Pl} = 1.1 \times 10^{-23} eV, \qquad (4)$$

with $m_{Pl} \equiv G^{-1/2}$ the Planck mass.

3. Scalar field dark matter in galaxies

The formation of galaxies through gravitational collapse of dark matter is not an easy problem to understand. A good model for galaxy formation has to take into account all the observed features of real galaxies. There are some ideas in this respect when dealing with a scalar field. Our main aim in this section is to present a plausible scenario for galaxy formation under the scalar field dark matter (SFDM) hypothesis. Through a gravitational cooling process [23, 24, 25], a cosmological fluctuation of the scalar field collapses to form a compact oscillaton by ejecting part of the field. The key idea consists precisely in assuming that such final object could distribute as galactic dark matter does. The final configuration then should consist of a central object (a core), i.e. an oscillaton, surrounded by a diffuse cloud of scalar field, both formed at the same time due to the same collapse process. The idea follows the standard idea of galaxy formation, namely that scalar field (dark matter) fluctuations are the responsible for the origin of the galaxies. In the case of the scalar field potential $\cosh(\lambda\Phi)$, we have used in [17] scalar field fluctuations of the form $\cosh(\lambda\Phi) \to \cosh(\lambda(\Phi + \delta\Phi)) \sim \exp(\lambda\delta\Phi) = \exp(\alpha\phi)$ for the regions where the scalar field fluctuation dominates $\delta\Phi > \Phi$. Of course, as in the standard theory of galaxy formation, the dark matter fluctuations are of different size in different regions of the Universe, for different galaxies. Therefore, at galactic level, we have a scalar field potential which depends on the local variable α. Thus, the exponential potential approximates the cosh potential in some regimes of the scalar field and we could develop some interesting aspects of a scalar dark matter halo in galaxies and although all this work is fully relativistic, it was done assuming staticity in galaxies [12, 26].

The question we are facing now is whether there is a dynamical mechanism that could provide a realistic scenario of galaxy formation using the scalar field dark matter hypothesis. First of all, a complete evolution of galactic and under galactic fluctuations belong to the non–linear regime of perturbations. The right answer would be provided by numerical evolutions of Einstein's equations in order to see the galaxy formation from the cosmological context. Fortunately, a partial answer is given in numerical research on Einstein's equations developed since 1990. In particular, the collapse of a scalar field has been studied deeply in [23, 24, 25] and it was found that there are final equilibrium and stable configurations for collapsed scalar field particles: boson stars (when the scalar field is complex) and oscillatons (when the scalar field is real and time-dependent), both of them being formed through a process called gravitational cooling[24]. Let us draw a possible physical picture of a galaxy collapse with dark matter being a scalar field endowed with the potential (1).

4. Scalar field collapse

In a realistic model the metric and fields should depend on space and time, thus a complete study would involve numerical calculations within and beyond General Relativity. One alternative is to study the behavior of the galaxy numerically with all the hypotheses stated above. In this section we will adopt this alternative, i.e. we will perform numerical simulations and in the next section we will perform semi–analytical calculations.

If a galaxy is an oscillaton, i.e. an oscillating soliton object, it must correspond to coherent scalar oscillations around the minimum of the scalar potential (1). For the scalar field collapse, the critical value for the mass of an oscillaton (the maximum mass for which a stable configuration exists) will depend on the mass of the boson. Roughly speaking, if we take $m_\Phi = 1.1 \times 10^{-23} eV$, and use the formula for the critical mass of the oscillaton corresponding to a scalar field with a Φ^2 potential (i.e. just a mass term), we expect the critical mass to be [23, 24]

$$M_{crit} \sim 0.6 \frac{m_{Pl}^2}{m_\Phi} \sim 10^{12} M_\odot. \tag{5}$$

This is a surprising result: the critical mass of the model shown in [17] is of the order of magnitude of the dark matter content of a standard galactic halo.

In order to study this situation for the case of a potential of the form (1), we present a numerical simulation of Einstein's equations in which the energy momentum tensor is that of a real scalar field. The scenario of galactic formation we assume is as follows: a sea of scalar field particles fills the Universe and forms localized primordial fluctuations that could collapse to form stable objects, which we will interpret as the dark matter halos of galaxies.

The numerical simulations suggest that the critical mass for the case considered here, using the scalar potential (1), is approximately [25, 28]

$$M_{crit} \simeq 0.1 \frac{m_{Pl}^2}{\sqrt{\kappa_0 V_0}} = 2.5 \times 10^{13} M_\odot. \tag{6}$$

The results of the numerical simulations are as follows. Essentially, we have found three different types of behavior for the scalar field collapse. In the first case, a generic feature is that scalar field distributions with an initial mass slightly larger than the critical mass collapse very violently and form a black hole. In the second type of behavior, fluctuations with an initial mass significantly smaller than the critical mass can not form stable oscillatons: the scalar field is completely ejected out as the system evolves [28]. The third behavior corresponds to a case where a fraction of the initial density is spread out, leaving an oscillating object that appears to be stable. This situation happens in a narrow window of initial conditions, between $0.05 - 1 \times M_{crit}$ [28].

From the cosmological point of view, the narrow window of initial conditions means that not all fluctuations will collapse into stable objects. Moreover, the collapsed objects will have masses of the same order of magnitude $M_{final} \sim 10^{12} M_\odot$, as it seems to be precisely the case for galaxies.

Summarizing, from the results of the numerical simulations of the collapse of the real scalar field with a cosh potential we find many similarities with the structure of the halos of galaxies. The scalar field density profile is not singular at the center. This fact, and the values of the final masses obtained using the cosmological values (3) and (4) for the parameters of the self-interaction potential, could correspond to objects like realistic galaxies. Moreover, it is in agreement with the observational constraints related to the phenomenological maximum galactic mass pointed out by Salucci and Burkert [29]. Therefore, we expect that fluctuations of this scalar field, due to Jeans instabilities, will in general collapse to form objects of the order of the mass of the halo of a typical galaxy.

We have shown before [17] that the SFDM model could be a good model for the universe at cosmological level, here we see that the scalar field could also be a good candidate for the dark matter content of individual galaxies (as suggested in [12, 26, 30]).

5. The galactic model with scalar field dark matter

5.1. A long exposition photograph of a galaxy

Summarizing, we have considered two working hypothesis up to now. First, we identify the formation of a central compact object and a halo with the gravitational collapse of a scalar field. The compact object could be a) an oscillaton (since we are dealing only with a real scalar field) or b) a Bose condensate. Second, we identify the ejected scalar field with the halo of this galaxy.

In this section we will adopt the second alternative, i.e. we will build a toy model for case a) stated above in purely geometrical terms and considering only the final stage of the collapse. To start with, we support our toy model on the numerical results studied in the previous section (see also [23, 24, 25]). First, since the time–dependence of the metric in an oscillaton is quite small [23], we suppose then that the center of this toy galaxy is an oscillaton which oscillates coherently but considering a *static* metric. This is an approximation because neither the galactic nuclei nor the oscillaton are expected to be static. However, for the purposes of this analytic work, we suppose that the dynamics of the oscillaton can be frozen in time in a way we explain below. Second, we do not expect the scalar halo to possess the same properties than the collapsed oscillaton; in some sense, they must be different. Thus, we will consider the scalar halo as another scalar field. Third, baryonic matter is considered to lie in the galaxy. Thus, we suppose that baryonic matter, which is part of the central object, (and only to the central object) will contribute to the curvature of the space–time of the galaxy.

This matter can be modelled as dust very well. As in previous works [12, 26], we let the luminous matter around the galaxy as test particles, i.e. they do not essentially contribute to the curvature of the space-time. Thus, we will take the baryonic matter surrounding the halo of the galaxy as test particles.

5.2. The analytical solution

Then, starting from a spherically symmetric space–time, using the *harmonic maps ansatz* [31] we were able to find a solution of the system. The exact solution of the averaged Einstein's field equations is

$$ds^2 = -B_0(r^2 + b^2)^{v_a^2}\left(1 - \frac{2M}{r}\right)dt^2 + \frac{A_0}{(1 - \frac{2M}{r})}dr^2 + r^2 d\Omega^2 \qquad (7)$$

with $d\Omega^2 = d\theta^2 + \sin^2\theta d\varphi^2$ and M is a constant with the interpretation discussed below. This metric is singular at $r = 0$, but it has an event horizon at $r = 2M$. This metric does not represent a black hole because it is not asymptotically flat. Nevertheless, for regions where $r \ll b$ but $r > 2M$ the metric behaves like a Schwarzschild black hole. Inside of the horizon the pressure of the perfect fluid is not zero anymore, thus our toy model is valid only in regions outside of the horizon, where it could be an approximation of the galaxy. Metric (7) in not asymptotically flat, but it has a natural cut off when the dark matter density equals the intergalactic density as mentioned in [26].

5.3. Physical features of the model

In order to understand the other parameters of the metric, let us proceed in the following way. It is believed that in a standard galaxy, the central object has a mass of $M \sim 2-3\times 10^6 M_\odot$ ~some $a.u.$ Far away from the center of the galaxy, say from $1pc$ up, the term $2M/r \ll 1$. In this limit metric (7) becomes

$$ds^2 = -B_0(r^2+b^2)^{v_a^2} dt^2 + A_0 dr^2 + r^2 d\Omega^2. \qquad (8)$$

This space–time is very similar to metric (18) of reference [26], but now with the potential

$$U(\psi(r)) = \frac{2v_a^2}{\kappa_0(1-v_a^2)} \frac{1}{(r^2+b^2)} \qquad (9)$$

being both solution the same in the limit $r \to \infty$. This implies that $A_0 = 1 - v_a^4$, recovering in this way the asymptotic results shown in [26]. Parameter b is related to parameter b of metric (21) in reference [12], where it acts as a gauge parameter. Of course this metric is only valid far away from the center of the galaxy. With parameter b it is now possible to fit quite well the rotation curves of spiral galaxies. Therefore metric (7) could not only represent the exterior part of the galaxy, but it could be a good approximation for the core part of it as well. Let us see this point.

The rotation curves v^{rot} seen by an observer at infinity for a spherically symmetric metric are given by $v^{rot} = \sqrt{rg_{tt,r}/(2g_{tt})}$ [26]. For metric (7) such result reads

$$v^{rot}(r) = \sqrt{\frac{v_a^2(r-2M)r^2 + M(r^2+b^2)}{(r-2M)(r^2+b^2)}}$$

formula that allows one to fit observational curves. Now let us explore an ADM–like concept of mass associated to our scenario. This mass can be calculated using the standard form of the metric

$$ds^2 = -e^{2\delta} dt^2 + \frac{dr^2}{(1-\frac{2m}{r})} + (1-\alpha)r^2 d\Omega^2 \qquad (10)$$

where $m = m(r)$ is interpreted as the mass function and $\delta = \delta(r)$ as the gravitational potential. This form of the metric is convenient because in this coordinates $m_{,r} = 4\pi r^2 \rho_T$, where ρ_T is the total density of the object. This interpretation is correct in regions where the space–time is almost flat, i.e. far away from the horizon $r = 2M$. Close to the horizon or inside of it, function m is a quantity that should be similar to the mass of the object, but it is not since it contains the contribution of all the components together; in this region, where the curvature of the space-time is huge, the volume element is different from $4\pi r^2 dr$. Furthermore, inside of the horizon we are not able to know the real physics of the object. On the other side, far away from the center of the toy galaxy, this function can be interpreted as the mass of an infinitesimal shell at radius r. Anyway we will call function m the mass function everywhere. Thus, the ADM-like mass is obtained at infinity by $M_{ADM} = \lim_{r \to \infty} m$. We perform the coordinate transformation $\sqrt{A_0} r \to r$, $\sqrt{A_0} b \to b$ in order to compare metrics (7) and (10). We obtain

$$ds^2 = -\frac{B_0}{A_0^{v_a^2}} \left(r^2 + b^2\right)^{v_a^2} \left(1 - \frac{2M\sqrt{A_0}}{r}\right) dt^2 + \frac{dr^2}{(1 - 2M\sqrt{A_0}/r)} + \frac{r^2}{A_0} d\Omega^2 \tag{11}$$

with $A_0 = 1 - v_a^4$. Thus, $M_{ADM} = \sqrt{1 - v_a^4} M$. Probably an observer at infinity would see the mass M_{ADM} at the center of the galaxy. Observe that for $r \gg b$ it follows the line element

$$ds^2 = -B_0 r^l dt^2 + A_0 dr^2 + r^2 d\theta^2 + r^2 \sin^2 \theta d\varphi^2 \tag{12}$$

where $l = 2(v^\varphi)^2$. This result is not surprising. Remember that the Newtonian potential ψ is defined as $g_{00} = -exp(2\psi) = -1 - 2\psi - \cdots$. On the other side, the observed rotational curve profile in the dark matter dominated region is such that the rotational velocity v^φ of the stars is constant, the force is then given by $F = -(v^\varphi)^2/r$, which respective Newtonian potential is $\psi = (v^\varphi)^2 \ln(r)$. If we now read the Newtonian potential from the metric (11), we just obtain the same result. Metric (11) is then the metric of the general relativistic version of a matter distribution, which test particles move in constant rotational curves. Observe that

$$V = -\frac{l}{\kappa_0 (2-l)} \frac{1}{r^2} \tag{13}$$

Thus, we recover the very important result, namely the scalar potential goes always as $1/r^2$ for a spherically symmetric metric with the *flat curve condition*. It is remarkable that this behavior of the stress tensor coincides with the expected behavior of the energy density of the dark matter in a galaxy. The effective density depends on the velocities of the stars in the galaxy, $\rho = (v^\varphi)^4/(1-(v^\varphi)^4) \times 1/(\kappa_0 r^2)$ which for the typical velocities in a galaxy is $\rho \sim 10^{-12} \times 1/(\kappa_0 r^2) \sim 1/3 10^{-12} H_0^{-2} \rho_{crit}/r^2$, while the effective radial pressure is $|P| = (v^\varphi)^2((v^\varphi)^2 + 2)/(1-(v^\varphi)^2) \times 1/(\kappa_0 r^2) \sim 10^{-6} \times 1/(\kappa_0 r^2)$, i.e., six orders of magnitude greater than the scalar field density. This is the reason why it is not possible to understand a galaxy with Newtonian dynamics. Newton theory is the limit of the Einstein theory for weak fields, small velocities but also for small pressures (in comparison with densities). A galaxy fulfills the first two conditions, but it has pressures six orders of magnitude bigger than the dark matter density, which is the dominating density in a galaxy. This effective pressure is the responsible for the behavior of the flat rotation curves in the dark matter dominated part of the galaxies.

Metric (12) is not asymptotically flat, it could not be so. An asymptotically flat metric behaves necessarily like a Newtonian potential provoking that the velocity profile somewhere decays, which is not the observed case in galaxies. Observe also that the matter density around a galaxy is smaller than the critical density, say $\rho_{around} \sim 0.06\rho_{crit}$, then $r_{crit} \approx 14 Kpc$, which correspond to a typical size for galaxies. At infinity, the observer will only measure M_{ADM}, i.e. it will see a Black-Hole-like metric at the center of the galaxy which horizon lies at $r = 2M_{ADM}$.

In Figure 2 of [30] the fit of the curves is done using the observed rotation curves of some dwarf galaxies, whose dark matter contribution is extremely dominating and therefore are considered as the *test of fire* for a dark matter model in galaxies. In general, for disc galaxies, the fit of the rotation curves using this metric is analogous as in reference [12]. It seems then that metric (7) is a good approximation for some late stadium of the space-time of a spiral galaxy; it is a good approximation of a "long exposition photograph" of a galaxy

Acknowledgments

I would like to thank L. A. Ureña, F. Siddhartha Guzmán and Dario Núñez for many very helpful discussions and comments, Erasmo Gómez, Aurelio Espíritu and Kenneth Smith for technical support. This work was partly supported by CONACyT México, Grant 34407-E.

References

[1] S. Perlmutter *et al. Astrophys. J.* **517** (1999) 565.

[2] A. G. Riess *et al.*, *Astron. J.* **116** (1998) 1009.

[3] P. de Bernardis *et al.*, *Nature* (London) **404** (2000) 955.

[4] A. Balbi *et al.*, *Astrophys. J.* **545** (2000) L1; Erratum–ibid. **558** (2001) L145. E–print *astro–ph/0005124*.

[5] Anthony H. Gonzalez, Dennis Zaritsky, Julianne J. Dalcanton, Amy Nelson, *Astrophys J.*, in press. E–print *astro–ph/0106055*; Kevin A. Pimbblet, Ian Smail, Alastair C. Edge, Warrick J. Couch, Eileen O'Hely, *Ann I. Zabludoff. MNRAS*, in press. E–print *astro–ph/0106258*

[6] Navarro, J. F. and Steinmetz, M. *Astrophys. J* **528** (2000) 607.

[7] Liming Wang, R. R. Caldwell, J. P. Ostriker and Paul J. Steinhardt, *Astrophys.J.* **530** (2000) 17.

REFERENCES

[8] D. N. Schramm, In "Nuclear and Particle Astrophysics", ed. J. G. Hirsch and D. Page, *Cambridge Contemporary Astrophysics, 1998*. Shi, X., Schramm, D. N. and Dearborn, D. *Phys. Rev.* **D50** (1995) 2414.

[9] Peebles, P. J. E. *Principles of Physical Cosmology*, (Princeton University Press, Princeton, 1993).

[10] Persic, M., Salucci, P. and Stel, F. *MNRAS* **281** (1996) 27.

[11] B. Moore, F. Governato, T. Quinn, J. Stadel and G. Lake, *Astrophys J.* **499** (1998) L15; Y. P. Jing and Y. Suto, *Astrophys J.* **529** (2000) L69.

[12] F. S. Guzmán and T. Matos, *Class. Quant. Grav.* **17** (2000) L9. E–print *gr-qc/9810028.*; T. Matos and F. S. Guzmán, Ann. Phys. (Leipzig) **9** (2000) S133.

[13] P.J.E. Peebles and B. Ratra, *Astrophys. L. Lett.* **325** (1988) L17; B. Ratra and P.J.E. Peebles, *Phys. Rev.* **D37** (1988) 3406.

[14] P.J.E. Peebles and A. Vilenkin, *Phys. Rev.* **D60** (1999) 103506. P.J.E. Peebles. E–print *astro-ph/0002495*.

[15] M. S. Turner, *Phys. Rev.* **D 28** (1983) 1253; L.H. Ford, *Phys. Rev.* **D35** (1987) 2955.

[16] Varun Sahni and Liming Wang, E–print *astro-ph/9910097*.

[17] T. Matos and L. A. Ureña-López, *Class. Quantum Grav.* **17** (2000) L75; T. Matos and L. A. Ureña–López, *Phys. Rev.* **D63** (2001) 63506.

[18] J. Goodman, E–print *astro-ph/0003018*.

[19] T. Matos and L. A. Ureña–López, *Nucl. Phys.* B (2002), in press. E–print *astro-ph/0010226*.

[20] D. Merritt, L. Ferrarese, and C. L. Joseph, *Science* **293** (2001) 1116.

[21] B. Moore, Nature **370** (1994) 629. A. Burkert, *Astrophys J.* **477** (1995) L25; J. A. Tyson, G. P. Kochanski and I. P. Dell'Antonio, *Astrophys J.* **498**(1998) L107.

[22] L. Ferrarese and D. Merritt, *Astrophys J. Lett.* **539**(2000) L9. K. Gebhardt et al., *Astrophys J. Lett.* **539** (2000) L13.

[23] E. Seidel and W. Suen, *Phys. Rev. Lett.* **66** (1991) 1221.

[24] E. Seidel and W. Suen, *Phys. Rev. Lett.* **72**(1994) 2516.

[25] L. A. Ureña-López. To be published.

[26] T. Matos, F. S. Guzmán,and D. Núñez, *Phys. Rev.* **D62** (2000) 061301.

[27] A. Riotto and I. Tkachev, *Phys. Lett.* **B484**(2000) 177.

[28] M. Alcubierre, F. S. Guzmán, T. Matos, D. Núñez, L. A. Ureña, and P. Wiederhold. To be published.

[29] P. Salucci and A. Burkert. E–print *astro-ph/0004397*.

[30] T. Matos and F. S. Guzmán, *Class. Quantum Grav.* **18** (2001) 5055. E–print *gr-qc/0108027*.

[31] T. Matos, *Ann. Phys.* (Leipzig) **46** (1989) 462; T. Matos, *J. Math. Phys.* **35** (1994) 1302; T. Matos, *Math. Notes* **58**(1995) 1178.

SCALAR SOLITON MODELLING DARK MATTER HALOS

Eckehard W. Mielke, Humberto H. Peralta
Departamento de Física,
Universidad Autónoma Metropolitana–Iztapalapa,
Apartado Postal 55-534, C.P. 09340, México, D.F., México
ekke@xanum.uam.mx, estocastico@msn.com

Franz E. Schunck
Institut für Theoretische Physik, Universität zu Köln,
50923 Köln, Germany
fs@thp.uni-koeln.de

Abstract   Within *standard* Newtonian gravity, galactic *dark matter* is modelled by a scalar field in order to *effectively* modify Kepler's law. In particular, we show that a *solvable* toy model with a *self-interaction* $U(\Phi)$ borrowed from non-topological solitons produces already qualitatively correct *rotation curves*. Although relativistic effects in the halo are very small, we indicate corrections arising from the generally relativistic formulation.

Keywords: Dark matter, rotation curves, scalar field models.

1. Introduction

In 1933 Zwicky [42], while investigating the Coma cluster, suggested the existence of *dark matter* in galaxy clusters. The total mass needed to gravitationally bind this cluster exceeds the amount of the luminous matter by roughly one order of magnitude. In the beginning of the 1970's one was able to extend the measurements of the rotation curves of galaxies so that one could find also there a higher mass to luminosity relation: after some radius one can see from the rotation curves that there is more mass than luminous matter. From these investigations, a halo radius of more than 220 kpc (the half–mass radius) is inferred for our Galaxy [7], and more recent results from satellite galaxies of a set of spiral galaxies indicate even more than 400 kpc [40]. By investigating data of

967 spirals, Persic et al. [24] confirmed that the structural properties of dark and visible matter are linked together. Accordingly, a galaxy with low luminosity is stronger dominated by a dark matter halo than a spiral one with high luminosity.

According to Newton's theory, the rotation curves for galaxies or galaxy clusters should show a Keplerian decrease $v^2 \propto 1/x$ at the rim of the luminous matter. Instead one observes *rather flat* rotation curves [29, 37]. More precisely, dwarf irregular and low surface brightness galaxies, whose halos are considered to be dark matter dominated, universally exhibit a logarithmic modification $v^2 \propto \ln x/x$ far from the center, according to the phenomenological fit of Burkert [3, 4]. On the other hand, bright galaxies with a supermassive center have rotation curves decreasing even more steeply at spatial infinity.

For several classes of gravitational theories, it has been argued that the introduction of dark matter is necessary [41]. Massive compact halo objects, so–called MACHOs, consisting of baryonic matter or, alternatively, of *boson stars* (BS) [20] do not seem to be sufficient to resolve this problem completely [1].

An intriguing possibility is a *scalar field* model of the halo first proposed by Schunck [30, 31, 32, 33] using a complex massless field coupled to the Einstein equation, an idea which was later taken up by several other authors. More recently, Spergel and Steinhardt [38] proposed that dark matter is self–interacting, cf. also Ref. [27].

Here we consider a model where dark matter is modelled by a primordial scalar field with a *self–interaction* $U(\Phi)$. Since the observed[1] rotation velocities are roughly bounded by $v_\varphi/c \leq \cdot 10^{-3}$, i.e. are non–relativistic, a Newtonian type approximation would be sufficient. In particular, a toy model with a Φ^6 repulsive self–interaction is known [16] to allow in the limiting case of *flat* spacetime an exact spherically symmetric *soliton* solution of the corresponding the nonlinear Klein–Gordon equation, as is shown in Sec. I. Simulating the halo by such a *non–topological soliton* (NTS) for the *positive* range of the potential, yields a Newtonian mass distribution which provides a qualitatively rather good fit to the rotation curve data of dwarf irregular and low surface brightness galaxies, see Sec. II and the comparison with observations in Sec. III. In Sec. IV, we indicate the extension to the generally relativistic formulation.

2. Approximate NTS solution

In the following, let us consider as a solvable toy model, a Φ^6 type potential

$$U(|\Phi|) = m^2 |\Phi|^2 \left(1 - \chi |\Phi|^4\right), \qquad \chi |\Phi|^4 \leq 1, \tag{1}$$

where m is the 'bare' mass of the boson and χ a coupling constant. The self–interaction in the radial Klein–Gordon equation takes the form $dU(P)/dP^2 = m^2 - 3m^2 \chi P^4$. In flat spacetime, such a model was first considered in Ref. [16] for constructing *non–topological soliton* (NTS) solutions.

For a spherically symmetric configuration, the corresponding nonlinear Klein–Gordon equation simplifies to an *Emden type equation* familiar from the astrophysics of gaseous spheres. It has the completely *regular* exact[2] solution

$$P(r) = \pm \chi^{-1/4} \sqrt{\frac{A}{1 + A^2 x^2}}, \tag{2}$$

where we introduced the *dimensionless* radial coordinate $x := mr$, and $A = \sqrt{\chi} P^2(0)$ in terms of the initial value. The solution depends essentially on the non-linear coupling parameter χ, since the limit $\chi \to 0$ would be singular. This feature is rather characteristic for *soliton solutions*. Already 1978, it has been generalized [16] to a NTS with (quantized) angular momentum l.

In the following, it suffices to restrict ourselves to the above given range for which the potential $U(|\Phi|)$ remains *positive*. If effects of self–gravitation are taken into account, this scalar potential needs not to be bounded from below: In the case of generally relativistic *boson stars* (BS) [17, 9, 12, 10, 34, 11] with a self–interaction $\lambda |\Phi|^4$, it has been proven numerically [35] that NTS exist even for negative values of λ. The extension to *soliton stars*[3] have been considered in Ref. [13] by Lee and Pang as well as Gleiser [5]. For related Q-stars, cf. [14].

The canonical energy–momentum tensor of a relativistic spherically symmetric scalar field is diagonal, i.e. $T_\mu{}^\nu(\Phi) = \text{diag}\,(\rho, -p_r, -p_\perp, -p_\perp)$ with

$$\begin{aligned} \rho &= \frac{1}{2}\left(\omega^2 P^2 + P'^2 + U\right), \\ p_r &= \rho - U, \\ p_\perp &= p_r - P'^2. \end{aligned} \tag{3}$$

where $' = d/dr$. The form (3) is familiar from perfect fluids, *except* that the radial and tangential pressure generated by the scalar field are in general different, i.e. $p_r \neq p_\perp$. In Ref. [15], e.g. this *anisotropy* of scalar matter has cavalierly been ignored, although it holds, as shown above, even in flat spacetime, or in the Newtonian approximation.

From (3) we find in flat spacetime the energy-density

$$\begin{aligned} \rho &= \frac{m^2}{2}\left[2P^2 + P'^2 - \chi P^6\right] \\ &= \frac{Am^2}{2\sqrt{\chi}(1+A^2x^2)}\left[2 + \frac{A^4x^2 - A^2}{(1+A^2x^2)^2}\right]. \end{aligned} \qquad (4)$$

(If we would consider a real scalar field instead, there is merely the change of the first proportionality constant 2 in the bracket to 1.) For dwarf galaxies, such a behavior of the density is observed. There, one gets good results by using the empirical isothermal density profile

$$\rho(r) \simeq \frac{\rho_0 r_c^2}{r_c^2 + r^2} = \rho_0[1 - (r/r_c)^2 + O(r^4)] ; \qquad (5)$$

comparing with our model, the central density is

$$\rho(0) = \frac{Am^2}{2\sqrt{\chi}}(2 - A^2) > 0 , \qquad (6)$$

and the core radius is $r_c^2 = (2-A^2)/[2m^2A^2(2A^2-1)]$. The requirement of the positivity of $U(\Phi)$ at the origin yields the constraint $A < 1$, which we will adopt in the following.

From observations we know that the central density of the halo is nearly independent of the total mass of the galaxy with an average value of $\rho(0) \simeq 0.02 M_\odot/\text{pc}^3$. Thus we expect the mass m and the coupling constant χ of the scalar field to be rather universal.

This non-linearly coupled scalar field exerts the following radial and tangential pressures:

$$\begin{aligned} p_\text{r} &= \frac{m^2}{2}\left[\chi P^6 + P'^2\right] = \frac{A^3 m^2}{2\sqrt{\chi}(1+A^2x^2)^2} \simeq \frac{A^3 m^2}{2\sqrt{\chi}} , \\ p_\perp &= \frac{m^2}{2}\left[\chi P^6 - P'^2\right] = \frac{A^3 m^2(1-A^2x^2)}{2\sqrt{\chi}(1+A^2x^2)^3} \simeq \frac{A^3 m^2}{2\sqrt{\chi}} , \\ \rho + p_\text{r} &= m^2[P^2 + P'^2] \\ &= \frac{Am^2}{\sqrt{\chi}(1+A^2x^2)}\left[1 + \frac{A^4x^2}{(1+A^2x^2)^2}\right]. \end{aligned} \qquad (7)$$

Thus, at the center of the NTS, we have $p_\text{r}(0) = p_\perp(0)$. Asymptotically, we find at radial infinity

$$p_\text{r}, -p_\perp \to \frac{m^2}{2\sqrt{\chi}Ax^4} . \qquad (8)$$

Dark matter halos

It is a crucial feature of our model that the energy of a thin spherical shell

$$\rho r^2 dr \to \frac{dr}{A\sqrt{\chi}} \tag{9}$$

tends for $x \to \infty$ to a constant. The resulting *mass function* $M(r) := \int_0^r \rho y^2 \, dy$ can be obtained by straightforward integration. With the aid of REDUCE) we find:

$$\begin{aligned} M(r) &= \frac{1}{mA\sqrt{\chi}} \left[x + \frac{A^2 - 8}{8A} \arctan(Ax) - \frac{A^2 x}{8} \frac{1 + 3A^2 x^2}{(1 + A^2 x^2)^2} \right] \\ &\simeq \frac{A}{6\sqrt{\chi}} m^2 \left(2 - A^2 \right) \cdot r^3 \\ &\to \frac{r}{A\sqrt{\chi}} \end{aligned} \tag{10}$$

3. Rotation curves

The tangential velocity v_φ of stars moving like "test particles" around the center of a galaxy is not directly measurable, but can be inferred from the redshift z_∞ observed at spatial infinity, for which

$$(1 + z_\infty)^2 \simeq \frac{(1 \pm v_\varphi)^2}{1 - v_\varphi^2} \tag{11}$$

holds. Due to their non–relativistic velocities in galaxies bounded by $v_\varphi/c \leq \cdot 10^{-3}$, we observe $z_\infty \simeq v_\varphi$ (as first part of a geometric series)[4]

In general, for the static spherically symmetric situation, an observer at rest on the equator of the galactic coordinate system measures the following tangential velocity[5] squared as a point particle (a star) flies past him in its circular orbit

$$v_\varphi^2 \simeq \frac{\kappa}{2} \left[\frac{M(r)}{r} + p_r r^2 \right] . \tag{12}$$

Without radial pressure for a weak gravitational field, (12) reduces to the Newtonian form $v_{\varphi,\text{Newt}}^2 \simeq M(r)/r$. As is well–known [39], a naive application of the Newtonian limit to general relativity (GR) would have led us to geodesics in flat spacetime, i.e. as if gravity would not affect the motion of test bodies like our stars moving in the dark matter halo. Thus it is mandatory to go beyond, as is indicated by our approximation of the generally relativistic formula in Ref. [22]. Then, also the pressure component $p_r \neq 0$ of an anisotropic 'fluid' contributes, our prime example being the case of scalar fields. For our NTS model, we will consider both terms, although, due to the fast decrease of p_r, cf. Eq. (8), the contribution of p_r to the asymptotic value of the rotation velocity is almost negligible.

From the shell energy (9) of the Newtonian NTS solution (2) and its radial pressure (6), we find for the rotation velocities

$$v_\varphi^2/v_\infty^2 = 1 + \left(\frac{A^2}{8} - 1\right)\frac{\arctan(Ax)}{Ax} + \frac{A^2}{8}\frac{A^2x^2 - 1}{(1 + A^2x^2)^2}, \quad (13)$$

for which the following approximations

$$v_\varphi^2 \simeq \frac{\kappa A(1 + A^2)m^2}{6\sqrt{\chi}} r^2$$

$$v_\varphi^2 \to v_\infty^2 = \frac{\kappa}{2A\sqrt{\chi}} = \frac{\kappa}{2\chi P^2(0)} \leq 10^{-6}, \quad (14)$$

hold near the center and at the far field, respectively. Together with (9) we conclude that asymptotically $M = 2v_\infty^2 r$.

Observationally, there is the rough restriction $v_\varphi/c \leq \cdot 10^{-3}$ of the rotation velocities of galaxies, which can be used to constrain the mass m and the coupling constant χ of our NTS model.

4. Comparison with observations

Already in 1995 Burkert [3, 28] noted the *universality* of galactic rotation curves and proposed the empirical fitting formula

$$\begin{aligned}v_{\varphi B}^2/v_0^2 &= \frac{1}{2x}\left\{\ln[(1 + x)^2(1 + x^2)] - 2\arctan(x)\right\} \\ &\simeq 1 - \frac{\arctan(x)}{x} = \frac{2}{3}x^2 - \frac{1}{2}x^3 + O(x^4) \\ &\to 2\frac{\ln x}{x}\end{aligned} \quad (15)$$

which has a maximum at $x = 3.3$ in dimensionless units $x = r/r_c$. It amounts to a *logarithmic modification* of the Kepler law at spatial infinity.

For 10 irregular dwarf and 7 low surface brightness galaxies, Kravtsov et al. [8] motivated the empirical density function $\rho/\rho_0 = x^{-0.2}(1 + x^2)^{-1.4}$ which implies the velocity profile $v_\varphi/v_0 = x^{0.9}/(1 + x^{3/2})^{2.48/3}$ which is fitting rather well to this set of observational data, cf. Fig. 3 of Ref. [4]. Since the universal Burkert profile (15) is nearly identical, we use it for the comparison of our NST curve with the data.

Our velocity curve (13) deduced from the NTS scalar model has a leading term which resemble those of another recent model based on a

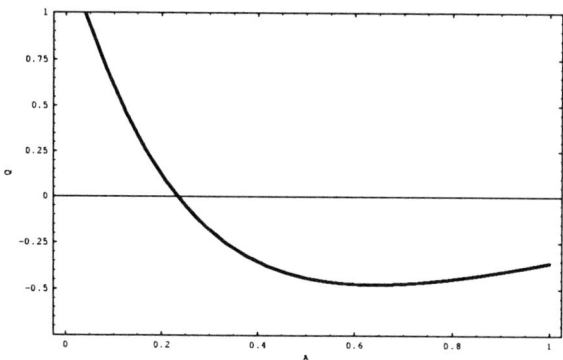

Figure 1. The quadratic deviation $Q(A)$ depending on the initial parameter A of the NTS rotation curve.

particular solution of the *Lane–Emden equation* [36] for a *T*runcated, nonsingular, *I*sothermal *S*phere (TIS), cf. [6]. This model agrees up to 1% with the empirical Burkert fit (15), whereas our result still has a considerable but astrophysically tolerable residue of up to 20 % within the observed range of 3 times the core radius. We notice that the form of rotation curve (13) depends on

the central value of the scalar field as well as the coupling constant χ due to parameter A. For $v_\infty = v_0$ a comparison of the expansions in (13) and (15) at the origin reveal that $A \simeq 1$.

In order to obtain a better fit to the observations, we calculate the *quadratic deviation*

$$Q(A) := \frac{1}{v_\infty^2} \int_0^4 (v_\varphi - v_{\varphi B})^2 \, dx = \frac{1}{v_\infty^2} \int_0^4 \left(v_\varphi^2 + v_{\varphi B}^2 - 2\sqrt{v_\varphi^2 v_{\varphi B}^2} \right) dx \tag{16}$$

of our rotation curve (13) from the Burkert fit (15) as a function of the initial parameter A of our exact NTS solution (2). The first and second term can be calculated analytically with MATHEMATICA, whereas the third term has been approximated by the iterated Simpson formula $S_n = \frac{h}{3} \left[f(a) + 2 \sum_{k=1}^{n-1} f(a + 2kh) + 4 \sum_{k=1}^{n} f(a + (2k-1)h) + f(b) \right]$ for the spacing $h = (b-a)/2n = 0.005$ of the intervals. This approximation of $Q(A)$ is drawn in Figure 1.

For the best empirical fit we use the standard condition

$$\frac{dQ(A)}{dA} = 0 \quad \text{and} \quad \frac{d^2 Q(A)}{dA^2} \geq 0 \tag{17}$$

of a minimum, which turns out to be $A \simeq 0.65$.

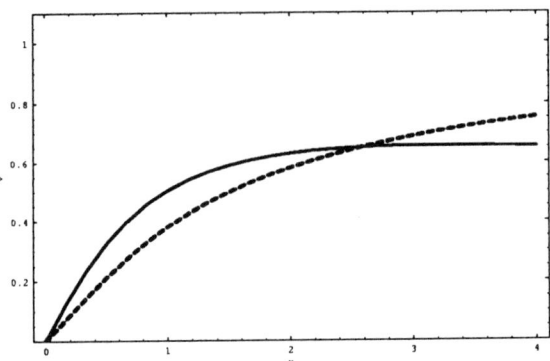

Figure 2. Comparison of the NTS rotation curve (- - -) for $A = 0.65$ with the Burkert fit (—). This choice of A gives the best overall agreement with the observations.

Since for the minimum of $Q(A)$ at $A \simeq 0.65$ the scalar potential $U(\Phi)$ is *still positive*, we can use this value of A in order to plot the NTS rotation curve (13) in comparison with the empirical Burkert fit (15), cf. Figure 2.

5. Outlook: Towards generally relativistic rotation curves

Although we have restrict ourselves to a Newtonian approximation, we may indicated the full *generally-relativistic framework* which departs from the coupled Einstein–Klein–Gordon equations

$$G_{\mu\nu} := R_{\mu\nu} - \frac{1}{2}g_{\mu\nu}R = -\kappa T_{\mu\nu}(\Phi), \qquad (18)$$

$$\left(\Box + \frac{dU}{d|\Phi|^2}\right)\Phi = 0. \qquad (19)$$

Here R is the curvature scalar, $\kappa = 8\pi G$, the gravitation constant ($\hbar = c = 1$), g the determinant of the metric $g_{\mu\nu}$, and Φ a *complex* scalar field. Moreover, $\Box := \left(1/\sqrt{|g|}\right)\partial_\mu\left(\sqrt{|g|}g^{\mu\nu}\partial_\nu\right)$ denotes the generally covariant d'Alembertian.

The stationarity ansatz

$$\Phi(r,t) = P(r)e^{-i\omega t} \qquad (20)$$

describes a spherically symmetric bound state of the scalar field with frequency ω. (The alternative case of a real scalar field could readily be accommodated in our formalism by putting $\omega = 0$ in our Ansatz.)

In the case of spherical symmetry, the line–element reads
$$ds^2 = e^{\nu(r)}dt^2 - e^{\lambda(r)}dr^2 - r^2\left(d\theta^2 + \sin^2\theta d\varphi^2\right), \tag{21}$$

in which the functions $\nu = \nu(r)$ and $\lambda = \lambda(r)$ depend on the Schwarzschild type radial coordinate r.

The decisive non–vanishing components of the Einstein equation are
$$\nu' + \lambda' = \kappa(\rho + p_r)re^\lambda, \tag{22}$$
$$\lambda' = \kappa\rho re^\lambda - \frac{1}{r}(e^\lambda - 1), \tag{23}$$

i.e. the 'radial' equations. Two further components are identically fulfilled because of the contracted Bianchi identity $\nabla^\mu T_\mu{}^\nu \equiv 0$ which is equivalent to the equation
$$\frac{d}{dr}p_r = -\nu'\left(\rho + p_r - \frac{2}{r}(p_r - p_\perp)\right) \tag{24}$$

of 'hydrostatic' equilibrium for an anisotropic fluid, a generalization of the Tolman–Oppenheimer–Volkoff equation, see Ref. [19]. In comparison, Eq. (18) of Ref. [23] seems to be misprinted[6].

The general solution of Eq. (23) is
$$e^{-\lambda} = 1 - \kappa\frac{M(r)}{r} \to 1 - 2v_\infty^2. \tag{25}$$

Asymptotically, according to (9) of the NTS model, the radial metric component $e^{-\lambda}$ approaches the value $1 - 2v_\infty^2 < 1$. After a redefinition of the radial coordinate $r \to \tilde{r} := r/\sqrt{1 - 2v_\infty^2}$, the asymptotic space has a *deficit solid angle*. The area of a sphere of radius r is not $4\pi r^2$, but $4\pi(1 - 2v_\infty^2)r^2$; cf. analogous results for global monopoles and global textures [2, 32; 23]. Thus, in GR, one could use the redefined radial coordinate in order to avoid a linear increase of the mass function which otherwise would cause a problem for the asymptotics. In the case of the more *realistic* phenomenological Burkert fit (15) the velocity tends to zero at spatial infinity, with the consequence that *no* such deficit angle is to be expected.

So far we have considered a solvable model in flat spacetime. However, when the tangential velocity $v_\varphi = v_\varphi(x)$ is known, we readily find from (12) that the metric components are in general given by
$$e^\nu = \exp\left\{2\int v_\varphi^2 d\ln x\right\} \to x^{2v_\infty^2} + K, \tag{26}$$
$$e^\lambda = \frac{1 + 2v_\varphi^2}{1 + \kappa p_r r^2} \to 1 + 2v_\infty^2 \simeq (1 - 2v_\infty^2)^{-1}, \tag{27}$$

where K is a constant of integration; in the last line we used (25). The approximation for e^λ is valid for $p_r r^2 \simeq 0$, i.e. for non-singular radial pressure at the origin or for sufficiently fast decreasing pressure at infinity. For example, in our NTS model, this condition is satisfied at the origin as well as at infinity, cf. the asymptotic function (8).

For our NTS rotation curve (13), we have derived the scalar field by setting the metric components to one in zeroth order approximation. Then, we calculated the mass function and, consequently, the rotation curve. Now, we can do better and determine the metric potentials in first order approximation. We find

$$e^\nu = \exp\left\{\frac{v_\infty^2}{2\chi^{3/2}}\left[\ln\left\{x^7\sqrt{1+x^2}\left(\frac{1+A^2x^2}{A^2x^2}\right)^{7/2}\right\}\right.\right.$$
$$\left.\left. -\frac{1}{1+x^2} + \frac{7A}{x}\arctan(Ax)\right]\right\}. \quad (28)$$

The shift function e^λ is influenced by both, the rotation velocity (13) and the radial scalar field pressure (6).

A more ambitious approach would be the *reconstruction* of the scalar potential $U(|\Phi|)$ from the empirical rotation curves on the basis of the Einstein equations, similarly as in the case of inflation [18] In view of the occurrence of scalar fields, like the dilaton or Kalb–Ramond axion, in effective superstring models, cf. [21] and the Refs. therein, the concordance of theory and observations in a viable self-interaction potential deserves further study,

Acknowledgments

We would like to thank Carl Brans and Roberto Sussman for helpful discussions and comments. This work was partially supported by CONACyT, grant No. 28339–E. One of us (F.E.S.) acknowledges research support provided by a personal fellowship. Moreover, (E.W.M.) thanks Noelia and Markus Gérard Erich for encouragement.

Notes

1. The maximum velocity within a rotation curve of almost 400 km/s is found in a Sa galaxy [25]; the velocity of 367 km/s at the farthest measured point belongs also to a Sa galaxy [26].
2. As is well-known within the Lane–Emden approach, there exists the further solution $P(r) = \pm(12\chi)^{-1/4}/\sqrt{x}$ which, in that classical context, corresponds to singular isothermal matter. Since it is singular at the origin and gives rise to a divergent shell energy $\rho r^2 dr \to (m/12\sqrt{\chi})rdr$, it will be discarded in the following. (Similarly, the corresponding singular solutions of [15] are rather unphysical.)

3. An example of a bounded potential is $U(|\Phi|) = m^2|\Phi|^2 \left(1 - |\Phi|^2/|\Phi_0|^2\right)^2$ typical for soliton stars, but then simple analytical expressions are not available and one needs to deal with asymptotic solutions like $P(r) \to \pm\sqrt{A/(1 + B^2 e^{2x})}$ for large radius.

4. This criterion is not met in Ref. [35] due to the high rotation velocities involved which appear near strong relativistic sources.

5. This is based on the generally relativistic relation $\tilde{v}_\varphi^2 = rN'/N$ for the tangential velocity, where $N = e^\nu/2$ is the lapse function. It should be noted, however, that for the Newtonian approximation $ds^2 \simeq N^2 dt^2 - d\vec{x}^2 = (1 - 2M/r)dt^2 - d\vec{x}^2$ of the metric this does not reproduce the usual Keplerian limit. Since $ds \simeq \sqrt{N^2 dt^2 - v^2 dt}$ in this approximation, the identification of the velocity in the Lorentz type factor $1/(N\sqrt{1 - rN'/N}) = 1/\sqrt{N^2 dt^2 - v^2}$ in the redshift formula (10) of Ref. [23] would have led us back to the alternative expression $v_\varphi^2 := \frac{1}{2}r\nu'e^\nu = \frac{1}{2}\left[1 - e^{-\lambda} + \kappa p_r r^2\right]e^{\lambda+\nu}$. Anyhow, outside of matter ($\nu = -\lambda$) for a weak gravitational field, both forms reduce to the Keplerian relation $v_{\varphi,\text{Newt}}^2 \simeq M(r)/r$.

6. Nucamendi et al. [23] have mistakenly criticized Ref. [32]: Actually, the rotation curves in [32] use a dark matter component for which the gravitational influence over the gravitational redshift is at least two order of magnitudes lower than the influence of the velocity of the stars, cf. Sec. IV; so their argument is wrong. Additionally, those authors doubt the *correct* formula of Ref. [32] for the composition of velocities.

References

[1] C. Alcock et al., *Phys. Rev. Lett.* **74** (1995) 2867.

[2] M. Barriola and A. Vilenkin, *Phys. Rev. Lett.* **63** (1989) 341; N. Turok and D. Spergel, *Phys. Rev. Lett.* **64** (1990) 2736.

[3] A. Burkert, *Astrophys. J.* **447** (1995) L25.

[4] A. Burkert and J. Silk, astro–ph/9904159.

[5] M. Gleiser, *Phys. Rev. Lett.* **63** (1989) 1199.

[6] I.T. Iliev and P.R. Shapiro, *Astrophys. J.* **546** (2001) L5.

[7] C.S. Kochanek, *Astrophys. J.* **457** 228 (1996).

[8] A.V. Kratsov, A.A. Klypin, J.S. Bullock and J.R. Primack, *Astrophys. J.* **502** (1998) 48.

[9] F.V. Kusmartsev, E.W. Mielke and F.E. Schunck, *Phys. Rev.* **D 43** (1991) 3895.

[10] F.V. Kusmartsev and F.E. Schunck, *Physica* **B178** (1992) 24.

[11] F.V. Kusmartsev and F.E. Schunck, in: *Classical and Quantum Systems — Foundations and Symmetries*, Proceedings of the 2nd International Wigner Symposium (Goslar, 16–20 July 1991), H.D. Doebner, W. Scherer, and F.E. Schroeck (eds.) (World Scientific Publ., Singapore, 1993), pp. 766.

[12] F.V. Kusmartsev, E.W. Mielke and F.E. Schunck, *Phys. Lett.* **A157** (1991) 465.

[13] T.D. Lee and Y. Pang, *Phys. Rep.* **221** (1992) 251.

[14] B.W. Lynn, *Nucl. Phys.* **B321** (1989) 465.

[15] T. Matos, F.S. Guzman and D. Nuñez, *Phys. Rev.* **D62** (2000) 061301.

[16] E.W. Mielke, *Phys. Rev.* **D18** (1978) 4525; *Lett. Nuovo Cim.* **25** (1979) 424.

[17] E.W. Mielke and R. Scherzer, *Phys. Rev.* **D24** (1981) 2111.

[18] E.W. Mielke and F.E. Schunck, *Phys. Rev.* **D52** (1995) 672.

[19] E.W. Mielke and F.E. Schunck, in: *Gravity, Particles and Space–Time*, ed. by P. Pronin and G. Sardanashvily (World Scientific, Singapore, 1996), pp. 391.

[20] E.W. Mielke and F.E. Schunck, *Nucl. Phys.* **B564** (2000) 185.

[21] E.W. Mielke and F.E. Schunck, *Gen. Rel. Grav.* **33** (2001) 805.

[22] E.W. Mielke and F.E. Schunck, preprint 2001.

[23] U. Nucamendi, M. Salgado and D. Sudarsky, *Phys. Rev.* **D 63** (2001) 12016.

[24] M. Persic, P. Salucci and F. Stel, *Mont. Not. R. Astr. Soc.* **281** (1996) 27.

[25] V.C. Rubin, W.K. Ford, Jr., and N. Thonnard, *Astrophys. J.* **238** (1980) 471.

[26] V.C. Rubin, D. Burstein, W.K. Ford, Jr., and N. Thonnard, *Astrophys. J.* **289** (1985) 81.

[27] A. Riotto and I. Tkachev, astro–ph/0003388v2.

[28] P. Salucci and A. Burkert, *Astrophys. J.* **537** (2000) L9.

[29] D.W. Sciama: *Modern cosmology and the dark matter problem* (Cambridge, Cambridge University Press, 1993).

[30] F.E. Schunck: *Selbstgravitierende bosonische Materie*, Ph. D. thesis, University of Cologne (Cuvillier Verlag, Göttingen 1996).

[31] F.E. Schunck, in: *Aspects of Dark Matter in Astro- and Particle Physics*, H.V. Klapdor-Kleingrothaus and Y. Ramachers, eds. (World Scientific, Singapore 1997), pp. 403.

[32] F.E. Schunck, astro–ph/9802258.

[33] F.E. Schunck, in: *Proc. of the Eigth Marcel Grossman Meeting on General Relativity*, Jerusalem, 1997, T. Piran and R. Ruffini, eds. (World Scientific, Singapore, 1999) pp. 1447.

[34] F.E. Schunck, F.V. Kusmartsev and E.W. Mielke, in: *Approaches to Numerical Relativity*, R. d'Inverno ed. (Cambridge University Press, Cambridge, 1992), pp. 130.

[35] F.E. Schunck and A.R. Liddle, *Phys. Lett.* **B404** (1997) 25.

[36] P.R. Shapiro, I.T. Iliev and A.C. Raga, *Mont. Not. R. Astr. Soc.* **307** (1999) 203.

[37] R. Smith: *Observational Astrophysics* (Cambridge, Cambridge University Press, 1995).

[38] D.N. Spergel and P.J. Steinhardt, *Phys. Rev. Lett.* **84** (2000) 3760.

[39] R.M. Wald: *General Relativity* (University of Chicago Press, Chicago 1984), p. 78.

[40] D. Zaritzky, R. Smith, C.S. Frenk and S.D.M. White, astro–ph/9611199.

[41] V.V. Zhytnikov and J.M. Nester, *Phys. Rev. Lett.* **73** (1994) 2950.

[42] F. Zwicky, *Helv. Phys. Acta* **6** (1933) 110.

THE DE SITTER/ANTI- DE SITTER BLACK HOLES PHASE TRANSITION?

Shin'ichi Nojiri
Department of Applied Physics
National Defence Academy, Hashirimizu Yokosuka 239-8686, JAPAN
nojiri@cc.nda.ac.jp, snojiri@yukawa.kyoto-u.ac.jp

Sergei D. Odintsov
Tomsk State Pedagogical University
Tomsk, RUSSIA and Instituto de Fisica de la Universidad de Guanajuato,
Lomas del Bosque 103, Apdo. Postal E-143, 37150 Leon,Gto., MEXICO
odintsov@ifug5.ugto.mx, odintsov@mail.tomsknet.ru

Abstract We investigate the Schwarzschild–Anti–deSitter (SAdS) and SdS BH thermodynamics in 5d higher derivative gravity. The interesting feature of higher derivative gravity is the possibility for negative (or zero) SdS (or SAdS) BH entropy which depends on the parameters of higher derivative terms. The appearance of negative entropy may indicate a new type instability where a transition between SdS (SAdS) BH with negative entropy to SAdS (SdS) BH with positive entropy would occur or where definition of entropy should be modified.

Keywords: Black hole thermodynamics, (anti-)de Sitter space, higher derivative gravity

BH thermodynamics is quite attractive, as it provides the understanding of gravitational physics at extremal conditions. Moreover, it has been realized recently that AdS BH may be relevant in the study of AdS/CFT correspondence. Hence, there appears nice way to describe strong coupling gauge theories via their gravitational duals. In the present work we discuss the thermodynamics of dS and AdS BHs in higher derivative gravity. The fundamental issue of entropy for such objects leads to some interesting conclusions.

We start from the following action of d dimensional R2–gravity with cosmological constant. The action is given by:

$$S = \int d^{d+1}x\sqrt{-g}\left\{aR2 + bR_{\mu\nu}R^{\mu\nu} + cR_{\mu\nu\xi\sigma}R^{\mu\nu\xi\sigma} + \frac{1}{\kappa^2}R - \Lambda\right\}. \quad (1)$$

We discuss the relation between SdS and SAdS BHs based on entropy considerations. For simplicity, we consider $c = 0$ case in (1) for most results. When $c = 0$, Schwarzschild–anti de Sitter space is an exact solution:

$$ds2 = -e^{2\nu(r)}dt2 + e^{-2\nu(r)}dr2 + r2\sum_{i,j=1}^{d-1}\tilde{g}_{ij}dx^i dx^j,$$

$$e^{2\nu} = e^{2\nu_0} \equiv \frac{1}{r^{d-2}}\left(-\mu + \frac{kr^{d-2}}{d-2} + \frac{r^d}{l^2}\right). \quad (2)$$

Here $\tilde{g}_{ij}$ is the metric of the Einstein manifold, which is defined by $\tilde{R}_{ij} = kg_{ij}$. $\tilde{R}_{ij}$ is the Ricci curvature given by $\tilde{g}_{ij}$ and k is a constant. For example, one has $k = d - 2$ for $d - 1$–dimensional unit sphere, $k = -(d - 2)$ for $d - 1$–dimensional unit hyperboloid, and $k = 0$ for flat surface. The curvatures have the following form: $\hat{R} = -\frac{d(d+1)}{l^2}$ and $\hat{R}_{\mu\nu} = -\frac{d}{l^2}\hat{G}_{\mu\nu}$. In (1), μ is the parameter corresponding to mass and the scale parameter l is given by solving the following equation:

$$0 = \frac{d2(d+1)(d-3)a}{l^4} + \frac{d2(d-3)b}{l^4} - \frac{d(d-1)}{\kappa^2 l^2} - \Lambda. \quad (3)$$

When the solution for l^2 (3) is positive (negative), the spacetime is asymptotically anti–de Sitter (de Sitter) space and especially if $\mu = 0$, the solution expresses the pure anti–de Sitter or de Sitter space. If $\Lambda = 0$, there is asymptotically flat (Minkowski) solution, where $\frac{1}{l^2} = 0$. In the following we concentrate on the case of $d = 4$.

The calculation of thermodynamical quantities like free energy F, the entropy S and the energy E may be done with the help of the following method: After Wick-rotating the time variable by $t \to i\tau$, the free energy F can be obtained from the action S (1) where the classical solution is substituted: $F = -TS$. Substituting Eq.(3) into (1) in the case of $d = 4$ with $c = 0$, one gets

$$S = -\int d^5x\sqrt{-G}\left(\frac{8}{l^2\kappa^2} - \frac{320a}{l^4} - \frac{64b}{l^4}\right)$$

$$= -\frac{V_3}{T}\int_{r_H}^{\infty}drr^3\left(\frac{8}{l^2\kappa^2} - \frac{320a}{l^4} - \frac{64b}{l^4}\right) \quad (4)$$

Here V_3 is the volume of 3d sphere and we assume τ has a period of $1/T$. The expression of S contains the divergence coming from large r. In order to subtract the divergence, we regularize S (3) by cutting off the integral at a large radius r_m and subtracting the solution with $\mu = 0$:

$$S_{reg} = -\frac{V_3}{T}\left\{\int_{r_H}^{r_m} drr^3 - \frac{e^{\nu(r=r_m)}}{e^{\nu(r=r_m;\mu=0)}}\int_0^{r_m} drr^3\right\}$$
$$\times \left(\frac{8}{\kappa^2 l^2} - \frac{320a}{l^4} - \frac{64b}{l^4}\right) \quad (5)$$

The factor $e^{\nu(r=r_m)}/e^{\nu(r=r_m;\mu=0)}$ is chosen so that the proper length of the circle which corresponds to the period $\frac{1}{T}$ in the Euclidean time at $r = r_m$ coincides with each other in the two solutions. Taking $r_m \to \infty$, one finds, as found in [1],

$$F = -V_3\left(\frac{l^2\mu}{8} - \frac{r_H 4}{4}\right)\left(\frac{8}{l^2\kappa^2} - \frac{320a}{l^4} - \frac{64b}{l^4}\right) \quad (6)$$

The horizon radius r_h is given by solving the equation $e^{2\nu_0(r_H)} = 0$ in (1):

$$r_H^2 = -\frac{kl^2}{4} + \frac{1}{2}\sqrt{\frac{k^2}{4}l^4 + 4\mu l^2} . \quad (7)$$

The Hawking temperature T_H is

$$T_H = \frac{k}{4\pi r_H} + \frac{r_H}{\pi l^2} \quad (8)$$

where $'$ denotes the derivative with respect to r. From the above equation (8), r_H can be rewritten in terms of T_H as

$$r_H = \frac{1}{2}\left(\pi l^2 T_H \pm \sqrt{(\pi l^2 T_H)^2 - kl^2}\right) \quad (9)$$

In (9), the plus sign corresponds to $k = -2$ or $k = 0$ case and the minus sign to $k = 2$ case.[1] One can also rewrite μ by using r_H as $\mu = \frac{r_H^4}{l^2} + \frac{kr_H^2}{2}$. Then we can rewrite F using r_H as

$$F = -\frac{V_3}{8}r_H^2\left(\frac{r_H^2}{l^2} - \frac{k}{2}\right)\left(\frac{8}{\kappa^2} - \frac{320a}{l^2} - \frac{64b}{l^2}\right) . \quad (10)$$

Then the entropy $S = -\frac{dF}{dT_H}$ and the thermodynamical energy $E = F + TS$ can be obtained as follows [1]:

$$S = 4V_3\pi r_H 3\left(\frac{1}{\kappa^2} - \frac{40a}{l^2} - \frac{8b}{l^2}\right), \quad E = 3V_3\mu\left(\frac{1}{\kappa^2} - \frac{40a}{l^2} - \frac{8b}{l^2}\right),$$
$$(11)$$

This seems to indicate that the contribution from the $R2$–terms can be absorbed into the redefinition:

$$\frac{1}{\tilde{\kappa}^2} = \frac{1}{\kappa^2} - \frac{40a}{l^2} - \frac{8b}{l^2}, \tag{12}$$

although this is not true for $c \neq 0$ case.

On the other hand, by using the surface counter term method [2], one gets the following expression for the conserved mass M:

$$M = \frac{3l^2}{16} V_3 \left(\frac{1}{\kappa^2} - \frac{40a}{l^2} - \frac{8b}{l^2} - \frac{4c}{l^2} \right) \left(k2 + \frac{16\mu}{l^2} \right). \tag{13}$$

One can also start from the expression for M with $c = 0$ as the thermodynamical energy E:

$$E = 3V_3 \left(\frac{1}{\kappa^2} - \frac{40a}{l^2} - \frac{8b}{l^2} \right) \left(\frac{k^2 l^2}{16} + \frac{\mu}{l^2} \right) \tag{14}$$

The expression of energy E (14) is different from that in (11) by a first μ–independent term, which comes from the AdS background. Using the thermodynamical relation $dS = \frac{dE}{T}$, we find

$$\begin{aligned} S &= \int \frac{dE}{T_H} = \int dr_H \frac{dE}{d\mu} \frac{d\mu}{dr_H} \frac{1}{T_H} \\ &= \frac{V_3 \pi r_H 3}{2} \left(\frac{8}{\kappa^2} - \frac{320a}{l^2} - \frac{64b}{l^2} \right) + S_0. \end{aligned} \tag{15}$$

Here S_0 is a constant of the integration. Up to the constant S_0, the expression (14) is identical with (11). We should note that the entropy S (11) becomes negative, when

$$\frac{8}{\kappa^2} - \frac{320a}{l^2} - \frac{64b}{l^2} < 0.. \tag{16}$$

This is true even for the expression (14) for the black hole with large radius r_H since S_0 can be neglected for the large r_H.

We now investigate in more detail what happens when Eq.(16) is satisfied. First we should note l^2 is determined by (3), which has, in case of $d = 4$, the following form:

$$0 = \frac{80a + 16b}{l^4} - \frac{12}{\kappa^2 l^2} - \Lambda, \tag{17}$$

There are two real solutions for l^2 when $\frac{6}{\kappa^2} + (80a + 16b)\Lambda \geq 0$ and the solutions are given by

$$\frac{1}{l^2} = \frac{\frac{6}{\kappa^2} \pm \sqrt{\frac{6}{\kappa^2} + (80a + 16b)\Lambda}}{80a + 16b}. \tag{18}$$

Suppose $\kappa^2 > 0$. Then if

$$(80a + 16b)\Lambda > 0, \tag{19}$$

one solution is positive but another is negative. Therefore there are both of the asymptotically AdS solution and asymptotically dS one. Let us denote the positive solution for l^2 by l^2_{AdS} and the negative one by $-l^2_{\text{dS}}$:

$$l^2 = l^2_{\text{AdS}}, \ -l^2_{\text{dS}}, \quad l^2_{\text{AdS}}, \ l^2_{\text{dS}} > 0. \tag{20}$$

Then when the asymptotically AdS solution is chosen, the entropy (14) has the following form:

$$\mathcal{S}_{\text{AdS}} = \frac{V_3 \pi r_H 3}{2} \left(\frac{8}{\kappa^2} - \frac{320a + 64b}{l^2_{\text{AdS}}} \right). \tag{21}$$

Here we have chosen $\mathcal{S}_0 = 0$. On the other hand, when the solution is asymptotically dS, the entropy (14) has the following form:

$$\mathcal{S}_{\text{dS}} = \frac{V_3 \pi r_H 3}{2} \left(\frac{8}{\kappa^2} + \frac{320a + 64b}{l^2_{\text{dS}}} \right). \tag{22}$$

When

$$\frac{8}{\kappa^2} - \frac{320a + 64b}{l^2_{\text{AdS}}} < 0, \tag{23}$$

the entropy $\mathcal{S}_{\text{AdS}}$ (21) is negative!

There are different points of view to this situation. Naively, one can assume that above condition is just the equation to remove the non-physical domain of theory parameters. However, it is difficult to justify such proposal. Why for classical action on some specific background there are parameters values which are not permitted? Moreover, the string/M-theory and its compactification would tell us what are the values of the theory parameters.

From another side, one can conjecture that classical thermodynamics is not applied here and negative entropy simply indicates to new type of instability in asymptotically AdS black hole physics. Indeed, when Eq.(23) is satisfied, since $80a + 16b > 0$ (same range of parameters!), the entropy $\mathcal{S}_{\text{dS}}$ (22) for asymptotically dS solution is positive. In other words, may be the asymptotically dS solution would be preferable?

On the other hand, when

$$\frac{8}{\kappa^2} + \frac{320a + 64b}{l^2_{\text{dS}}} < 0, \tag{24}$$

the entropy $\mathcal{S}_{\text{dS}}$ in (22) is negative and the asymptotically dS solution is instable (or does not exist). In this case, since $80a + 16b < 0$, the entropy $\mathcal{S}_{\text{AdS}}$ in (21) for asymptotically AdS solution is positive and the asymptotically AdS solution would be preferable. Expression for the AdS black hole mass in (14) tells that when $\frac{8}{\kappa^2} - \frac{320a+64b}{l_{\text{Ads}}^2} = 0$, the AdS black hole becomes massless then there would occur the condensation of the black holes, which would make the transition to the dS black hole. On the other hand, when $\frac{8}{\kappa^2} + \frac{320a+64b}{l_{\text{ds}}^2} = 0$, the dS black hole becomes massless then there would occur the condensation of the black holes and the AdS black hole would be produced. Note that above state with zero entropy (and also zero free energy and zero conserved BH mass) is very interesting. Perhaps, this is some new state of BHs. As we saw that is this state which defines the border between physical SAdS (SdS) BH with positive entropy and SdS (SAdS) BH with negative entropy.

Hence, there appeared some indication that some new type of phase transition (or phase transmutation) between SdS and SAdS BHs in higher derivative gravity occurs. Unfortunately, we cannot suggest any dynamical formulation to describe explicitly such phase transition (it is definitely phase transition not in standard thermodynamic sense).

Let us consider now the entropy for Gauss-Bonnet case, where $a = c$ and $b = -4c$ in (1). For this purpose, we use the thermodynamical relation $dS = \frac{dE}{T}$. For the Gauss-Bonnet case, the energy (13) has the following form [2]:

$$E = M = \frac{3l^2}{16} V_3 \left(\frac{1}{\kappa^2} - \frac{12c}{l^2} \right) \left(k2 + \frac{16\mu}{l^2} \right) \tag{25}$$

We also found [2]

$$\mu = \frac{1}{2l^2} \left(k\epsilon - \frac{1}{2} \right)^{-1} \left\{ (2\epsilon - 1)r_H 4 - kr_H 2l^2 \right\} . \tag{26}$$

Here $\epsilon \equiv \frac{c\kappa^2}{l^2}$. Then using (25), (26), and the expression of the Hawking temperature,

$$4\pi T_H = \frac{1}{2} \left(ck + \frac{r_H 2}{2\kappa^2} \right)^{-1} \left[\frac{kr_H}{\kappa^2} - \frac{8cr_H 3}{l^4} + \frac{4r_H 3}{\kappa^2 l^2} - \frac{2Q2}{3g2r_H 3} \right] , \tag{27}$$

the entropy can be obtained as

$$S = \int \frac{dE}{T_H} = \int dr_H \frac{dE}{d\mu} \frac{\partial \mu}{\partial r_H} \frac{1}{T_H} \tag{28}$$

$$= \frac{V_3}{\kappa^2} \left(\frac{1 - 12\epsilon}{1 - 4\epsilon} \right) (4\pi r_H 3 + 24\epsilon k\pi r_H) + \mathcal{S}_0 . \tag{29}$$

Here S_0 is a constant of the integration, which could be chosen to be zero if we assume $S = 0$ when $r_H = 0$. When $\epsilon = 0$ ($c = 0$), the expression reproduces the standard result

$$S \to \frac{4\pi V_3 r_H{}^3}{\kappa^2} \,. \tag{30}$$

The entropy (28) becomes negative (at least for the large black hole even if $S_0 \neq 0$) when

$$\frac{1}{12} < \epsilon < \frac{1}{4} \,. \tag{31}$$

Therefore the unitarity might be broken in this region but it might be recovered when $\epsilon > \frac{1}{4}$. Even in case $\epsilon < 0$ ($k = 2$), the entropy becomes negative when

$$r_H{}^2 < -12\epsilon \,, \tag{32}$$

if $S_0 = 0$. Then the small black hole might be unphysical.

The fact discovered here–that entropy for S(A)dS BHS in gravity with higher derivatives terms may be easily done to be negative by the corresponding choice of parameters is quite remarkable. It is likely that thermodynamics for black holes with negative entropies should be reconsidered. In this respect one possibility would be to redefine the gravitational entropy for higher derivative gravity.

Acknowledgments

The authors are grateful to M. Cvetič for collaboration. SDO would like to thank the organizers of First Mexican Meeting on Mathematical and Experimental. Physics for hospitality. The work by SN is supported in part by the Ministry of Education, Science, Sports and Culture of Japan under the grant n. 13135208.

Notes

1. When $k = 2$, as we can see from (7) and (8), r_H, and also T_H, are finite in the limit of $l \to \infty$, which corresponds to the flat background. Therefore we need to choose the minus sign in (9) for $k = 2$ case.

References

[1] S. Nojiri, S.D. Odintsov and S. Ogushi, hep–th/0105117, to appear in *Int.J.Mod.Phys.* **A**; hep—th/0108172, to appear in *Phys.Rev.* **D**.

[2] M. Cvetič, S. Nojiri and S.D. Odintsov, hep–th/0112045.

UNDERSTANDING HAIRY BLACK HOLES WITH THE ISOLATED HORIZON FORMALISM

Daniel Sudarsky
Instituto de Ciencias Nucleares
Universidad Nacional Autónoma de México
A. Postal 70-543, México D.F. 04510, México
sudarsky@nuclecu.unam.mx

Abstract   We present a novel application of the Isolated Horizon formalism, to the study of various properties of hairy black hole solutions that exist in various theories of non–lineal matter fields coupled with Einstein's gravity. These solutions have been found with the use of numerical techniques and their properties have been left mostly unexplained. The use of this approach has allowed the construction of a phenomenological model that provides an easy understanding of most of the qualitative properties of the existing results, as well as to the prediction of numerical relations between the different numerical values of several of the parameters characterizing these solutions. Particularly interesting are the relations between parameters describing the black hole solutions, with the regular solitonic solutions that often appear in association with the former in these types of theories. The available numerical calculations support all these predictions.

Keywords: Hairy Black Holes, Isolated Horizon.

1.   Introduction

The discovery, using numerical techniques, of solutions representing nontrivial regular and static solutions black holes in various simple theories of matter fields interacting with gravitation, came in the early 90's as a big surprised to a large segment of the general relativistic community, which was for the most part persuaded by the arguments behind of Wheeler's "no hair" conjecture, and the theorems supporting it in a large set of specific situations. Perhaps, due to the numerical nature of their discovery, or simply because there didn't seem to be a way to address the issue, the specific data describing such solutions has for the most part remain unexplained and almost no attempt was made for a long time to try to look for, and to understand regularities in the numerical results.

A development coming seemingly from a completely unconnected direction has changed the situation. The search for a method to evaluate the entropy of a black hole on the basis of the loop quantum gravity program, lead Ashtekar and his collaborators [1], to develop a formalism that allows to treat in a fully hamiltonian setting the description of certain types of Black Holes: The Isolated Horizon Formalism. The basic idea is that in order to have a well defined hamiltonian treatment of the " exterior of the Black Hole" one need to impose conditions that allow the writing of appropriate surface terms to the naive hamiltonian so the full hamiltonian is differentiable. The conditions at infinity correspond to the requirement of asymptotic flatness, and boundary term at infinity is as usual the ADM mass. Similarly the conditions at the inner boundary, correspond to the Isolated Horizon conditions, and the boundary term corresponds essentially to a "Horizon Mass". Basic properties of the resulting Hamiltonian, allows us, when considering this formulation for the case of theories containing Hairy Black Holes, to establish a series of relations among the parameters describing them. This resulted in testable predictions supported the existing numerical data, and which can be extended to theories and situations in which results are yet to be obtained. The work presented here has been carried out in collaborations with A. Ashtekar and A. Corichi and U. Nucamendi and has been reported in several articles [2, 3]. I have kept references to a minimum due to space constraints and so I apologize to the many authors whose relevant work is not cited explicitly. A more complete referencing has been included in the above mentioned articles.

1.1. Isolated horizon and horizon mass

The notion of Isolated Horizon is a sort of generalization of the definition of a Killing horizon. The idea is to avoid constraining the space-time away from the horizon, while preserving just those properties that are needed to obtain a consistent Hamiltonian framework. In essence, an Isolated horizon, or more precisely a weakly isolated horizon, Δ is an expansion–free, null, 3–dimensional sub–manifold Δ of the 4-dimensional space–time (M, g_{ab}), equipped with an equivalence class of future–directed null normals ℓ^a (where two such null normals are equivalent is one is a rescaling of the other) which Lie–drag (certain components of) the intrinsic connection on Δ. For details see [4]. In fact it turns out that in order to have a consistent Hamiltonian formulation for the evolution of the external region we can not choose the evolution vector field as in an arbitrary way on phase space, and the consistency condition turns out to be equivalent to the requirement that the horizon energy satisfies first law of black hole mechanics.

To define the horizon mass M_{hor}, one proceeds as follows. Recall that, in the standard asymptotically flat context, the ADM energy E^t_{ADM} arises as the boundary term at spatial infinity in the expression of the Hamiltonian generating evolution along an asymptotic time–translation t^a. Now the idea is to use as phase space the sector of general relativity consisting of space–times which are asymptotically flat at spatial infinity *and* admit an isolated horizon Δ as *internal boundary*. The Hamiltonian generating an appropriate time-translation then has two surface terms, one at infinity, $E^{(t)}_\infty$, and one at the internal horizon boundary, $E^{(t)}_\Delta$. The first is precisely the ADM energy $E^t_\infty = E^t_{ADM}$. The second, $E^{(t)}_\Delta$, is interpreted as the horizon energy associated with the time translation t^a at the horizon. Then, the Hamiltonian H_t generating evolution along an acceptable (one that is compatible with the first law) vector field t^a is given by:

$$H^{(t)} = \int_\Sigma (\text{constraints})\, d^3x + E^{(t)}_{ADM} - E^{(t)}_\Delta \qquad (1)$$

where Σ is any partial Cauchy slice extending from the isolated horizon Δ to spatial infinity.

At infinity, one uses the asymptotic flatness and choose a canonical normalization for t^a there, thus leading to a well defined prescription to define M_{ADM}: i.e. the value of $E^{(t)}_\infty$ corresponding to a unit time translation that is normal to the hypersurface in question. At the internal boundary, there is in general no canonical choice for such normalization. In the case of theories in which there is no hair, namely in theories where a unique static black hole solution, is fully determined by the charges at infinity, one can use the normalization of t^a at the horizon to coincide with of the properly normalized Killing Field of the static solution that has the same horizon parameters (i.e. Charge and Area in the case of Einstein–Maxwell Theory). In the cases where there is hair, the situation is more complicated (and in general there might no be such a global canonical choice). However in the static cases which are the ones of interest in this paper we have a preferred normalization: the one given by the static Killing field that is appropriately normalized at infinity. We defined the Horizon mass as the surface term in the hamiltonian associated with the internal boundary, corresponding to the above chosen normalization of t^a at the horizon;

$$M_{\text{hor}} = E_\Delta^{(t_0)} \qquad (2)$$

where t_0^a is the static Killing field that is normalized to unity at infinity. In practice, we can use the first law to evaluate the Horizon mass of a static solution that lies in a branch of static solutions connecting to the solution with zero horizon area (i.e. the regular solitonic solution) when such solution exist. In this way we obtain for the case of Einstein–Yang–Mills Theory (EYMT) and the nth branch of static spherically symmetric colored black holes, the expression for the horizon mass:

$$M_{\text{hor}}^{(n)}(r_H) = (1/2) \int_0^{r_H} \beta^{(n)}(r) dr \qquad (3)$$

where, the horizon radius is defined by $r_H = \sqrt{A/4\pi}$ with A the horizon area, and $\beta^{(n)}(r)$ is the surface gravity of the nth colored black hole of radius r, multiplied by $2r$.

The main relation between black holes and solitons. There is a general argument that states that, within each connected component of the phase space the value of the Hamiltonian H remains constant. The argument goes as follows: The Hamilton equations of motion can be written as $\delta H = \Omega(\delta, X_H)$, where δ is an arbitrary variation and X_H is the Hamiltonian vector field. A static solution is one at which the Hamiltonian vector field either vanishes or generates pure gauge evolution. In either case, the symplectic structure evaluated on X_H and *any* arbitrary vector field δ vanishes. Therefore, for this point of the phase space, $\delta H = 0$ for any direction δ. In particular $\delta H = 0$ for variations relating two static solutions. Thus H is constant on any connected component of static solutions (actually the result only requires stationarity rather than staticity).

The question now is to find the value of the Hamiltonian for one such component. Lets concentrate for simplicity on the EYMT and more specifically on the space of static spherically symmetric solutions (SSS) within this theory. This set of black holes is one–dimensional (parameterized by r_H), and spaces corresponding to distinct values of n belong to disconnected components. That is, the space SSS in EYMT has a countable number of connected components. Thus, for the nth branch we have:

$$H^{(n)} = M_{ADM}^{(n)}(r_H) - M_{hor}^{(n)}(r_H)] = c^{(n)} \qquad (4)$$

This in particular implies that its value is independent of the radius r_H of the horizon. Thus one is allowed to take the limit $r_H \to 0$. Now, it is well known that the colored black holes converge point-wise to the Bartnik–McKinnon soliton solutions [5] and that the ADM mass satisfies $M_{\text{ADM}}^{(n)} \mapsto M_{\text{BK}}^{(n)}$ when $r_H \mapsto 0$. Furthermore, the mass of the black hole goes to zero in this limit, so we can conclude that $H^{(n)} = M_{\text{BK}}^{(n)}$ that is, the total value of the Hamiltonian equals the ADM mass of the nth Bartnik–McKinnon soliton solution. Thus we have arrived at the unexpected relation,

$$M_{\text{ADM}}^{(n)}(r_H) = M_{\text{BK}}^{(n)}(r_H) + M_{\Delta}^{(n)}(r_H) \tag{5}$$

The formula above suggests that we think of a colored black hole as a 'bound state' of an ordinary, 'bare' black hole and a 'solitonic residue'. Indeed, we can rewrite the total mass of the colored black hole as a sum of three parts, the soliton's mass, the mass of the 'bare' black hole, and the binding energy:

$$M_{\text{ADM}}^{(n)}(r_H) = M_{\text{sol}}^{(n)} + M_{\text{hor}}^{(0)}(r_H) + E_{\text{binding}}(r_H), \tag{6}$$

where $M_{\text{hor}}^{(0)}(r_H) = r_H/2$ is the horizon mass of the Schwarzschild black hole of radius r_H and $E_{\text{binding}}(r_H) = (M_{\text{hor}}^{(n)} - M_{\text{hor}}^{(0)})(r_H)$ is the binding energy.

We can now use the physically expected properties of the gravitational binding energy to make predictions. In particular, since E_{binding} is the gravitational binding energy must be negative, we are lead to the expectation that: i) $M_{\text{hor}}^{(n)}(r_H) < M_{\text{hor}}^{(0)}(r_H)\ \forall n > 0\ and\ \forall r_H$. Using the expression (3) of the horizon mass, we conclude that ii) $(\beta_{(n)} - \beta_{(0)})(r_H) < 0$, i.e. $(\kappa_{(n)} - \kappa_{(0)})(r_H) < 0$. Next, since on physical, grounds one expects the absolute value of the gravitational binding energy increases as the mass of either of the two bound objects grows. Then, let us first look at the situation where r_H is fixed and we vary n. The mass $M_{\text{hor}}^{(0)}$ of the bare black hole is fixed but, as is well-known, the soliton mass $M_{\text{sol}}^{(n)}$ increases with n. Thus $|E_{\text{binding}}|$ must increase with n. This implies iii) *For any fixed value of r_H both the horizon mass $M_{\text{hor}}^{(n)}$ and the surface gravity $\kappa_{(n)}$ are monotonically decreasing functions of n.*

Next, let us consider the opposite situation and keep $n > 0$ fixed and vary r_H. Now the soliton mass $M_{\text{sol}}^{(n)}$ is fixed and the bare black hole mass $M_{\text{hor}}^{(0)}(r_H) = r_H/2$ increases with r_H. Hence, again (the absolute value of) the gravitational binding energy must increase with r_H. Thus, we conclude that iv) *For any fixed n, $(r_H/2 - M_{\text{hor}}^{(n)})(r_H)$ is a monotonically increasing function of r_H whence, from (3), $\beta^{(n)}(r_H) < 1$ for all $n > 0$ and r_H.*

Next, we use the fact that, for any fixed n the ADM mass $M^{(n)}_{\text{ADM}}(r_H)$ is a monotonically increasing function of r_H. Since $M^{(n)}_{\text{sol}}$ is fixed on the nth branch of static solutions, we conclude that: **v)** $M^{(n)}_{\text{hor}}(r_H)$ *is also a monotonically increasing function of* r_H. This implies that, for all n and r_H, $\beta^{(n)}$ and $\kappa^{(n)}$ are positive functions of r_H. Hence $M^{(n)}_{\Delta}$ is also positive for all n and r_H, except for $r_H = 0$ where it vanishes.

Finally, using the fact that $M^{(n)}_{\text{ADM}}(r_H) > M^{(0)}_{\text{ADM}}(r_H) = r_H/2$, we conclude that $M^{(n)}_{\text{hor}}(r_H) = M^{(n)}_{\text{ADM}}(r_H) - M^{(n)}_{\text{sol}} > \frac{r_H}{2G} - M^{(n)}_{\text{sol}}$. Thus, we conclude that, for any given n, the curve $M^{(n)}_{\text{hor}}(r_H)$ lies between the two parallel lines $f_1(r_H) = r_H/2 - M^{(n)}_{\text{sol}}$ and $f_2(r_H) = r_H/2$. Furthermore, since $M^{(n)}_{\text{hor}}(r_H)$ is a monotonically increasing function of r_H, it follows that: **vi)** *For any* n, *the curve showing the dependence of* $M^{(n)}_{\text{hor}}(r_H)$ *on* r_H *becomes asymptotically parallel to the two lines which bound it, i.e., has slope* $1/2$ *for large* r_H. Hence, the curves showing the dependence of $\beta^{(n)}$ on r_H asymptotically approach the curve $\beta^{(0)}(r_H) = 1$ and the curves showing the functional dependence of $\kappa_{(n)}$ of r_H asymptotically approach the curve $\kappa^{(0)}(r_H) = 1/2r_H$.

These properties account for all qualitative features of the dependence of $M^{(n)}_{\text{hor}}$ and $\beta^{(n)}$ on n and r_H. This qualitative understanding of the horizon properties, relied on the model and on two known qualitative properties of the ADM mass: $M^{(n)}_{\text{sol}}$ increases with n and, for any given n, $M^{(n)}_{\text{ADM}}$ increases monotonically with r_H.

More complicated theories and crossing phenomena. Let us now turn to more general theories with hair, by allowing Higgs or Proca fields in addition to the Yang–Mills fields, or considering Einstein–Skyrme theories [6, 7, 8]. These theories have additional dimensionfull constants which trigger new phenomena.

We will begin with a qualitative argument to elucidate the 'origin' of the crossing phenomenon. Let us first ignore gravity and consider these theories in Minkowski space–time. The coupling constants and other parameters (such as the vacuum values of fields) in these theories provide a length scale which determines the 'size' $R^{\text{Mink}}_{\text{sol}}$ of the solitonic solution. Let us now couple these theories to general relativity. Gravitational effects, being attractive, can only reduce this size. Hence, if a black hole is to exhibit any hair at all, its horizon radius r_H must be *less than* $R^{\text{Mink}}_{\text{sol}}$. (In fact, using more detailed arguments involving the pressure of matter fields near the horizon and at infinity, one can give a sharper bound: $r_H < (2/3)R^{\text{Mink}}_{\text{sol}}$ [10].) Thus, in these theories which have an in-built length scale already in Minkowski space–time, the horizon radius r_H of hairy black holes is bounded above by that length scale which is independent of Newton's constant. If there is an upper bound to the horizon radius, it is natural to ask what happens in the phase diagram to the individual, connected branches corresponding to static black holes. The most natural expectation is for disconnected branches in the Einstein–Yang–Mills theory to merge smoothly in these theories. Then, one would see two branches which cross at the point corresponding to the maximum value of r_H. This is indeed what is found in numerical investigations of the spherically symmetric, static black hole sectors of Einstein–Skyrme, Einstein–Yang–Mills–Proca, and Einstein–Yang–Mills–Higgs theories [6, 7, 9] and, more recently, in the static axially symmetric black hole sector of Einstein–Yang–Mills-Higgs theory [11]. Our discussion suggests that this crossing phenomenon will occur in any theory that contains a length scale independent of G (which we have set as 1) and allows hairy black holes.

Let us start by analyzing the way two branches merge at the cross–over point. For simplicity, let us focus on static black holes that carry no other charge *at infinity* beyond the ADM mass. As long as the energy momentum tensor in the theory satisfies appropriate energy conditions, the static black holes can be assumed to admit a Killing horizon with a bifurcation two surface. For such black holes, arbitrary perturbations are known to obey the first law of black hole dynamics: $\delta M_{\text{ADM}} = \frac{1}{8\pi} \kappa \, \delta a_\Delta$. Now consider the plot of the ADM mass as a function of the horizon radius, for this family and focus on the intersection point between the two branches. If the two curves intersect at a finite angle, the first order change in the mass for a given infinitesimal change in the horizon area would depend on the branch used to approach the intersection point. This would contradict the fact that the surface gravity has a well defined value at the solution corresponding to the intersection point. Hence, we conclude: **vii)** *at any cross–over point, the two branches must have the same tangent vector*. Explicit calculations have confirmed this behavior in all theories where crossing phenomena has been seen to occur.

Next, let us examine how the crossing phenomenon affects the notion of the horizon mass. As before, given any permissible evolution vector field t^a we are again led to a horizon energy $E_\Delta^{(t)}$ and the corresponding first law . The delicate point again is the selection of a preferred evolution vector field. As in the Einstein–Yang–Mills case, we can again focus on any one branch, say the nth, of static solutions to select the evolution vector field $t^a_{(n)}$ and construct the horizon mass $M_{\text{hor}}^{(n)}$ of any black hole on this branch. This strategy is tenable except at the cross–over points where evidently there is an ambiguity. This is a reflection of the fact that, in general, the notion of the horizon mass refers to the phase space, and thus the value of the horizon mass of a black hole is not determined simply by its space–time geometry.

However, in spite of — or rather, because of — this ambiguity, one can extract useful information from this definition of horizon mass. Let the nth and $(n+1)^{th}$ branches intersect at the horizon radius r_H^*. Then, using (5) it follows that:

$$M_{\text{ADM}}(r_H^*) - M_{\text{hor}}^{(n)}(r_H^*) = M_{\text{sol}}^{(n)}, \quad M_{\text{ADM}}(r_H^*) - M_{\text{hor}}^{(n+1)}(r_H^*) = M_{\text{sol}}^{(n+1)}. \tag{7}$$

This yields:

$$M_{\text{sol}}^{(n+1)} - M_{\text{sol}}^{(n)} = M_{\text{hor}}^{(n)}(r_H^*) - M_{\text{hor}}^{(n+1)}(r_H^*) = \frac{1}{2} \oint \beta(r)\,dr \tag{8}$$

where the last closed counter integral is performed by first moving along the nth branch from $r_H = 0$ to $r_H = r_H^*$, then moving back along the $(n+1)^{th}$ branch to $r_H = 0$ and finally sliding down the "y axis" to the point $r_H = 0$ on the nth branch.

This equation is striking in that it provides a quantitative relation between the horizon properties of hairy black holes and masses of solitons, even though the two belong to completely different sectors of the theory. The equality was first checked numerically in certain static, axisymmetric hairy black hole solutions in the Einstein–Yang–Mills–Higgs theory [11]. Given the generality of our considerations, one expects this relation to hold also for more general matter sources which lead to the crossing phenomena, such as Yang–Mills–Higgs, Yang–Mills–Proca and Skyrme fields.

Finally we note that it is possible to extend the above arguments to make quantitative predictions on the behavior of the ADM mass of *solitons* in the *Einstein–Yang–Mills theory*, despite the fact that in this theory there is no crossing phenomena at all. The idea is to consider the static sector of Einstein–Yang–Mills case as the limit, of the static sector in Einstein–Yang–Mills–Higgs (EYMH) Theory, when the vacuum expectation value η of the Higgs field goes to zero. This is justified, by the following argument: Let us now, consider the static sector of a family of EYMH theories, parameterized by η. As η decreases, the second mass parameter of the theory becomes smaller and the upper bound on the area of the hairy black holes increases. Thus we expect that the point of intersection of the nth and $(n+1)^{th}$ branches would move towards larger values of R as η decreases. Moreover in the limit $\eta \to 0$ we would expect this intersection point to move towards infinity leading to a situation where the different branches do not intersect, as is in fact observed to happen in pure EYM theory. We will support this argument by explicitly proving that for the case $\eta = 0$ the static, purely magnetic solutions of EYMH theory have vanishing Higgs field. The proof is a simple generalization of a Bekenstein "No hair Theorem" [12]:

Consider an EYMH theory with scalar field Φ with potential $V(\Phi) = \lambda(\Phi^*\Phi - \eta^2)^2$, where λ, and η are constants. Consider a static black hole solution with timelike Killing field ξ^a. The equation of motion for the scalar field is:

$$D^a D_a \Phi - \frac{\partial V}{\partial \Phi^*} = 0 \qquad (9)$$

where D_a stands for the gauge and metric covariant derivative $D_a = \nabla_a - ieA_a^i T^i$ with ∇_a is the metric compatible derivative operator, A_a^i stand for the gauge fields, T^i for the lie algebra generators, and e is gauge coupling constant. For non–extremal black holes the space-time has a bifurcate Killing Horizon with bifurcation surface S. Let t be the Killing parameter and consider M the region of space–time bounded by Σ_1 and Σ_2 hypersurfaces of constant t, S and asymptotic infinity. Multiplying the field equation by Φ^* and integrating over M we obtain:

$$0 = \int_{\partial M} \Phi^* D_a \Phi d\Sigma^a - \int_M (D_a\Phi^* D_b\Phi g^{ab} + \Phi^* \frac{\partial V}{\partial \Phi^*})\sqrt{-g} d^4 x \qquad (10)$$

The boundary integral consists of four terms: The integrals over Σ_1 and Σ_2 that are equal in magnitude and opposite in sign due to the time translation invariance, the integral at infinity that vanishes because of the fall off conditions on Φ required from asymptotic flatness, and the terms associated with S which vanish as S has zero measure. Next we write the inverse metric $g^{ab} = -N^2 \xi^a \xi^b + h^{ab}$, where N is the norm of the Killing Filed ξ^a and h^{ab} is the inverse of spatial metric Σ. Using the fact that $\xi^a D_a \Phi$ vanishes as long as the solution is static and that $A_0^i \equiv \xi^a A_a^i = 0$ corresponding to the purely magnetic sector, we obtain:

$$\int_M (D_a\Phi^* D_b\Phi h^{ab} + \Phi^* \frac{\partial V}{\partial \Phi^*})\sqrt{-g} d^4 x \qquad (11)$$

The first term in the integrand is positive semidefinite in general and the second term becomes positive semidefinite when $\eta = 0$. Thus in this situation the only possibility is $D_a \Phi = 0$ and $\Phi^* \frac{\partial V}{\partial \Phi^*} = 0$ which for $\eta = 0$ implies $\Phi = 0$. Thus by considering the limit of the static and purely magnetic sector of EYMH theory in the limit $\eta \to 0$ we expect to obtain the static and purely magnetic sector of EYM theory, and thus, by considering Eq. 8 in this limit we obtain a formula for the mass of the nth soliton in EYM theory:

$$M_{sol}^{(n)} = \frac{1}{2} \oint \beta(r) dr \qquad (12)$$

where the integral is performed first along the Schwarzschild branch to $r = \infty$ and back to $r = 0$ along the nth branch. That is, we are regarding Minkowski space–time as the 0th (degenerated) soliton, where the Schwarzschild branch (the 0th branch) begins. This relation has been checked directly and verified by the numerical data [3].

2. Discussion

The subject of hairy black holes has became an important point in the mathematical physics literature. By now, static hairy black hole solutions have been found numerically for a variety of matter sources coupled to general relativity. Their horizon attributes, such as the dependence of the surface gravity on the horizon area, have been plotted and their stability has been studied in some detail. The literature in the field has grown steadily over the last decade with discoveries of new families of solutions made by combining analytic and numerical techniques. Therefore, a long list of facts regarding hairy black holes is now available. The heuristic model we have introduced can account for many of qualitative features of the accumulated data ass well as for some quantitative relations amongst them.

Acknowledgments

This work was supported in part by the CONACYT grant 32272–E and by the DAGPA–UNAM grant 112401.

References

[1] A. Ashtekar, C. Beetle, and S. Fairhurst, *Class. Quantum Grav.* **16** (1999) L1; A. Ashtekar, C. Beetle and S. Fairhurst, *Class. Quantum Grav.* **17** (2000) 253; A. Ashtekar and A. Corichi, *Class. Quantum Grav.* **17** (2000) 1317.

[2] A. Corichi and D. Sudarsky, *Phys. Rev.* **D61** (2000) 101501; A. Corichi, U. Nucamendi and D. Sudarsky, *Phys. Rev.* **D 62** (2000) 044046; A. Ashtekar, A. Corichi and D. Sudarsky, *Class. Quantum Grav.* **18** (2001) 919.

[3] A Corichi, U. Nucamendi, and D. Sudarsky, *Phys. Rev.* **D64** (2001) 107501.

[4] A. Ashtekar, S. Fairhurst, and B. Krishnan, *Phys. Rev.* **D62** (2000) 104025.

[5] R. Bartnik and J. McKinnon, *Phys. Rev. Lett.* **61** (1988) 141.

[6] P. Bizon and T. Chmaj, *Phys. Lett.* **B297** (1992) 55.

[7] T. Torii and K. Maeda, *Phys. Rev.* **D48** (1993) 1643.

[8] For a recent review on all known properties of solitons and hairy black hole solutions see M. S. Volkov and D. V. Gal'tsov, *Physics Reports* **319** (1999) 1.

[9] K. Maeda, T. Tachizawa, T. Torii, and T. Maki, *Phys. Rev. Lett.* **72** (1994) 450; T. Torii, K. Maeda, and T. Tachzawa, *Phys. Rev.* **D51** (1995) 1510.

[10] D. Nuñez, H. Quevedo, and D. Sudarsky, *Phys. Rev. Lett.* **76** (1996) 571.

REFERENCES

[11] B. Kleihaus and J. Kunz, *Phys. Lett.* **B494** (2000) 130.

[12] J. D. Bekenstein, *Phys. Rev. Lett.* **28** (1972) 452; *Phys. Rev.* **D5** (1972) 1239; 2403.

SKETCHING THE INFLATON POTENTIAL

César A. Terrero-Escalante
Instituto de Física, UNAM, Apdo. Postal 20-364, 01000, México D.F., México
cterrero@fis.cinvestav.mx

Abstract Based on solutions of the Stewart–Lyth inverse problem it is argued that in the CMB data analysis the parametrization of the primordial spectra from inflation must include the 'running' of both, scalar and tensorial, spectral indices if information beyond the exponential potential model is wanted to be detected.

Keywords: Inflation, inflaton potential, CMB.

1. Introduction

Usually, to understand a grownup behavior it is necessary to look back into his childhood. This seems to be also the case in Cosmology. (See Refs. [1] for reviews and references on the topics mentioned in this introduction).

The eldest picture we have of our universe comes from the times when it was about 10^5 years old. In that epoch matter became cold enough for radiation to decouple and almost freely travel across the space up to the present. Since cosmic time ticks logarithmically and the current age of the universe is estimated to be about 10^{10} years, this nearly uniform cosmic background of radiation is a tidy picture of the young universe.

Nuclear physics allow us to trace the cosmic history back to the times when the lightest chemical elements were synthesized. The predicted abundances of these elements can be matched with current surveys, providing evidence about the universe when it was a three seconds old kid.

High–energies physics as described by the Standard Model of Particles allow us to glance a little bit further into the past, but the description of the events happening immediately after the birth of our universe still remains highly speculative. The main problem here is the lack of a consistent and tested theory, and the practical impossibility of reproducing the relevant events in laboratory conditions because of the very high energies involved. It is, therefore, quite stimulating that the inflationary paradigm links the physics in the very early universe to the current cosmological state.

A period of rapid accelerated expansion of the universe just after its birth can explain several features of the currently observed universe like its age, its size and its topology. Perhaps the more attractive feature that can be explained in this framework are the initial conditions for our own existence. In a perfectly homogeneous universe there are not chances for such complex structures like we are to arise. At the epoch when non–relativistic matter dominated over the relativistic one (or radiation), inhomogeneities acting like perturbations in the gravitational potential were required to seed the formation of galaxies through gravitational instability. Ultimately, this process led to the formation of solar systems and planets like the Earth. Those perturbations produced when the pressure of the radiation was still high enough to compensate gravity attraction induced an oscillatory motion of expansions and contractions. This motion led to inhomogeneities in the background temperature which we observe today as anisotropies in the cosmic microwave background (CMB) radiation. Therefore, to describe the formation of galaxies and the CMB anisotropies spectrum it is necessary to set the spectrum of the corresponding initial perturbations of the gravitational potential. Given the very special characteristics of this primordial spectrum, the inflationary scenario is the most widely accepted mechanism for setting these initial conditions [1, 2]. In an inflationary universe quantum fluctuations of matter and space–time are stretched by the expansion up to cosmological scales well beyond the distance within which causal interaction can take place. After the end of inflation these fluctuations reenter the causal horizon producing perturbations in the gravitational potential.

It is remarkable that the kind of primordial perturbation spectra generated in the simplest version of the inflationary scenario, the single scalar field scenario, fits quite well as the seeds for the CMB anisotropies [1, 2]. In this scenario the expansion is driven by the potential energy of the, so called, inflaton field. This way, if it is assumed, as we do here, that the inflaton physics closely correspond to the actual scenario of the very early universe, then recalling the earliest memories of our universe is equivalent to sketching the inflationary potential.

The aim of this contribution is to discuss how the Stewart–Lyth inverse problem [3] can be used as a powerful tool for drawing a photo-robot of the inflaton potential.

2. Stewart–Lyth inverse problem

To describe the inflationary dynamics and the corresponding perturbations it has proved to be useful to introduce the set of horizon flow functions [4]:

$$\epsilon_0 \equiv \frac{d_H(N)}{d_{Hi}}, \qquad (1)$$

and,

$$\epsilon_{m+1} \equiv \frac{d\ln|\epsilon_m|}{dN}, \quad m \geq 0. \qquad (2)$$

where $d_H \equiv 1/H \equiv a/\dot{a}$ denotes the Hubble distance, with d_{Hi} evaluated at some initial time t_i and dot stands for differentiation with respect to cosmic time. The scale factor a measures the expansion of the spatial volume, and $N \equiv \ln(a/a_i)$ is the e-folds number. The first horizon flow function ϵ_1 can be written in several useful ways:

$$\epsilon_1 \equiv \frac{d\ln d_H}{dN} = \dot{d}_H = \frac{3}{2}\frac{\rho+p}{\rho} = 3\frac{T}{T+V} = \kappa\frac{T}{H^2}, \qquad (3)$$

where ρ and p are, respectively, the energy density and the pressure in a universe dominated by the potential energy, $V(\phi)$, of the inflaton field, ϕ. $T \equiv \dot{\phi}^2/2$, $\kappa = 8\pi/m_{Pl}^2$ is the Einstein constant and m_{Pl} is the Planck mass. Inflation happens for $0 \leq \epsilon_1 < 1$. For $m > 1$, ϵ_m may take any real value.

At the high energies corresponding to the inflationary period, the inflaton energy density and the space–time metrics undergo quantum fluctuations. These are the fluctuations that stretched by the expansion reentered the Hubble horizon, d_H, close to the point of equal density of matter and radiation, to seed the formation of large scale structure and the CMB anisotropies. The spectra of these seeds can be parametrized by the series:

$$\ln A_S^2(k) = \ln A_S^2(k_*) + \Delta(k_*)\ln\frac{k}{k_*} + \frac{1}{2}\frac{d\Delta(k)}{d\ln k}\Big|_{k=k_*}\ln^2\frac{k}{k_*} + \cdots \qquad (4)$$

$$\ln A_T^2(k) = \ln A_T^2(k_*) + \delta(k_*)\ln\frac{k}{k_*} + \frac{1}{2}\frac{d\delta(k)}{d\ln k}\Big|_{k=k_*}\ln^2\frac{k}{k_*} + \cdots, \qquad (5)$$

where A_S and A_T stand respectively for the normalized amplitudes of the density (scalar) and metrics (tensor) perturbations, and k_* is the wavenumber corresponding to a pivotal length scale. Functions $\Delta(k)$ and $\delta(k)$ will be called here the scalar and tensorial spectral indices respectively. Their derivatives with respect to $\ln k$ are known as the 'running' of the spectral indices. In the procedure of fitting the CMB data the order where series (4) and (5) are truncated is determined by the precision of the observations. It is worthy to mention that a careful analysis of the perturbations generated during inflation shows that the spectra of such perturbations while reentering the Hubble horizon are generically almost scale–invariant (see Ref. [1] for details). With this regard, one can assume that each higher term in series (4) and (5) is smaller than the corresponding lower order term.

Typically, most of CMB data analyses have neglected the possible effects of the primordial gravitational wave spectrum [2]. The tensor contribution to the CMB spectrum can be parametrized in terms of the quantity

$$r \equiv \alpha \frac{A_T^2}{A_S^2}, \qquad (6)$$

representing the relative amplitudes of the tensor and scalar perturbations, where the constant, α, depends on the particular normalization of the spectral amplitudes that is chosen. In the last few years, however, there has been a growing recognition that the role of the tensor perturbations deserves more attention when determining the best–fit values of the cosmological parameters (For a recent review, see, e.g., Ref. [2]).

The Stewart–Lyth inverse problem (SLIP) was introduced in Ref. [3] as a method for finding the inflaton potential using information on the functional form of the spectral indices. With this aim, the following non–linear differential equations must be solved:

$$2C\epsilon_1 \hat{\hat{\epsilon}}_1 - (2C+3)\epsilon_1 \hat{\epsilon}_1 - \hat{\epsilon}_1 + \epsilon_1^2 + \epsilon_1 + \Delta = 0, \qquad (7)$$
$$2(C+1)\epsilon_1 \hat{\epsilon}_1 - \epsilon_1^2 - \epsilon_1 - \delta = 0, \qquad (8)$$

where $C = -0.7293$ and a circumflex accent denotes differentiation with respect to $\tau \equiv \ln H^2$. These equations are derived from next–to–leading order expressions for the spectral indices in terms of the horizon flow functions (2) [3]. In Ref. [5] it was shown that to this order $\delta \leq 0$. The validity of the conclusions to be drawn here are constrained by the next–to–leading order precision.

Then, having an expression for $\tau(\epsilon_1)$ (which is typically the form of the solutions to Eqs. (7) and (8)) the corresponding inflaton potential is given by the parametric function [5],

$$V(\phi) = \begin{cases} \phi(\epsilon_1), \\ V(\epsilon_1), \end{cases} \qquad (9)$$

where,

$$V(\epsilon_1) = \frac{1}{\kappa}(3 - \epsilon_1)\exp\left[\tau(\epsilon_1)\right], \qquad (10)$$

and,

$$\phi(\epsilon_1) = -\frac{2(C+1)}{\sqrt{2\kappa}} \int \frac{\sqrt{\epsilon_1} d\epsilon_1}{\epsilon_1^2 + \epsilon_1 + \delta} + \phi_0. \tag{11}$$

Here V_0 and ϕ_0 are integration constants. The SLIP solutions are constrained by the following conditions [5]:

$$\begin{cases} \hat{\epsilon}_1 \frac{d\phi}{d\epsilon_1} < 0, \\ \hat{\epsilon}_1 \frac{dV}{d\epsilon_1} > 0. \end{cases} \tag{12}$$

In order link the SLIP solution with the primordial spectra given by the series (4) and (5) it proved to be useful to rewrite Eqs. (7) and (8) by converting from derivatives with respect to τ to derivatives with respect to $\ln k$, with k corresponding to the wavelength crossing the Hubble horizon by the first time, i. e., $k = aH$ [6].

3. Constraining the inflaton potential

The Stewart–Lyth inverse problem has two strong drawbacks. First, the full power of this procedure can be used only when information on the functional forms of both spectral indices is available. Unfortunately, this information is rather difficult to be directly obtained from observations. In addition, simple functional forms of the spectral indices involve great difficulties while solving the SLIP. In spite of these drawbacks, there is an alternative way of using the SLIP related expressions. This method allows to test for the internal consistency in the procedure of fitting the CMB data by finding and describing the inflaton potential corresponding to the assumptions on the primordial perturbations used in that procedure. Thus, the SLIP can help in constraining the possible inflaton potentials by linking distinct features of the power spectra to the inflationary dynamics.

Historically, the first analytical calculations of the form of the initial conditions required for galaxy formation due to Harrison and Zel'dovich (see Ref. [1] for details) yield scale–invariant spectra, i.e., constant amplitudes ($\Delta = \delta = 0$). Current precision of the CMB measurements still allows for this kind of spectra to be used as initial conditions [2]. However, in the inflationary scenario some degree of scale dependence is necessarily present if a dynamical mechanism for ending the accelerated expansion acted in the very early universe in order to recover the success of the Hot Big Bang Theory as a sequence of a radiation and a matter dominated universes.

Since the contribution of the tensor modes to the CMB anisotropies is typically very small, one can wonder whether the actual inflaton potential could produce perturbations spectra where the scale dependence is entirely contained in the scalar component, i. e., $\delta = 0$ and $\Delta = \Delta(k)$.

The corresponding SLIP was solved in Ref. [5]. The inflaton potential has the form,

$$V(\phi) = V_0 \frac{3 - \tan^2\left[\frac{\sqrt{2\kappa}}{4(C+1)}(\phi - \phi_0)\right]}{\cos^{4(C+1)}\left[\frac{\sqrt{2\kappa}}{4(C+1)}(\phi - \phi_0)\right]}. \tag{13}$$

Nevertheless, the analysis of the behavior of the scalar index,

$$\Delta(k) = \frac{1}{8(C+1)^2}\left\{\left(\frac{k}{k_0}\right)^{1/(C+1)}\right.$$
$$- \sqrt{\left(\frac{k}{k_0}\right)^{1/(C+1)}\left[\left(\frac{k}{k_0}\right)^{1/(C+1)} - 4\right]}\left\{\left(\frac{k}{k_0}\right)^{1/(C+1)} + 2C\right.$$
$$\left.\left.- \sqrt{\left(\frac{k}{k_0}\right)^{1/(C+1)}\left[\left(\frac{k}{k_0}\right)^{1/(C+1)} - 4\right]}\right\}\right\}, \tag{14}$$

corresponding to the potential (13), shows that, for large k, Eq. (14) converges to $\Delta = 1/2(C+1) \approx 1.85$, which is too far from values allowed by theory and observations [1, 2]. Therefore, a correct parametrization of the inflationary spectra must take into account terms beyond the linear ones in both series (4) and (5). The next step is, then, to consider spectra slightly tilted from scale invariance, i.e., with constant and close to zero spectral indices. In that case the spectra take on a power–law form. It is well known that an exponential potential produces exactly such perturbation spectra. This scenario is called power–law inflation because $a \propto t^p$, where $p \gg 1$ is a constant [7]. Using Eqs. (3), it is easy to check that for this model, $\epsilon_1 = 1/p$.

We proved in Ref. [5] that, even to next–to–leading order, the only SLIP solution corresponding to these conditions on both spectra, tensorial and scalar, is power–law inflation. If, in the same mood as it was done while analyzing the case $\delta = 0$, the condition of a power–law spectrum is relaxed by allowing scale dependence only for the scalar index, then it will be obtained that, though power–law inflation is still a trivial solution, other SLIP solutions arise [5]. (Hereafter the involved formulas for the SLIP solutions are omitted. For details see the cited references). These solutions converge to power–law inflation; the scalar index converges to a constant realistic value from above or from below, depending on the initial conditions. Since the depart from power–law behavior takes place at large angular scales where the cosmic variance is dominant, then it will be very difficult from the observational point of view to distinguish between these SLIP solutions and power–law inflation.

This way, if one wants to move beyond the power–law scenario, it is necessary to include the running of both spectral indices in the parametrization given by Eqs. (4) and (5). It is reasonable, for instance, to consider that the scale dependence of the tensorial spectral index is distinctly perceived only up to next–to–leading order in terms of ϵ_1, consistently with the approximation used to derive Eqs. (7) and (8). We solved the SLIP with the ansatz [8],

$$\delta(\epsilon_1) = -\left[(1+a)\epsilon_1^2 + (1+b)\epsilon_1 + c\right], \quad (15)$$

where a, b, c are real numbers. It was found that for $b^2 > 4ac$, power–law inflation is again an attractor of the corresponding inflationary dynamics. For $b^2 < 4ac$, power–law inflation is no longer an attractor but a transient regime of the dynamics. Obviously, the spectra depart from a power law before and after the quasi power–law behavior. Those perturbations produced before the quasi power–law regime are imprinted in the very large angular scales, being the detection of their signatures difficult because of the cosmic variance. On the other hand, since inflation ends very fast after the quasi power–law regime is left behind, the scales crossing the horizon at that time were extraordinarily small and reentered it back immediately with not relevant effect. If the actual inflaton potential is close to this model, it will be also very difficult to observe any difference from the exponential potential. Then, even including the running of both spectral indices in the parametrization of the primordial spectra could be not sufficient to get ride of the power–law bias.

There are cases where power–law inflation is a repeller of the inflationary dynamics. In Ref. [9] the SLIP was solved with the condition of constant tensor to scalar ratio, r. This was partially motivated by the possibility, in the near future, of estimating a constant central value for r from the observation of the CMB polarization [1, 2]. From Eqs. (4), (5), (7) and (8), it is simple to show that the condition $r = constant$ is equivalent to $\Delta(k) = \delta(k)$. Using this, after adding Eqs. (7) and (8),

$$2C\epsilon_1\hat{\hat{\epsilon}}_1 - (\epsilon_1 + 1)\hat{\epsilon}_1 = 0. \quad (16)$$

Obviously, a trivial solution for this case is $\epsilon_1 = 1/p = constant$, corresponding to power–law inflation. The remaining SLIP solutions depart very quickly from this power–law solution. In one case δ grows and becomes positive, thus indicating a breakdown in the next–to–leading order analysis. For the other case, δ begins to evolve extremely rapidly, probably indicating that the running of the spectral index, $d\Delta/d\ln k$, becomes too large. Either way, observational constraints are difficult to satisfy. Thus, the potential in the region open to observation must be sufficiently close to the exponential (power–law inflation) model.

4. Conclusions

The quality of the photo–robot of the inflaton potential depends crucially on the order of the series used to parametrize the primordial spectra while fitting the CMB anisotropies spectrum. In turn, that order depends on the precision of the available observational data.

If the inflaton potential is suspected to differ from the exponential potential corresponding to power–law inflation, then quadratic terms of both scalar and tensor modes are necessary (but perhaps not sufficient) to observe the differences.

If, while fitting the CMB data, the running of both spectral indices happen to be distinctly different from zero, then the constant value of the tensor to scalar ratio as determined from the CMB polarization likely will not be characteristic of a wide range of scales.

When this contribution was ready to be submitted, a paper by Leach et al. [10] was posted at Los Alamos including a very interesting discussion on the parametrization of the primordial spectra from inflation.

Acknowledgments

The author wish to thank the organizers of the Meeting by their invitation to participate. The author was partially supported by the CONACyT grant 38495–E and the Sistema Nacional de Investigadores (SNI).

References

[1] A.D. Linde, *Particle Physics and Inflationary Cosmology*, (Chur: Harwood, 1990), pp. 362; A.R. Liddle and D.H. Lyth, *Cosmological Inflation and Large–Scale Structure*, (Cambridge: Univ. Press, 2000), pp. 400.

[2] W. Hu and S. Dodelson, arXiv:astro–ph/0110414 (2001).

[3] E. Ayon–Beato, A. Garcia, R. Mansilla and C. A. Terrero–Escalante, *Phys. Rev.* **D62** (2000) 103513. arXiv:astro–ph/0007477.

[4] D. J. Schwarz, C. A. Terrero–Escalante and A. A. Garcia, *Phys. Lett.* **B517** (2001) 243. arXiv:astro–ph/0106020.

[5] C. A. Terrero–Escalante, E. Ayon–Beato and A. A. Garcia, *Phys. Rev.* **D64** (2001) 023503. arXiv:astro–ph/0101522.

[6] E. Ayon–Beato, A. Garcia, R. Mansilla and C.A. Terrero–Escalante, arXiv:astro–ph/0009358 (2000).

[7] F. Lucchin and S. Matarrese, *Phys. Rev.* **D32** (1985) 1316.

REFERENCES

[8] C.A. Terrero–Escalante and A.A. Garcia, *Phys. Rev.* **D65** (2002) 023515. arXiv:astro–ph/0108188.

[9] C.A. Terrero–Escalante, J.E. Lidsey and A.A. Garcia, arXiv:astro–ph/0111128 (2001).

[10] S.M. Leach, A.R.Liddle, J.Martin, and D.J. Schwarz, arXiv:astro–ph/0202094 (2002).

III

EXACT SOLUTIONS

COSMOLOGICAL SOLUTIONS AND THEIR STABILITY IN SCALAR–TENSOR THEORIES

Jorge L. Cervantes-Cota
Departamento de Física, Instituto Nacional de Investigaciones Nucleares (ININ)
P.O. Box 18-1027, México D.F. 11801, México
jorge@nuclear.inin.mx

M. A. Rodriguez-Meza
Departamento de Física, Instituto Nacional de Investigaciones Nucleares (ININ)
P.O. Box 18-1027, México D.F. 11801, México.
Instituto de F´initialsica, Universidad Autónoma de Puebla
P.O. Box J-48, Puebla 72570, México.
mar@nuclear.inin.mx

Marcos Nahmad
Instituto de Física, Universidad Autónoma de Puebla
P.O. Box J-48, Puebla 72570, México.

Abstract   Using scaled variables we present a set coupled equations valid for isotropic and anisotropic Bianchi type I, V, IX models in scalar–tensor theories. This technique permits one to integrate the system for some cases of FRW, Bianchi I and V models; otherwise, we integrate the system numerically. In particular, for the Brans–Dicke theory we analyze the stability of solutions.

Keywords: Scalar–tensor theories, exact solutions, cosmology.

1.   Introduction

The universe at big scales is homogeneous and isotropic [1], and must also has been having these properties since, at least, the era of nucleosynthesis [2]. In order to explain the isotropy of the universe from theoretical anisotropic models, many authors have considered the Bianchi models that can in principle evolve to a Friedmann–Robertson–Walker (FRW) cosmology [3]. Motivated by these facts, we have been working in scalar–tensor theories to investigate if Bianchi universes are able to isotropize as the universe evolves [4], and if its evolution can be inflationary [5, 6]. In the present report we present some numerical solutions for Bianchi models, and analyze the stability of some of our solutions.

2. Equations for FRW and Bianchi models

We consider the following field equations of scalar–tensor theories:

$$R_{\mu\nu} - \frac{1}{2} R g_{\mu\nu} = -\frac{8\pi}{\phi} \left[T_{\mu\nu} + V(\phi) g_{\mu\nu} \right] - \frac{\omega}{\phi^2} \left[\phi_{|\mu} \phi_{|\nu} - \frac{1}{2} \phi_{|\lambda} \phi^{|\lambda} g_{\mu\nu} \right]$$
$$- \frac{1}{\phi} \left[\phi_{|\mu||\nu} - \phi^{|\lambda}{}_{||\lambda} g_{\mu\nu} \right] \quad (1)$$

and

$$\phi^{|\lambda}{}_{||\lambda} = \frac{8\pi}{3 + 2\omega} \left[T - 2\phi \frac{dV(\phi)}{d\phi} + 4 V(\phi) \right], \quad (2)$$

where ϕ is the scalar field, ω a coupling constant, the symbol $|$ partial derivative, $||$ the covariant derivative, $V(\phi)$ the potential, and T is the trace of the energy–momentum tensor, $T_{\mu\nu}$. The continuity equation (energy-momentum conservation law) reads $T_\mu{}^\nu{}_{||\nu} = 0$.

In previous investigations [4, 5, 6, 7] we have used scaled variables, in terms of which some solutions have been given, therefore following we use them: the scaled field $\psi \equiv \phi a^{3(1-\nu)}$, a new cosmic time parameter $d\eta \equiv a^{-3\nu} dt$, $()' \equiv \frac{d}{d\eta}$, the 'volume' $a^3 \equiv a_1 a_2 a_3$, and the Hubble parameters $H_i \equiv a_i'/a_i$ corresponding to the scale factors $a_i = a_i(\eta)$ for $i = 1, 2, 3$. We assume comoving coordinates and a perfect fluid with barotropic equation of state, $p = \nu\rho$, where ν is a constant. Using these definitions one obtains the cosmological equations for FRW and Bianchi type I, V and IX models:

$$(\psi H_i)' - \psi a^{6\nu} R_{ij} = \frac{8\pi a^{3(1+\nu)}}{3 + 2\omega} [[1 + (1-\nu)\omega] \rho + (1 + 2\omega)V$$
$$+ \psi \frac{\partial V}{\partial \psi}], \quad \text{for} \quad i = 1, 2, 3. \quad (3)$$

$$H_1H_2 + H_1H_3 + H_2H_3 + [1+(1-\nu)\omega](H_1+H_2+H_3)\frac{\psi'}{\psi}$$

$$- (1-\nu)[1+\omega(1-\nu)/2](H_1+H_2+H_3)^2 - \frac{\omega}{2}\left(\frac{\psi'}{\psi}\right)^2$$

$$- \frac{1}{2}a^{6\nu}R_j = 8\pi\frac{a^{3(1+\nu)}}{\psi}(\rho+V), \qquad (4)$$

$$\psi'' - (1-\nu)\psi a^{6\nu}R_j = \frac{8\pi\, a^{3(1+\nu)}}{3+2\omega}\left[[2(2-3\nu)+3(1-\nu)^2\omega]\rho\right.$$
$$\left.+[4+3(1-\nu)(1+2\omega)]V + (1-3\nu)\psi\frac{\partial V}{\partial \psi}\right] \qquad (5)$$

where a column sum is given by $R_j \equiv \Sigma_i R_{ij}$, with $j = $ I, V or IX and $R_{iI} = 0$ for Bianchi I, $R_{iV} = 2/a_1^2$ for Bianchi V, and $R_{iIX} = (a_i^4 - a_l^4 - a_m^4 + 2a_l^2 a_m^2)/(-2a^6)$ for Bianchi IX, where i, l, m are three cyclic indexes (no tensor notation) for the scales factors [4].

For the Bianchi type V model one has the additional constraint: $H_2 + H_3 = 2H_1$, implying that a_2 and a_3 are inverse proportional functions, $a_2 a_3 = a_1^2$; note that the mean Hubble parameter, $H \equiv \frac{1}{3}(H_1+H_2+H_3)$, is for this Bianchi type model $H = H_1$. FRW models are obtained if $a_1 = a_2 = a_3$, whereby the flat model corresponds to the Bianchi I curvature term, the open to Bianchi V, and closed model to Bianchi IX.

Additionally, the continuity equation yields: $\rho a^{3(1+\nu)} = \text{const.} \equiv M_\nu$, M_ν being a dimensional constant depending on the type of fluid present. The vacuum case is attained when $M_\nu = 0$. For the dust case ($\nu = 0$) M_0 has units of mass, whereas for the radiation case ($\nu = 1/3$) $M_{\frac{1}{3}}$ has units of mass×length.

The system of ordinary differential equations, Eqs. (2–4), can be recast in the following set:

$$\psi\psi'' + (1-\nu)\omega\,\psi'^2 - 2(1-\nu)[1+(1-\nu)\omega](H_1+H_2+H_3)\psi\,\psi'$$
$$+\frac{1}{3}(1-\nu)\left[2(2-3\nu)+3(1-\nu)^2\omega\right][(H_1+H_2+H_3)\psi]^2$$
$$+[2+(1-\nu)(1+3\nu)\omega]m_\nu\,\psi + (1-\nu)(h_1^2+h_2^2+h_3^2)$$
$$= \frac{8\pi}{3+2\omega}\psi a^{3(1+\nu)}\left[[1+3\nu+2(1-\nu)\omega]V + (1-3\nu)\psi\frac{\partial V}{\partial \psi}\right] \qquad (6)$$

and

$$\psi'' - (1-\nu)[(H_1 + H_2 + H_3)\psi]' - (1-3\nu)m_\nu = \frac{8\pi}{3+2\omega}a^{3(1+\nu)}$$
$$\times \left[4V - 2\psi\frac{\partial V}{\partial \psi}\right] \quad (7)$$

where $m_\nu \equiv \frac{8\pi M_\nu}{3+2\omega}$, and we have written the Hubble rates as follows [8]:

$$H_i = \frac{1}{3}(H_1 + H_2 + H_3) + \frac{h_i}{\psi}, \quad (8)$$

where the h_i's are some unknown functions of η that determine the anisotropic character of the solutions. Furthermore, the models obey the condition

$$h_1 + h_2 + h_3 = 0 \quad (9)$$

to demand consistency with Eq. (8). For the Bianchi type V model one has additionally that $h_1 = 0$, since $H_1 = H$ as mentioned above. For FRW models one has that $h_1 = h_2 = h_3 = 0$.

3. Solutions for the potential $V = V_o\psi^2$

The problem to find solutions of Bianchi and FRW models in the scalar–tensor theories is, firstly, to solve Eqs. (5) and (6), and, secondly, to solve for each scale factor through Eq. (2). This task is very complicated to achieve analytically, therefore this work have to be done, in general, numerically. There are, however, a few cases that can be analytically worked out. Let us the see these cases. We observe from Eq. (6) that if its right hand side is zero, that is for $V = V_o\psi^2$, then this equation can be once integrated to get that

$$H = \frac{\psi' - (1-3\nu)m_\nu \eta - \delta}{3(1-\nu)\psi}, \quad (10)$$

where δ is an integration constant. This equation expresses the average Hubble rate in terms of ψ, and substituting it into Eq. (8), gives us each of the Hubble rates. The point that is left is to have solutions for ψ. Let us substitute Eq. (10) into Eq. (5) to get the following:

$$\psi\psi'' - \frac{2}{3(1-\nu)}\psi'^2 - \frac{2(1-3\nu)}{3(1-\nu)}[(1-3\nu)m_\nu\eta + \delta]\psi'$$
$$+[2 + (1-\nu)(1+3\nu)\omega]m_\nu\psi + \frac{1}{3(1-\nu)}[2(2-3\nu) + 3(1-\nu)^2\omega] \times$$
$$[(1-3\nu)m_\nu\eta + \delta]^2 + (1-\nu)(h_1^2 + h_2^2 + h_3^2)$$
$$= \frac{8\pi}{3+2\omega}\psi a^{3(1+\nu)}\left[[1 + 3\nu + 2(1-\nu)\omega]V + (1-3\nu)\psi\frac{\partial V}{\partial \psi}\right] \quad (11)$$

where $V = V_o\psi^2$. Still Eq. (10) is too complicated since it is coupled to the other equations through the h_i's and to the source term. But, if $V_o = 0$, that is for Brans–Dicke (BD) theory, we can make some further progress. Then, let us consider this case. Accordingly, FRW models, for which $h_i = 0$, imply that Eq. (10) decouples, and the problem is reduced to solve this single, second order differential equation. For Bianchi type I and V models it turns out that the only possible solutions imply that the h_i's are constants (hence, the equations decouple again), whereas for type IX they are unknown functions of η [4, 5, 6, 7]. An explanation of this fact resides in the property that Bianchi type I and V models have curvature terms of FRW type, whereas type IX has a very much complicated form. So the things, it seems that the most general solution of Eq. (10) with h_i's constants would give general solutions for Bianchi models I and V, as well as for the FRW models. It turns out that the solution $\psi = A\eta^2 + B\eta + C$, with A, B and C constants, is the general solution for flat FRW and Bianchi I models, and it is a particular solution for non–flat FRW and Bianchi V models [4, 6].

For if $V_o \neq 0$, the work has to be done numerically. We have integrated the Bianchi V model, see figure 1, for the cases when $\omega = 2000$ (plots A and B) as it is demanded by the current values of the BD parameter, and when $\omega = -1$ (plot C), that is, for the case in which scalar–tensor theories mimic string effective theories. In all plots the potential plays initially a minor role, since we set the initial data accordingly. Later on, the potential dominates the dynamics and, hence, inflation takes place in all scale factors. The graceful exit problem is present.

4. Stability analysis

We want to analyze the stability of the solutions for the case of vanishing potential (BD case). In order to proceed we define the following transformation

$$\Omega \equiv \psi - (A\eta^2 + B\eta + C); \qquad \chi \equiv \frac{d\Omega}{d\eta}, \qquad (12)$$

where A, B, C are real constants, which are given in Ref. [4] for FRW, Bianchi type I, and V models. With these definitions, Eq.(10) transforms

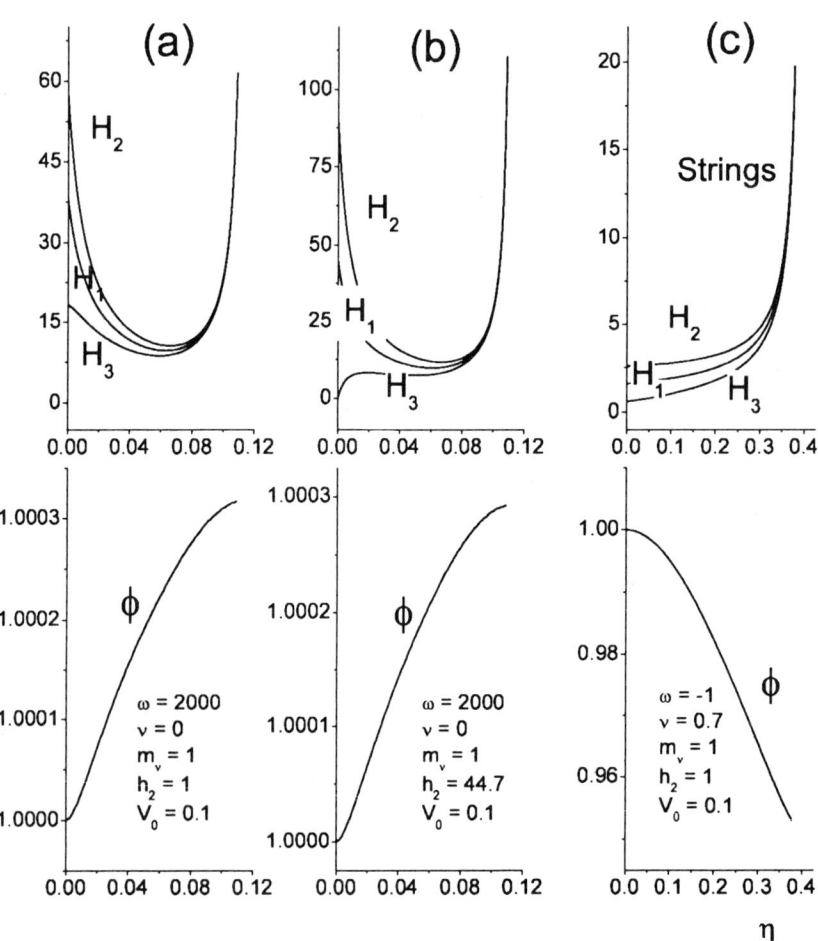

Figure 1. The Hubble parameters and scalar field as a function of η for $\omega = 2000$ with different anisotropic parameters (plots A and B), and for $\omega = -1$ (plot C).

to the following system:

$$\frac{d\Omega}{d\eta} = \chi$$

$$\frac{d\chi}{d\eta} = \frac{1}{(\Omega + A\eta^2 + B\eta + C)} \left[-K\Omega + \frac{2\chi^2}{3(1-\nu)} \right.$$
$$\left. + \frac{2\chi}{3(1-\nu)} \left[(4A + m_\nu(1-3\nu)^2)\eta + 2B + \delta(1-3\nu) \right] \right], \quad (13)$$

where $K = 2A + 2m_\nu + (1-\nu)(1+3\nu)m_\nu \omega$ is a constant. We want to check the stability of this system at the origin of the plane (Ω, χ), i.e., the stability around the typical solution $\psi = A\eta^2 + B\eta + C$.

The equilibrium points are obtained by setting $d\Omega/d\eta = d\chi/d\eta = 0$, which implies that the origin $\Omega = \chi = 0$ is an equilibrium point. This, in turn, implies through Eqs. (12) that the quadratic polynomial solution is an equilibrium solution for all Bianchi models.

In order to analyze the type of stability of the system above, we use the Lyapunov theorem on the stability of motion [9]. It states that if one is able to find a function, say $V(\Omega, \chi, \eta)$, associated to a perturbed differential equation system, as the given by Eqs. (12), with the following properties:

i) $V(\Omega, \chi, \eta)$ being a positive definite function. Accordingly, one requires the existence of another function $W(\Omega, \chi)$ such that:

$$V(\Omega, \chi, \eta) \geq W(\Omega, \chi) \quad \text{and} \quad W(\Omega, \chi) = 0 \Leftrightarrow (\Omega, \chi) = (0, 0), \quad (14)$$

moreover $V(0, 0, \eta) = W(0, 0) = 0$, for any η.

ii) Given $\frac{\partial \Omega}{\partial \eta}$ and $\frac{\partial \chi}{\partial \eta}$ by Eqs. (12) the condition is

$$\frac{dV(\Omega, \chi, \eta)}{d\eta} = \frac{\partial V(\Omega, \chi, \eta)}{\partial \Omega} \frac{\partial \Omega}{\partial \eta} + \frac{\partial V(\Omega, \chi, \eta)}{\partial \chi} \frac{\partial \chi}{\partial \eta} + \frac{\partial V(\Omega, \chi, \eta)}{\partial \eta} \leq 0 \quad (15)$$

then, the unperturbed system is stable.

To proceed to apply the Lyapunov theorem, we need to find $V(\Omega, \chi, \eta)$ and $W(\Omega, \chi, \eta)$ for the system. Although some methods to construct this type of functions can be found [10], it is a non–trivial task. In our case we were able to find such functions for the case in which $K = 0$. For this case, we have found the following functions:

$$V(\Omega, \chi, \eta) = \chi^2 \exp(-\Omega^3 \chi) + \frac{\Omega^2}{a\eta + b}, \quad (16)$$

and

$$W(\Omega, \chi) = \chi^2 \exp(-\Omega^3 \chi),$$

where a, b are positive, real constants. Since $K = 0$ all points of the form $(\Omega, 0)$ are equilibrium points for any value of Ω. In this way, one has that $W(\Omega, \chi) = 0 \iff \chi = 0$ for any finite value of Ω, that is, the origin is not an isolated equilibrium point, but the whole Ω–axis, $(\Omega, 0)$, represents an equilibrium line, in contrast with the condition given through Eq. (14) that implies isolated equilibrium points; then, in this case the stability theorem of Lyapunov cannot be directly applied. Nevertheless, one may extend the proof of theorem 7.1 of Ref. [9] (see Ref. [11]) such that the functions Eqs. (16) and (17) are sufficient, and in this way to prove the stability of our system at the origin. Observe that this proof does not prove the non–stability at the origin for the case that $K \neq 0$.

According with condition *ii)*, one has to demand that

$$\frac{dV(\Omega, \chi, \eta)}{d\eta} = \frac{\partial V(\Omega, \chi, \eta)}{\partial \Omega}\frac{\partial \Omega}{\partial \eta} + \frac{\partial V(\Omega, \chi, \eta)}{\partial \chi}\frac{\partial \chi}{\partial \eta} + \frac{\partial V(\Omega, \chi, \eta)}{\partial \eta} \leq 0$$

$$= -3\chi^4 \Omega^2 \exp(-\Omega^3 \chi) + \frac{2\Omega\chi}{(a\eta + b)} - \frac{a\Omega^2}{(a\eta + b)^2}$$

$$+ \frac{(2 - \chi\Omega^3)\chi \exp(-\Omega^3 \chi)}{(\Omega + A\eta^2 + B\eta + C)}\left[-K\Omega + \frac{2\chi^2}{3(1-\nu)}\right.$$

$$+ \left. \frac{2\chi}{3(1-\nu)}\left[(4A + m_\nu(1-3\nu)^2)\eta + 2B + \delta(1-3\nu)\right]\right]. \quad (17)$$

Note that all terms, except for the first, depend on η and they tend to zero as η tends to infinity. This means that there exists a number $\eta_+ \geq 0$, such that if $\eta \geq \eta_+$, then the dominant term is the first one, that always has non–positive sign. This property, according to theorem 7.1 [9] and its extension [11], proves the asymptotic stability at the origin.

For Bianchi I model, $K = 0$ implies that $\omega = -3/2$ for any value of ν, whereas for Bianchi V model, the following relationship must be valid: $\omega = \frac{-18\nu}{(1+3\nu)^2}$, which is a singular value of this model since the constant C_V becomes infinity [7]. For the FRW model given in Ref. [7] with $\kappa_2 = 0$, which corresponds to a second order polynomial solution for ψ, the solution is stable.

5. Conclusions

We have presented a set of differential equations written in rescaled variables that are valid for Bianchi type I, V and IX models, as well as for FRW models within general scalar–tensor theories. It seems to be very difficult to solve analytically this system given by Eqs. (2), (5) and (6). However, for the special case in which the potential vanishes, or when matter terms dominate over the potential [5], that is for the BD case, one is able to find solutions. Accordingly, one can integrate Eq. (10) for the cases of FRW models, and Bianchi type I and V models, because in these cases the anisotropic parameters (h_i's) are constants.

For the BD case, $V_o = 0$, the expression $\psi = A\eta^2 + B\eta + C$, with A, B and C constants, represents the general solution for flat FRW and Bianchi I models, and it is a particular solution for non–flat FRW and Bianchi V models [4, 6]. For the case that $V = V_o\psi^2$ we presented numerical integrations for the Bianchi V model. These show initially the anisotropic dynamics, then the potential dominates and inflation takes place in all scale factors. However, the graceful exit problem is present.

Further solutions of Eq. (10) are in order, which can be either within FRW cosmologies or Bianchi type V or IX models. In particular, the general solution of this equation with h_i's constants should provide the general solution of both curved FRW cosmologies and Bianchi type V model.

We have analyzed, following Lyapunov, the stability of our system at the origin for the specific case when $K = 0$. This case implies stability of the FRW solution with $\kappa_2 = 0$ [7], and of Bianchi I model with $\omega = -3/2$. For Bianchi V one finds a singular model, and nothing about the stability can be said.

Acknowledgments

This work has been financially supported by Conacyt, grant number 33278-E and DFG–Conacyt bilateral project.

References

[1] C.L. Bennett et al., *Astrophys. J.* **464** (1996) L1; K. M. Górski et al., *ibid.* L11; G. Hinshaw, *ibid.* L17.

[2] S.W. Hawking and R. J. Taylor, *Nature* (London) **209** (1966) 1278; J. D. Barrow, *Mon. Not. R. Astron. Soc.* **175** (1976) 359.

[3] C. B. Collins and S. W. Hawking, *Astrophys. J.* **180** (1973) 317; J. D. Barrow and D. H. Sonoda, *Phys. Rep.* **139** (1986) 1. J. D. Barrow, *Phys. Rev.* **D51** (1995) 3113.

[4] P. Chauvet and J. L. Cervantes–Cota, *Phys. Rev.* **D52** (1995) 3416; J. P. Mimoso and D. Wands, *Phys. Rev.* **D52** (1995) 5612.

[5] J. L. Cervantes–Cota and P. A. Chauvet, *Phys. Rev.* **D59** (1999) 043501.

[6] J. L. Cervantes–Cota, *Class. Quant. Grav.* **16** (1999) 3903.

[7] J. L. Cervantes–Cota and M. Nahmad, *Gen. Rel. Grav.* **33** (2001) 767.

[8] V. A. Ruban and A. M. Finkelstein, *Gen. Rel. Grav.* **6** (1975) 601.

[9] A. M. Lyapunov, *Probleme General de la Stabilité du Movement*, (Princeton University Press, 1949), translation by Taylor & Francis (1992); D. Merkin, *Introduction to the Theory of Stability*, (TAM Springer Verlag, 1992) Chapter 7, 226.

[10] V. Lakshimikantham, V.M Matrosov and S. Sivasundaram, *Vector Lyapunov Functions and Stability Analysis of Nonlinear Systems*, (Kluwer Academic Publishers,1991), Sec. 1.6; L. Yu Anapolsky, V.D Igertov, M.V. Matrosov, *Methods of Constructing Lyapunov Functions*, (Itogi Nauki i Tekhniki Ser. Obscaya Mecanika, 1975) Vol. 2, VINITI; D.G Schultz and J.E Gibson, *Trans AIIE*, 81(II) (1962) 203.

[11] M. Nahmad, J. L. Cervantes-Cota, M. A. Rodr´initialguez-Meza, *Extended Lyapunov theorem on the stability of motion*, (2002) submitted.

CONSIDERATIONS ON ACCELERATED EXPANSION IN A SCALAR–TENSOR COSMOLOGY

Pablo Chauvet–Alducin
Departamento de Física,
Universidad Autónoma Metropolitana–Iztapalapa
P. O. Box. 55-534, México D. F.
C. P. 09340 MEXICO
pcha@xanum.uam.mx

Abstract   An accelerated, or at least a non–decelerating one, expansion in scalar –tensor theories for isotropic models in the presence of barotropic fluids only, without the aid of quintessence matter is possible because of the given, natural properties, that the postulated scalar fields have. So the addition of matter fields minimally coupled to gravity with the sole purpose of speeding up its expansion may be superfluous. These special field characteristics can be more clearly examined when the scalar is sourceless. To this end it is therefore fundamental, and fortunately, possible to exhibit all the analytic solutions for at least the simplest scalar–tensor theory, the Brans–Dicke cosmology with a sourceless scalar.

Keywords: accelerated expansion, isotropic, barotropic, quintessence, scalar–tensor, cosmology, vacuum, radiation, dark matter, dark energy.

1. Introduction

The results of present measurements regarding certain supernovas[1] are thought to favor an actual acceleration of the Universe's expansion. Yet, one must bear in mind that certain details concerning the physics of Type Ia supernovae, currently used as cosmological distance indicators remain unclear. For instance it is possible that an evolution of the sources or other systematic uncertainties may have not yet been properly considered, and still cast reasonable doubt to prevent the full acceptance of the aforementioned accelerated expansion possibility.

On the other hand, many other independent observational data, like the galaxy clustering statistics, the peculiar velocities, the baryon mass fraction in clusters of galaxies, and the spectrum of cosmic microwave background radiation anisotropies, make investigators entertain the idea that the density of clumped, bright, and dark exotic matter as well, is not nearly enough to make the Universe spatially flat, specially when the evidence obtained from BOOMERANG-98 and MAXIMA-I [2] experiments to determine the position of the first acoustic peak in the angular power spectrum of the cosmic microwave background, indicates that the Universe is close to being spatially flat, and favors the "inflation proposal" which maintains that just after its big bang beginning the Universe underwent a short duration inflationary expansion, and in consequence its spatial geometry ends up being very nearly flat, and helps explain other cosmological observations as well, like the horizon problem.

All these issues, and several other related considerations, have lead to speculate that in addition to ordinary and dark matter, a considerable portion of the energy density that constitutes the Universe must take the form of a smooth –instead of clumped on cosmological scales, and would therefore not interfere with the formation of galaxies or clusters of galaxies– yet unknown ingredient: so it is also "dark". Moreover, concerning the accelerating expansion phenomenon this energy must have a repulsive character similar in its effects to those that are produced when a positive "cosmological constant", designated in Einstein's equations by the Greek letter lambda (Λ). This factor is now related to the energy of the quantum vacuum which is filled with virtual particles and antiparticles that constantly, but for a brief instance, come into existence to disappear again in the form of pure energy. The trouble with a constant Λ is that its value is at odds with those ones predicted by most particle physics models. Anyhow, briefly stated this time all kinds of matter, ordinary as well as dark, generate through gravitational attraction a considerable quantity of the familiar, partially visible, material cosmic structure. Meanwhile, a repulsive dark energy smoothly pervades the Universe. Tentatively, together dark energy and the dark matter conform around 95 percent of the cosmos.

So, assuming that the above statements are correct, in addition to ordinary and dark matter –both of which have an attractive character– Einstein's general relativity cosmology (GR) demands the presence of extra, repulsive matter fields, capable of generating an accelerated expansion. Such dynamical fields with self interaction are referred to as "quintessence matter" or Q–matter. This Q–matter can be phenomenologically treated as scalar fields with a potential able to generate nowadays a sufficient negative pressure, and with a slowly varying energy density it acts as an effective cosmological constant. Similar to the inflation process but with the difference that its energy scale is much lower. Nowadays when those fields' energy densities are dominant, in contrast to their early stage magnitude which then should be almost negligible, implies they must have grown enough to turn consonant with the present matter density, which sets numerous constraints and the fine tuning of parameters for the quintessence potential's adequate evolution [3]. None of the known potentials are entirely problem free.

Within GR the quintessence scalar fields are minimally coupled, and for the Friedmann–Lemaître–Robertson–Walker models (FLRW) this type of fields do their job only when the space is flat which, some think is a result that is little dull.

Non minimally coupled scalar fields are the mark of scalar–tensor gravity theories pondered to be natural alternatives to, classic GR. Beside other considerations, solutions that have the possibility of producing late time accelerated expansion have been examined when extra scalar fields minimally coupled to gravity, and with self interaction terms are included into their original design. The proposed solutions for these models always have a power law expansion.

Because the most simple scalar–tensor cosmology, Brans–Dicke theory (BD) [4], offers additional possibilities that can be made to conform with observations several authors, Banerjee and Pavón [5] in particular, have explored the outcome of a model which in addition to ordinary matter represented by "dust" a barotropic, perfect fluid, includes a second extra scalar field with a potential –Q-matter– which is minimally coupled to gravity, and show that its curved spaces can exhibit a non–decelerating solution where in this context the *deceleration parameter q*, has the value $q = 0$, and beside the solution may even explain other cosmological questions.

This paper exhibit the behavior of models for all kinds of space curvature available in FLRW, for a BD cosmology when its native scalar field is sourceless: First by itself, with a vacuous, matter energy–momentum tensor for it is well known that such cosmologies exist and, in contrast to GR, are non trivial, and next the corresponding solutions for the other similar cosmologies, this time in the presence of radiation because they also support a sourceless scalar field; are examined.

From the variation of the action for the BD Lagrangian this paper presents first, its well known field equations for two dependent variables which after proper combination to form a new, single variable, is made to depend on a redefined time. With their help the ensuing equations are easily solved and all its consequences can be grasped. Because of its transparency the vacuum case will be treated first, and next the radiation case. The remaining barotropic fluids can also be handled in

2. Scalar–tensor field equations

The Lagrangian of scalar–tensor gravity theories [6], but with no self interaction, which can be expressed as

$$L = \phi R - \frac{\omega(\phi)}{\phi}\phi_{;\alpha}\phi^{;\alpha} + 16\pi L_M, \qquad \alpha = 0,...,3 \qquad (1)$$

where R is the spacetime Ricci curvature scalar, ϕ is the scalar field, $\omega(\phi)$ a dimensionless coupling function, and the units I use are such that G Newton's gravitational constant, together with c the speed of light, obtain $8\pi G = c = 1$. Finally, L_M is here the Lagrangian for ordinary matter fields. The variation of the action gets BD equations when $\omega = const.$. The variation with respect to $g_{\mu\nu}$, and ϕ generate the following field equations

$$R_{\mu\nu} - \frac{1}{2}g_{\mu\nu}R = 8\pi\frac{T_{\mu\nu}}{\phi} + \frac{1}{\phi}[\phi_{;\mu\nu} - g_{\mu\nu}\phi_{;\alpha}\phi^{;\alpha}] + \frac{\omega}{\phi}[\phi_{;\mu}\phi_{;\nu} - \frac{1}{2}g_{\mu\nu}\phi_{;\alpha}\phi^{;\alpha}], \qquad (2)$$

and

$$\phi_{;\mu}{}^{;\mu} = \left(\frac{8\pi}{3+2\omega}\right)T. \qquad (3)$$

Just for completeness, the energy–momentum tensor for a perfect fluid where p is the pressure, and ρ its energy density is

$$T^{\mu\nu} = (\rho + p)u^\mu u^\nu + pg^{\mu\nu}$$

where u^β is its four velocity, and $T \equiv T_\mu{}^\mu$ its trace. The vacuum is simply represented by a null energy–momentum tensor. Barotropic fluids are described by the equation $p = n\rho$ with $-1 < n < 1$.

The Robertson–Walker (RW) metric in (t, r, θ, φ) coordinates

$$ds^2 = -dt^2 + a(t)^2[(1-kr^2)^{-1}dr^2 + r^2(d\theta^2 + sin^2\theta d\varphi^2)], \qquad (4)$$

where a is the scale factor, and $k = +1, 0 \, or -1$ for the positive, vanishing or negative space curvature respectively.

From

$$T^\nu_{\mu;\nu} = 0 \qquad (5)$$

one obtains

$$\dot\rho = -3(\rho + p)(\ln a)\dot{}$$

which yields
$$\rho = M_n a^{-3(1+n)}, \qquad M_n = \text{const.} \tag{6}$$
and for a single, perfect fluid, in a FLRW cosmology its expansion is determined by two unknown dependent variables: $a = a(t)$, and $\phi = \phi(t)$ that satisfy the following ordinary, but nonlinear field equations.

$$\frac{\ddot{\phi}}{\phi} + 3\frac{\dot{a}\dot{\phi}}{a\phi} = \frac{8\pi}{3+2\omega} \frac{(\rho - 3p)}{\phi} \tag{7}$$

is the wave, ϕ field equation, together with

$$\frac{\ddot{a}}{a} + 2\frac{\dot{a}^2}{a^2} + \frac{\dot{a}\dot{\phi}}{a\phi} + 2\frac{k}{a^2} = \frac{8\pi}{3+2\omega}\frac{[(1+\omega)\rho - \omega p]}{\phi}, \tag{8}$$

and

$$3\frac{\ddot{a}}{a} + \omega\frac{\dot{\phi}^2}{\phi^2} + \frac{\ddot{\phi}}{\phi} = \frac{-8\pi}{3+2\omega}\frac{[(2+\omega)\rho + 3(1+\omega)p]}{\phi} \tag{9}$$

where the over-dots stand for time derivatives.

3. The vacuum

The BD field equations for a vacuous stress–energy tensor written in terms of the corresponding reduced variable $v \equiv \phi a^3$, are obtained noting that by integrating the original wave equation one has

$$\frac{\dot{\phi}}{\phi} = \frac{\nu}{v}, \qquad \nu = const. \tag{10}$$

from which the relation between a and v

$$a^3 = v \exp\left(\frac{-\nu}{3}\int \frac{dt}{v}\right). \tag{11}$$

is obtained from

$$3\frac{\dot{a}}{a} = \frac{\dot{v}}{v} - \frac{\nu}{v}, \tag{12}$$

The dynamic equation is

$$\frac{\ddot{v}}{v} = -6\frac{k}{a^2}, \qquad k = 0, \pm 1 \tag{13}$$

and its constriction, where the curvature term in it has been replaced by Eq.(13)

$$3\frac{\ddot{v}}{v} - 2\left(\frac{\dot{v}}{v}\right)^2 - 2\left(\frac{\nu}{v}\right)\frac{\dot{v}}{v} + (4+3\omega)\left(\frac{\nu}{v}\right)^2 = 0. \tag{14}$$

The above equations are directly integrable: For $k = 0$

$$v = ct + t_0, \qquad c, t_0 = const. \tag{15}$$

where

$$2c = \left[-1 \pm (3(3+2\omega))^{\frac{1}{2}}\right]\nu. \tag{16}$$

For $k \neq 0$, the use of $dt = v d\tau$ hands out

$$v = \left[\cosh\left(\frac{(-\Delta)^{\frac{1}{2}}}{6}\tau\right)\right] e^{-\frac{\nu}{2}(\tau+\tau_0)}, \quad \omega > -3/2, \tag{17}$$

for $k = 1$ and

$$v = \left[\cos\left(\frac{(\Delta)^{\frac{1}{2}}}{6}\tau\right)\right] e^{-\frac{\nu}{2}(\tau+\tau_0)}, \quad \omega < -3/2, \tag{18}$$

for $k = -1$, with $\Delta \equiv -12(3+2\omega)\nu^2$.

The corresponding scale factors for the above two spaces are

$$a = \left[e^{-\frac{\nu}{2}(\tau+\tau_0)} \cosh\left(\frac{(-\Delta)^{\frac{1}{2}}}{-6}(\tau+\tau_0)\right)\right]^{\frac{1}{2}}, \tag{19}$$

when $k = 1$, where $\omega > -3/2$ and

$$a = \left[e^{\frac{\nu}{2}(\tau+\tau_0)} \cos\left(\frac{(\Delta)^{\frac{1}{2}}}{6}(\tau+\tau_0)\right)\right]^{\frac{1}{2}}, \tag{20}$$

when $k = -1$, where $\omega < -3/2$.

Cosmic time t, is obtained from

$$t = \int \left[\cosh\left(\frac{(-\Delta)^{\frac{1}{2}}}{6}\tau\right)\right] e^{-\frac{\nu}{2}(\tau+\tau_0)} d\eta, \tag{21}$$

for $k = 1$ with $\omega > -3/2$, or from

$$t = \int \left[\cosh\left(\frac{(-\Delta)^{\frac{1}{2}}}{6}\tau\right)\right] e^{-\frac{\nu}{2}(\tau+\tau_0)} d\eta, \tag{22}$$

for $k = 1$ with $\omega > -3/2$. So that the general vacuum model solutions are given in parametric form[7] (see also [8]).

4. Radiation

Incoherent radiation or ultra–relativistic fluids characterized by the relation $p = \frac{1}{3}\rho$, are the only ones for which its field equations have been fully solved [9], [10] independent of the space curvature. For the present Eq.(6) implies

$$\rho = M_r a^{-4}. \qquad M_r = const.$$

Such fluids cannot generate a scalar field, which if it exists must be sourceless. The ϕ field equation for it is

$$\dot{\phi} a^3 = \gamma = const \neq 0. \qquad (23)$$

When $\gamma = 0$, the solutions that are obtained correspond to GR solutions.

This case is smoothly solved when $r \equiv \phi a^2$ is used as the dependent variable on an "η" time which is defined by $dt \equiv a d\eta$. With its help the following equations are got

$$2\frac{a'}{a} = \frac{r'}{r} - \frac{\gamma}{r}, \qquad (24)$$

where

$$\frac{\phi'}{\phi} = \frac{\gamma}{r}, \qquad (25)$$

and so

$$a^2 = r \exp\left(-\int \frac{\gamma}{r} d\eta\right). \qquad (26)$$

The dynamic equation for r is

$$\frac{r''}{r} = \frac{2}{3}\frac{m_r}{r} - 4k, \quad k = 0, \pm 1, \qquad (27)$$

and its constriction is

$$\left(\frac{r'}{r}\right)^2 - \frac{3+2\omega}{3}\left(\frac{\gamma}{r}\right)^2 = \frac{4m_r}{3r} - 4k. \quad m_r \equiv 8\pi M_r \qquad (28)$$

These last two equations do not contain the scale factor "a" that in the equivalent equations for other fluids is always present when $k \neq 0$. The absence of the aforementioned term makes unnecessary to recourse, as in the next section, to a third order differential equation in order to eliminate it and almost effortlessly find some model's solutions.

For $k = 0$ Eq.(27), and Eq.(28) obtain

$$r = R\eta^2 + Q\eta + P, \qquad (29)$$

with $R = \frac{1}{3}m_r$, and $3Q^2 = 4m_rP + (3+2\omega)\gamma^2$ while P remains free (again, P and therefore Q get determined in a Bianchi I model). The discriminant is
$$\Delta_0 = -\frac{(3+2\omega)}{3}\gamma^2.$$
For the LFRW flat space ($k = 0$) model, three different cosmic solutions with its possible subcases exist distinguished by the discriminant whose sign depends on the sign of ω. These solutions have been analyzed in [10].

Particular solutions, with
$$r = r_0 = const., \qquad (30)$$
where $r_0 = \frac{1}{6k}m_r$ for $k \neq 0$ are such that
$$a = \left(\frac{1}{6k}m_r\right)^{\frac{1}{2}} \exp\left(-\frac{3k\gamma}{m_r}\eta\right). \qquad (31)$$
This last expression given in terms of the metric "cosmic time" t turns into a linear expansion given by
$$a = -\left(\frac{3k\gamma}{m_r}\right)t = \left(-\frac{3k}{3+2\omega}\right)^{\frac{1}{2}}t, \qquad (32)$$
with the help of the constriction equation because from it one gets
$$m_r^2 = -3(3+2\omega)k\gamma^2, \qquad (33)$$

For $k \neq 0$ Eq.(27) and Eq.(28) have the following, general solutions

$$2(\eta + \eta_0) = \begin{cases} \ln\left[2\left(16k^2r^2 + \frac{16}{3}m_rr + \frac{(3+2\omega)}{3}\gamma^2\right)^{\frac{1}{2}} - 8kr + \frac{4}{3}m_r\right], \\ \operatorname{arcsh}\frac{-8kr + \frac{4}{3}m_r}{\Delta^{\frac{1}{2}}}, \\ -\arcsin\frac{-8kr + \frac{4}{3}m_r}{(-\Delta)^{\frac{1}{2}}}, \\ \ln(-8kr + \frac{4}{3}m_r), \end{cases} \qquad (34)$$

the first with $\Delta = -\frac{16}{9}(m_r + 3(3+2\omega)\gamma^2$ is for $k = -1$, the second is for $\Delta > 0$ and $k = -1$ or for $\Delta < 0$, the third for $k = 1$, and the last is for $\Delta = 0$ and $k = -1$. Moreover, when $\Delta \neq 0$ for each type of determinant, depending on the sign of ω, one can get two solutions.

Acknowledgments

This work was supported by CONACyT Grant No. 5-3672-E9312.

References

[1] P. M. Garnavich et al., *Astrophys. J.* **509** (1998) 74; A G. Reiss et al., *Astron. J.* **116** (1998) 1009; S. Perlmutter et al., *Astrophys. J.* **517** (1999) 565.

[2] P. Bernadis et al., *Nature* **404**(2000) 955.

[3] P. J. Steinhardt, L. Wang and I. Zlatev, *Phys. Rev. Lett.* **59** (1999) 123504.

[4] C. Brans and R. H. Dicke, *Phys. Rev.* **124** (1961) 925.

[5] N. Banerjee and D. Pavón, *Class. Quantum Grav.* **1** (2001) 925.

[6] P. G. Bergmann, *Int. J. Theor. Phys.* **1** (1968) 25; R. V. Wagoner, *Phys. Rev.* **D1** (1970) 3209; K. Nordvedt, *Astrophys. J.* **161** (1970) 1059.

[7] P. Chauvet, *Astrophys. Space Sci.* **90** (1983) 51.

[8] J.P. Mimoso and D. Wands, *Phys. Rev.* **D51** (1995) 477; J. D. Barrow, *Phys. Rev.* **D47** (1993) 5329; P. Chauvet and O. Obregón, *Astrophys. Space Sci.* **66** (1979) 515; R. E. Dehnen and O. Obregón, *Astrophys. Space Sci.* **14** (1971) 454; R. E. Morganstern, *Phys. Rev.* **D4** (1971) 278.

[9] P. Chauvet, and J. L. Cervantes-Cota, *Phys. Rev.* **D52** (1995) 3416; P. Chauvet and O. Pimentel, *Gen. Rel. Grav.* **24** (1992) 243; P. Chauvet, J. L. Cervantes-Cota, and H. N. Núñez-Yépez, *Class. Quantum Grav.* **9** (1992) 1923.

[10] L. E. Gurevich, A. M. Finkelstein, and V. A. Ruban, *Astrophys. Space Sci.* **22** (1973) 231; V. A. Ruban, and A. M. Finkelstein, *Gen. Rel. Grav.* **6**(1975) 601.

ON CONFORMALLY FLAT STATIONARY AXISYMMETRIC SPACETIMES

Cuauhtemoc Campuzano and Alberto Garcia
Departamento de Fisica, Centro de Investigación y de Estudios Avanzados del IPN
Apdo. Postal 14-740, 07000 México D.F., MEXICO
ccvargas@fis.cinvestav.mx, aagarcia@fis.cinvestav.mx

Abstract   It is shown that within stationary axisymmetric spacetimes, there exists a class of metrics which is, at the same time, conformally flat. The freedom in the structure of the metric reduces to a single function depending on two variables.

Keywords: Conformally flat and stationary axisymmetric spacetimes, staticity.

1. Introduction

In 1976, Collinson [1] formulated the following theorem: Every conformally flat axisymmetric spacetime is necessarily static. If the source is a perfect fluid, then the spacetime can be reduced to the usual Schwarzschild interior metric, for short SIM. Later, in 1988, Garc´initiala demonstrated the existence of three branches of solutions described by SIM [2]. The main goal of this work is to review this problem and establish that the Collinson theorem fails to be true.

The starting point of our study is the axisymmetric stationary metric

$$ds^2 = e^{-2Q(z,\bar{z})}dz\,d\bar{z} + \frac{e^{-2G(z,\bar{z})}}{a+b}\left(a(z,\bar{z})d\phi + dt\right)\left(b(z,\bar{z})d\phi - dt\right). \quad (1)$$

The evaluation of the Newman–Penrose curvature coefficients, Weyl complex components, with respect to the null tetrad basis

$$e^1 = e^{-2Q}dz, \quad e^3 = \frac{e^{-2G}}{\sqrt{a+b}}(b\,d\phi - dt),$$
$$e^2 = e^{-2Q}d\bar{z}, \quad e^4 = \frac{e^{-2G}}{\sqrt{a+b}}(a\,d\phi + dt). \quad (2)$$

gives the following nonvanishing components

$$\Psi_0 = \frac{e^{2Q}}{a+b}\left(2\frac{\partial a}{\partial z}\left(\frac{\partial Q}{\partial z}-\frac{\partial G}{\partial z}\right)+\frac{\partial^2 a}{\partial z^2}-\frac{2}{a+b}\left(\frac{\partial a}{\partial z}\right)^2\right),$$

$$\overline{\Psi}_4 = \frac{e^{2Q}}{a+b}\left(2\frac{\partial b}{\partial z}\left(\frac{\partial Q}{\partial z}-\frac{\partial G}{\partial z}\right)+\frac{\partial^2 b}{\partial z^2}-\frac{2}{a+b}\left(\frac{\partial b}{\partial z}\right)^2\right),$$

$$6\Psi_2 = \frac{e^{2Q}}{(a+b)^2}\left(2(a+b)^2\left(\frac{\partial^2 Q}{\partial z\partial\bar{z}}-\frac{\partial^2 G}{\partial z\partial\bar{z}}\right)+5\frac{\partial a}{\partial z}\frac{\partial b}{\partial \bar{z}}-\frac{\partial a}{\partial \bar{z}}\frac{\partial b}{\partial z}\right). \quad (3)$$

If one demands the spacetime to be conformally flat, then the Weyl tensor has to vanish, which is equivalent to fulfill the requirement $\Psi_0 = \Psi_4 = \Psi_2 = 0$. Accordingly, one has:

$$\Psi_0 = 0 \implies 2\frac{\partial a}{\partial z}\frac{\partial P}{\partial z}+\frac{\partial^2 a}{\partial z^2}-\frac{2}{a+b}\left(\frac{\partial a}{\partial z}\right)^2 = 0, \quad (4)$$

where the function P stands for $Q - G := P$,

$$\Psi_4 = 0 \implies 2\frac{\partial b}{\partial z}\frac{\partial P}{\partial z}+\frac{\partial^2 b}{\partial z^2}-\frac{2}{a+b}\left(\frac{\partial b}{\partial z}\right)^2 = 0. \quad (5)$$

Subtracting equations (4) and (5) one obtains

$$2\left(\frac{\partial a}{\partial z}-\frac{\partial b}{\partial z}\right)\frac{\partial P}{\partial z}+\frac{\partial^2 a}{\partial z\partial z}-\frac{\partial^2 b}{\partial z\partial z}-\frac{2}{a+b}\left(\left(\frac{\partial a}{\partial z}\right)^2-\left(\frac{\partial b}{\partial z}\right)^2\right) = 0 \quad (6)$$

or, dividing by $\partial(a-b)/\partial z$

$$\frac{\partial}{\partial z}\ln\left[\frac{e^{2P}}{(a+b)^2}\left(\frac{\partial a}{\partial z}-\frac{\partial b}{\partial z}\right)\right] = 0 \implies \frac{\partial a}{\partial z}-\frac{\partial b}{\partial z} = \bar{g}(\bar{z})(a+b)^2 e^{-2P}. \quad (7)$$

Dividing Eq.(4) by $\partial a/\partial z$ and Eq.(5) by $\partial b/\partial z$, and subsequently adding the resulting equations one gets

$$\frac{\partial}{\partial z}\ln\left[\frac{e^{4P}}{(a+b)^2}\frac{\partial a}{\partial z}\frac{\partial b}{\partial z}\right] = 0 \implies \frac{\partial a}{\partial z}\frac{\partial b}{\partial z} = \bar{h}(\bar{z})(a+b)^2 e^{-4P}. \quad (8)$$

For real Ψ_2, one arrives at the condition

$$\frac{\partial a}{\partial z}\frac{\partial b}{\partial \bar{z}} = \frac{\partial a}{\partial \bar{z}}\frac{\partial b}{\partial z}. \quad (9)$$

Since the functions a, b, and P are real, therefore from (7) one has

$$g(z)\frac{\partial}{\partial \bar{z}}(a-b) = \bar{g}(\bar{z})\frac{\partial}{\partial z}(a-b). \quad (10)$$

Because of the remaining freedom in the choice of the variable z, introducing a new variable z, such that $g(z)\frac{\partial}{\partial z} \to \frac{\partial}{\partial z}$, one arrives at

$$\left(\frac{\partial}{\partial z} - \frac{\partial}{\partial \bar{z}}\right)(a - b) = 0. \tag{11}$$

Hence,

$$a - b = F(z + \bar{z}). \tag{12}$$

Substituting a from the above relation into Eq. (9) one obtains

$$\dot{F}\left(\frac{\partial}{\partial z} - \frac{\partial}{\partial \bar{z}}\right)b = 0, \tag{13}$$

where dots denote derivatives with respect to $z + \bar{z}$.
If $\dot{F} = 0$, then using linear transformations of the Killingian variables t and ϕ, one can achieve $a = b$, and consequently the metric is static.
If now $\dot{F} \neq 0$, then $b = b(z + \bar{z})$ and $a = a(z + \bar{z})$. By setting $g(z) = 1$, without loss of generality equations (7) and (8) rewrites as

$$\dot{a}\dot{b} = \epsilon k^2 (a + b)^2 \, e^{-4P} \tag{14}$$

and

$$\dot{a} - \dot{b} = (a + b)^2 \, e^{-2P} \tag{15}$$

Combining Eqs. (14) and (15), one obtains

$$\dot{a}\dot{b} = \epsilon k^2 \frac{(\dot{a} - \dot{b})^2}{(a + b)^2}. \tag{16}$$

Introducing new dependent functions X and Y on the variable $z + \bar{z}$ according with

$$\begin{aligned} a + b &= 2kY, \; a = k(Y + X), \\ a - b &= 2kX, \; b = k(Y - X), \end{aligned} \tag{17}$$

the equation (16) becomes

$$\dot{Y}^2 - \dot{X}^2 = \epsilon \frac{\dot{X}^2}{Y^2} \tag{18}$$

or equivalently

$$\left(\frac{dY}{dX}\right)^2 = 1 + \frac{\epsilon}{Y^2} \tag{19}$$

with general integral
$$(X - X_0)^2 = Y^2 + \epsilon. \tag{20}$$

In terms of the original functions a and b, one has
$$a = kX_0 - \frac{\epsilon k^2}{b + kX_0}, \tag{21}$$

where X_0 is an integration constant, at this stage the sign of the parameter $\epsilon = \pm 1$ remains free. Entering with (21) into Eq. (14) one gets
$$e^{-4P} = \frac{\dot{b}^2}{(b + kX_0)^2(a + b)^2}, \tag{22}$$

Using now the linear transformations of the Killingian variables
$$dt = \alpha dt' + \beta d\phi',$$
$$d\phi = \gamma dt' + \delta d\phi', \alpha\delta - \beta\gamma \neq 0, \tag{23}$$

one can achieve that the new metric component $g_{t'\phi'}$ vanishes if
$$\alpha = -\gamma k(X_0 \pm \sqrt{\epsilon}), \ \gamma = \gamma,$$
$$\beta = \delta k(-X_0 \pm \sqrt{\epsilon}), \ , \delta = \delta. \tag{24}$$

Hence, for $\epsilon = 1$ the stationary metric can be brought to the form of a static metric. Then, with $a = b$, one can reproduce the results derived by Collinson for conformally flat static metrics.

The case $\epsilon = -1$ deserves special attention.

2. Conformally flat axisymmetric stationary spacetimes, $\epsilon = -1$

This case escapes to previous considerations. It opens the possibility to consider conformally flat stationary axisymmetric spacetime structures in General Relativity. The corresponding metric remains stationary axisymmetric; there is no a real linear transformation of the Killingian variables t and ϕ, which could be used to cancel $g_{t\phi}$. Knowing that the functions $a = a(x)$, $b = b(x)$, and $P = P(x)$, where $2x = z + \bar{z}$, and $2y = z - \bar{z}$, it is more convenient to work with the metric in coordinates $\{x, y, \phi, t\}$. In these coordinates the metric amounts to

$$ds^2 = e^{-2G(x,y)}\left\{\left(-\frac{db}{dx}\right)\frac{dx^2 + dy^2}{(b + kx_0)^2 + k^2}\right.$$
$$\left. + \frac{b + kx_0}{(b + kx_0)^2 + k^2}\left[\left(kx_0 + \frac{k^2}{b + kx_0}\right)d\phi + dt\right](bd\phi - dt)\right\}. \tag{25}$$

Notice that the first derivative with respect to its argument x, db/dx, ought to be negative in its range of definition. The explicit form of this function $b(x)$, one determines by setting Ψ_2 equal to zero; this last quantity is given by

$$12\Psi_2 e^{-2G}\left(\frac{db}{dx}\right)^3 = +\left(k^2+(b+kx_0)^2\right)\left[\left(\frac{db}{dx}\right)\frac{d^3b}{dx^3}-\left(\frac{d^2b}{dx^2}\right)^2\right]$$
$$-2(kx_0+b)\left(\frac{db}{dx}\right)^2\frac{d^2b}{dx^2}+2\left(\frac{db}{dx}\right)^4, \qquad (26)$$

hence for $\Psi_2 = 0$ the general solution for vanishing Weyl tensor amounts to

$$b(x) = \frac{C_1 - \mu\tanh[\mu(x+C_3)]}{C_2}, \qquad (27)$$

where the positive constant $\mu := \sqrt{k^2 C_2^2 + (C_1+kx_0 C_2)^2}$, and C_1, C_2, and C_3 are integration constants.

An adequate choice of the constants allows one to fulfill the non–positive behaviour of the first derivative of function $b(x)$.

3. Conclusions

On the light of the present results, we conclude that the Collinson theorem fails. There exist branches of spacetimes, which are conformally flat and at the same time are stationary and axisymmetric. The conformal factor of this class of metrics depends on non–Killingian variables x and y; its explicit expression depend on the sources of the Einstein equations.

Acknowledgments

The authors thank Eloy Ayón–Beato for useful discussions. This work has been partially supported by CONACyT Grant 38495E.

References

[1] C.D. Collison, *Gen. Rel. Grav.* **7** (1976) 419.

[2] A. A. Garc'initiala, *Gen. Rel. Grav.* **20** (1988) 595.

QUANTUM COSMOLOGY IN SOME EXTENDED SCALAR–TENSOR THEORIES

Sergio del Campo
Instituto de Fisica, Facultad de Ciencias Básicas y Matemáticas,
Universidad Católica de Valparaiso, Avenida Brasil 2950, Valparaiso, Chile.
Departamento de Fisica, Facultad de Ciencia, Universidad de Santiago de Chile.
sdelcamp@ucv.cl

Samuel Lepe
Instituto de Fisica, Facultad de Ciencias Básicas y Matemáticas,
Universidad Católica de Valparaiso, Avenida Brasil 2950, Valparaiso, Chile.
Departamento de Fisica, Facultad de Ciencia, Universidad de Santiago de Chile,
Avda. Ecuador 3493, Santiago, Chile.
slepe@lauca.usach.cl

Luis O. Pimentel
Departamento de Fisica, Universidad Autónoma Metropolitana
Apartado Postal 55-534, C.P. 09340, México, D.F., México
lopr@xanum.uam.mx

Abstract The Wheeler–DeWitt equation is considered in a family of extended scalar–tensor theories (STT). We also include a Kaluza–Klein model. We find exact and/or approximated solutions to these equations. Their asymptotic forms (WKB wave function) are also specified.

Keywords: Quantum Cosmology, alternative theories of gravitation.

1. Introduction

In Quantum cosmology the description of the initial state of the Universe, via the wave function of the Universe, has become very important, since from it is possible to obtain specific initial values of the corresponding dynamical variables that are used for describing the evolution of the Universe.

In order to do so, it is necessary to impose some initial condition for the wave function of the Universe, which is obtained as a solution of a functional differential equation, the Wheeler–DeWitt (WDW) equation [1]. Perhaps, the most appealing boundary conditions that have been proposed up to day, are the Hartle–Hawking no–boundary[2] and the Vilenkin tunneling[3] boundary conditions. In these days have had a revival discussion about these boundary condition when they are applied to describe an emerging open universe[4]. It seems that we still need to know more before we can decide which boundary condition is the most appropriated for describing the initial state of the Universe.

In this paper we would like to present some solutions to the WDW equation in different cases, related mostly to scalar tensor theories. The solutions that we found are the most general possible. Their asymptotic forms are also described. Among the different cases that we deal with are the scalar tensor theories, non–linear gravity and multidimensional gravity.

2. Scalar–tensor theories

The action for scalar tensor theories is taken to be

$$S = \frac{1}{16\pi} \int \sqrt{-g} \left\{ \phi R - \frac{\omega(\phi)}{\phi} g^{\mu\nu} \phi_{,\mu} \phi_{,\nu} + 2\phi \lambda(\phi) \right\} d^4x. \tag{1}$$

In a FRW metric $ds^2 = dt^2 - a^2(t)[d\chi^2 + \sin^2\chi(d\theta^2 + \sin^2\theta d\phi^2)]$, and assuming that the scalar field ϕ is homogeneous, the action simplifies to

$$S = \frac{\pi}{8} \int \left[-6a\phi + 6a\dot{a}^2 \phi + 6a^2 \dot{a}\dot{\phi} - \omega_0 \frac{a^3 \dot{\phi}^2}{\phi} + 2a^3 \phi \lambda(\phi) \right] dt, \tag{2}$$

from which we could read the Lagrangian

$$L = \frac{\pi}{8} \left(6a\dot{a}^2 \phi + 6a^2 \dot{a}\dot{\phi} - 6a\phi - \omega_0 \frac{a^3 \dot{\phi}^2}{\phi} + 2a^3 \phi \lambda(\phi) \right), \tag{3}$$

and the canonical conjugate momenta corresponding to a and ϕ are

$$\pi_a = \frac{3}{4}\pi a^2 \phi \left(2\frac{\dot{a}}{a} + \frac{\dot{\phi}}{\phi} \right), \quad \pi_\phi = \frac{\pi}{4} a^3 \left(3\frac{\dot{a}}{a} - \omega_0 \frac{\dot{\phi}}{\phi} \right). \tag{4}$$

The Hamiltonian H of the system is

$$H = \pi_a \dot{a} + \pi_\phi \dot{\phi} - L = \frac{\omega_0}{6} a^2 \pi_a^2 - \phi^2 \pi_\phi^2 + a\phi \pi_a \pi_\phi +$$
$$\frac{3}{16}\pi^2(2\omega_0 + 3)a^4\phi^2 - \frac{\pi^2}{16}(2\omega_0 + 3)a^6\phi^2\lambda(\phi). \tag{5}$$

After canonical quantization

$$\pi_a^2 \to -\frac{1}{a^\alpha}\frac{\partial}{\partial a}\left(a^\alpha \frac{\partial}{\partial a}\right) \quad , \quad \pi_\phi^2 \to -\frac{1}{\phi^\beta}\frac{\partial}{\partial \phi}\left(\phi^\beta \frac{\partial}{\partial \phi}\right)$$

we get the WDW equation, $H\Psi(a,\phi) = 0$, for an arbitrary factor ordering, encoded in the α and β parameters,

$$\left\{\frac{\omega_0}{6}\left(a^2\partial_a^2 + \alpha a \partial_a\right) + a\phi \partial_a \partial_\phi - \left(\phi^2 \partial_\phi^2 + \beta \phi \partial_\phi\right) - U(a,\phi)\right\}\Psi(a,\phi) = 0, \tag{6}$$

with the superpotential given by

$$U(a,\phi) = \frac{3}{16}\pi^2(2\omega_0 + 3)a^4\phi^2\left(1 - \frac{a^2\lambda(\phi)}{3}\right). \tag{7}$$

With the following change of variables, $x = a^2\phi$, $y = \rho\ln\phi$, with $\rho^2 = (2\omega_0 + 3)/3$, the WDW equation becomes

$$\left\{x^2\partial_x^2 + (N(\alpha,\beta)+1)x\partial_x - \left(\partial_y^2 + \frac{\beta-1}{\rho}\partial_y\right) - U(x,y)\right\}\Psi(x,y) = 0, \tag{8}$$

where now the superpotential becomes

$$U(x,y) = \left(\frac{3\pi\rho}{4}\right)^2 x^2\left[1 - \frac{xG(y)}{3}\right], \tag{9}$$

with

$$G(y) = e^{-y/\rho}\lambda(e^{y/\rho}), \quad N(\alpha,\beta) = \frac{\omega_0(\alpha-1) - 3(\beta-1)}{2\omega_0 + 3}. \tag{10}$$

In the following we will assume that $G(y)$ is a slowly varying function in the interval of interest ($G'/G << 1$), with $G_0(y)$ the approximately constant value; then it is possible to separate the variables and the corresponding equation are,

$$x^2 X'' + (N(\alpha,\beta)+1)xX' + \left(k^2 - \frac{9\pi^2\rho^2}{16}x^2 + \frac{3\pi^2\rho^2 G_0(y)}{16}x^3\right)X = 0. \tag{11}$$

$$Y'' + \frac{\beta - 1}{\rho}Y' + k^2 Y = 0, \tag{12}$$

where k is a separation constant.

The equation for X can be solved by power series, but it is not very illuminating, instead we will solve a particular case in which the solution can be expressed in terms of standard functions. If we take the particular factor ordering characterized by $\alpha = -1$ and $\beta = 2$ and a vanishing separation constant, we have the following particular solution,

$$\Psi = [C_1 Ai(-\xi) + C_2 Bi(-\xi)]\left[Y_1 - Y_2 e^{-y/\rho}\right], \tag{13}$$

here $\xi = (3\pi\rho/4G_0)^{2/3}(3 - G_0 x)$, C_i and Y_i ($i = 1, 2$) are integration constants. Ai and Bi are the Airy functions.

By using the asymptotic forms of the Airy functions at large argument, it is found for the classical allowed region ($G_0 \phi a^2 > 3$)

$$\Psi \sim (G_0 \phi a^2 - 3)^{-1/4} e^{\pm i\left[\frac{\pi\rho}{2G_0}(G_0\phi a^2 - 3)^{3/2}\right]} (Y_1 - [Y_2/\phi]), \tag{14}$$

for the classical forbidden region ($G_0 \phi a^2 < 3$) we obtain

$$\Psi \sim (3 - G_0 \phi a^2)^{-1/4} e^{\pm\left[\frac{\pi\rho}{2G_0}(3 - G_0\phi a^2)^{3/2}\right]} (Y_1 - [Y_2/\phi]), \tag{15}$$

Note that expressions (14) and (15) are the WKB solutions, since they are of the form $\Psi(a, \phi) \to A(a, \phi)e^{\pm i S_{Cl}}$, $\Psi(a, \phi) \to A(a, \phi)e^{\pm S_E}$, respectively. Here S_{Cl} and S_E represent the Classical and Euclidean actions

3. Non–linear gravity

The action for the nonlinear theories that we will consider is

$$S = \frac{1}{16\pi}\int \sqrt{-g}\left\{\mathcal{R}^n\right\} d^4x, \tag{16}$$

where we have set the coupling constant equal to 1, and n is a constant that we will fix latter. Following Teyssandier and Tourrenc [5] we substitute the above action with one with an auxiliary scalar field, whose equation of motion identifies it with the scalar curvature,

$$S = \frac{1}{16\pi}\int \sqrt{-g}\left\{\phi^n + n\phi^{n-1}(\mathcal{R} - \phi)\right\} d^4x. \tag{17}$$

For FRW metric the action takes the form

$$S = \frac{1}{16\pi}\int dt \left\{(1-n)a^3\phi^n + 6n\phi^{n-1}[a - a\dot{a}^2 + (n-1)a^2\dot{a}\frac{\dot{\phi}}{\phi}]\right\}. \tag{18}$$

The resulting WDW equation for this theory is

$$\left\{(n-1)a\partial_a\phi\partial_\phi + \phi^2\partial_\phi^2 + \alpha\phi\partial_\phi - 6n(n-1)^3 a^6 \phi^{2n-1}\right.$$

$$\left.+36n^2(n-1)^2 a^4 \phi^{2(n-1)}\right\}\Psi(a,\phi) = 0. \qquad (19)$$

Here α takes care of the factor ordering in the variable ϕ. Setting $x = \ln a$ and $y = \ln \phi$ we have

$$\left\{(n-1)\partial_x\partial_y + \partial_y^2 + \alpha\partial_y - 6n(n-1)^3 e^{6x+(2n-1)y}\right.$$

$$\left.+36n^2(n-1)^2 e^{4x+2(n-1)y}\right\}\Psi(x,y) = 0. \qquad (20)$$

With a further change of variable and choosing n=5/7 and $\alpha = 0$ the equation can be made separable,

$$u = \frac{4\sqrt{5}}{7}\left(2x + \frac{1}{7}y\right), \quad v = -\frac{4\sqrt{5}}{7}\left(x - \frac{1}{7}y\right), \qquad (21)$$

and the WDW equation becomes

$$\left\{\partial_u^2 - \partial_v^2 + 15e^u + e^v\right\}\Psi(u,v) = 0. \qquad (22)$$

The general solution is given by

$$\Psi(u,v) = \int dk \; [\; C_1(k)H_{2k}^{(1)}(\sqrt{60}e^{u/2}) + C_2(k)H_{2k}^{(2)}(\sqrt{60}e^{u/2})]$$

$$[C_3(k)H_{2k}^{(1)}(2ie^{v/2}) + C_4(k)H_{2k}^{(2)}(2ie^{v/2})], \qquad (23)$$

where k is the separation constant, $H_{2k}^{(1,2)}$ are the Hankel functions, and we have made a superposition with $C_i(k)$ being the amplitude functions.

Using the asymptotic limits of Hankel functions for $z \to \infty$, and $z \to 0$ we have the asymptotic limit for $a \to \infty$ of the wave function

$$\Psi(a,\phi) \to A(a,\phi)e^{\pm iS_{Cl}}, \qquad (24)$$

where A is some amplitude and S_{Cl} represents the Classical action, and in that limit, it is given by $S_{Cl} = \sqrt{60}(a^2\phi^{1/7})^{2\sqrt{5}/7}$. In the other limit, $a \to 0$ the wavefunction is

$$|\Psi(a,\phi)| \to A_{\alpha,\beta}(a,\phi)e^{\pm 2(\phi^{1/7}/a)^{2\sqrt{5}/7}}, \qquad (25)$$

here $A_{\alpha,\beta}(a,\phi)$ is an amplitude that depends on the factor ordering being used. With our choice A happens to be a non singular function. We notice that $\Psi \to e^{\pm S_E}$, with S_E the Euclidean action.

4. Multidimensional gravity

There are some reasons for studying multidimensional spacetimes. Firstly, it is known that consistent theories that unify fundamental interactions require a multidimensional spacetime, such as superstring[6], supersymmetry[7], etc. Secondly, it is thought that extra dimensions might have played an important role early in the evolution of the universe[8]. At this point, a mechanism which could explain why we do not observe these extra dimension nowadays should be addressed.

In this paper we will consider a $4 + n$–dimensional Kaluza–Klein model, where n is an integer greater than one (associated with the number of extra–dimensions), and a topology governed by $R \times S^3(a) \times S^n(\phi)$, in which a and ϕ are the cosmic scale factors associated with the radii of the external space (three–sphere) and the internal space (n–sphere), respectively. The action of the model is dominated by a cosmological constant plus a magnetic monopole.

The effective action for our model is given by[9]

$$S = \int d^{n+4}x \sqrt{-g} \left[\frac{1}{16\pi G_{n+4}} (R + 2\Lambda) - \frac{1}{2n!} F_{M_1.....M_n} F^{M_1.....M_n} \right], \qquad (26)$$

where G_{4+n} represents the gravitational constant, Λ the cosmological constant, and $F_{M_1.....M_n}$ is a $(n-1)$-th rank antisymmetric tensor field.

In order to write down the corresponding field equations, we use the Robertson–Walker metric, i.e.,

$$g_{MN} = diag(-1, a^2(t)g_{lm}, \phi^2(t)g_{ij}), \qquad (27)$$

where $M, N = 0, ..., n+3$, $l, m = 1, 2, 3$, and $i, j = 4, ..., n+3$ for a given n. For $F_{M_1.....M_n}$ we take the ansatz

$$F_{M_1.....M_n} = \begin{cases} f\sqrt{12\bar{g}}\, \epsilon_{M_1.....M_n}, & \text{for internal space}, \\ 0, & \text{otherwise}. \end{cases}$$

where $\bar{g} = det(g_{ij})$ and $\epsilon_{M_1.....M_n}$ is the antisymmetric Levi–Civita tensor. At the quantum cosmological level, it has been shown[10] that the stability in an S^6 internal space can be achieved by introducing a supergravity rank–six antisymmetric tensor field.

The WDW equation is given by

$$\left\{ a^{-p}\partial_a(a^p\partial_a) + \frac{6}{n(n-1)}\left(\frac{\phi}{a}\right)^2 b^{-q}\partial_\phi(\phi^q\partial_\phi) \right.$$

$$\left. -\frac{6}{n-1}\left(\frac{\phi}{a}\right)\partial_a\partial_\phi - U_n(a,\phi)\right\}\Psi(a,\phi) = 0, \tag{28}$$

where the parameters p and q represent the factor-ordering ambiguity between a and ∂_a and ϕ and ∂_ϕ, respectively, and

$$U_n(a,\phi) = \beta_n^2(\phi)a^2(1 - V_n(\phi)a^2), \tag{29}$$

with

$$\beta_n^2(\phi) = \frac{1}{2}\frac{n+2}{n-1}\phi^{2n}, \tag{30}$$

and

$$V_n(\phi) = \lambda\left[1 + \frac{f^2}{\lambda\phi^{2n}} - \frac{n(n-1)}{6\lambda\phi^2}\right], \tag{31}$$

represents the superpotential.

The minisuperspace of the model is defined by $0 < a < \infty$ and $0 < \phi < \infty$. We shall solve equation (28) by assuming that we can neglect ϕ-derivatives[11], and by introducing in this case a new variable

$$x(a,\phi) = -\left[\frac{\beta_n(\phi)}{2V_n(\phi)}\right]^{2/3}(1 - V_n(\phi)a^2) \tag{32}$$

and with the choice $p = -1$, equation (28) reduces to

$$(\partial_x^2 + x)\Psi(x) = 0, \tag{33}$$

which presents exact solutions. The wave function corresponding to an outgoing wave in the classically allowed region ($U_n < 0$), up to a numerical coefficient, is given by[3])

$$\Psi(x) = [A_i(-x) + iB_i(-x)]/[A_i(-x_o) + iB_i(-x_o)] \tag{34}$$

here A_i and B_i are Airy functions and $x_o = x(0,b) = -[\beta_n(b)/2V_n(b)]^{2/3}$. The asymptotic form of equation (34) turns out to be

$$\Psi(a,\phi) \sim (V_n(\phi)a^2 - 1)^{-1/4}exp\left\{-\frac{\beta_n(\phi)}{3V_n(\phi)}\left[i(V_n(\phi)a^2 - 1)^{3/2} + 1\right]\right\}. \tag{35}$$

An analytic continuation of Eq. (35) enables us to obtain the wave function in the forbidden region ($U_n > 0$).

Acknowledgments

This work was supported by the international collaboration Chile–Mexico under grants Cooperación Cientifica International–Fundación Andes/CONICYT (95052)in Chile, and CONACYT E 120–243 in Mexico.

References

[1] J.A. Wheeler, in *Battele Rencontres*, edited by C. DeWitt and J.A. Wheeler (Benjamin, New York, 1968); B. S. DeWitt, *Phys. Rev.* **160** (1967) 1113.

[2] J. Hartle and S.W. Hawking, *Phys. Rev.* **D28** (1983) 2960.

[3] A. Vilenkin, *Phys. Rev.* **D30** (1984) 509; **D33** (1986) 3560; **D37** (1988) 888; **D50** (1994) 2581; A. Mac´initialas, *Gen. Rel. Grav.* **31** (1999) 653.

[4] S.W. Hawking and N.G. Turok, hep–th/9802030; A. Linde, gr-qc/9802038; A. Vilenkin, hep–th/9803084; W. G. Unruh, gr-qc/9803050.

[5] P. Teyssandier and P. Tourrenc, *J. Math. Phys.* **24** (1983) 2793.

[6] M.B. Green, J. H. Schwarz and E. Witten, *Superstring Theory* Vols. I and II.

[7] M.F. Sohnius, *Phys. Rep.* **128** (1985) 39.

[8] E. W. Kolb and M. S. Turner, *The Early Universe* (Frontries in Physics, Addison–Wesley Publ. Com., INC., 1988).

[9] P. G. O. Freud and M. A. Rubin, *Phys. Lett.* **B97** (1980) 233.

[10] U. Carow–Watamuno, T. Inami and S. Watamura, *Class. Quantum Grav.* **4** (1987) 23.

[11] The omission of phi–derivatives is justified if in the classical region, $(U_n < 0)$, $\phi\alpha \gg |\sqrt{6n(n-1)} - 2n|$ is satisfied, and in the forbidden region, $(U_n > 0)$, $V_n(\phi)a^2\phi\alpha \ll \sqrt{6n(n-1)} + 4n$, where $\alpha \equiv \partial_\phi[\ell n V_n(\phi)]$.

DEFORMATION QUANTIZATION OF SDIFF(Σ_2) SDYM EQUATION

M.Przanowski
Departamento de Fisica, CINVESTAV, 07000 Mexico D.F., Mexico.
Institute of Physics, Technical University of Łódź,
Wólczańska 219. 93-005 Łódź, Poland.
przan@fis.cinvestav.mx

J.F.Plebański
Departamento de Fisica, CINVESTAV, 07000 Mexico D.F., Mexico.
pleban@fis.cinvestav.mx

S.Formański
Institute of Physics, Technical University of Łódź,
Wólczańska 219. 93-005 Łódź, Poland.
sforman@ck-sg.p.lodz.pl

Abstract Deformation quantization (the Moyal deformation) of SDYM equation for the algebra of the area preserving diffeomorphisms of a 2-surface Σ_2, sdiff(Σ_2), is studied. Deformed equation we call the *master equation* (ME) as it can be reduced to many integrable nonlinear equations in mathematical physics. Two sets of conserved charges for ME are found. Then the linear systems for ME (the Lax pairs) associated with the conserved charges are given. We obtain the dressing operators and the infinite algebra of hidden symmetries of ME. Twistor construction is also done.

Keywords: Quantization.

1. Introduction

In 1994 V.Husain [1] was able to reduce the Ashtekar–Jacobson–Smolin equations describing the metric of self–dual complex vacuum spacetimes (the heavenly spacetimes) to one equation for one holomorphic function $\Theta_0 = \Theta_0(x,y,p,q)$

$$\partial_x^2 \Theta_0 + \partial_y^2 \Theta_0 + \{\partial_x \Theta_0, \partial_y \Theta_0\}_\mathcal{P} = 0 \tag{1}$$

where $\{\cdot,\cdot\}_\mathcal{P}$ stands for the Poisson bracket

$$\{f,g\}_\mathcal{P} = f \overleftrightarrow{\mathcal{P}} g, \qquad \overleftrightarrow{\mathcal{P}} := \frac{\overleftarrow{\partial}}{\partial q}\frac{\overrightarrow{\partial}}{\partial p} - \frac{\overleftarrow{\partial}}{\partial p}\frac{\overrightarrow{\partial}}{\partial q} \tag{2}$$

Eq. (1) is called the *Husain-Park heavenly equation* (*H-P equation*) as it has been also found by Q.H.Park [2] from another point of view. Namely, in Park's approach Eq. (1) is the principal chiral model equation for the algebra of the area preserving diffeomorphisms of a 2-surface Σ_2, sdiff(Σ_2), and it is obtained by a symmetry reduction of sdiff(Σ_2) SDYM equation

$$\partial_x \partial_{\tilde{x}} \Theta_0 + \partial_y \partial_{\tilde{y}} \Theta_0 + \{\partial_x \Theta_0, \partial_y \Theta_0\}_\mathcal{P} = 0 \tag{3}$$

where now $\Theta_0 = \Theta_0(x,y,\tilde{x},\tilde{y},p,q)$.

A natural generalization of Eq. (3) can be done when the Poisson bracket is changed by the Moyal one. Thus one arrives at the following equation [3, 4]

$$\partial_x \partial_{\tilde{x}} \Theta + \partial_y \partial_{\tilde{y}} \Theta + \{\partial_x \Theta, \partial_y \Theta\}_\mathcal{M} = 0 \tag{4}$$
$$\Theta = \Theta(\hbar; x,y,\tilde{x},\tilde{y},p,q)$$

where $\{\cdot,\cdot\}_\mathcal{M}$ denotes the Moyal bracket

$$\{f,g\}_\mathcal{M} := \tfrac{1}{i\hbar}(f*g - g*f) = f\tfrac{2}{\hbar}\sin(\tfrac{\hbar}{2}\overleftrightarrow{\mathcal{P}})g; \quad \hbar \epsilon \mathbf{R}$$
$$f*g := \sum_{n=0}^\infty \tfrac{1}{n!}(\tfrac{i\hbar}{2})^n \omega^{i_1 j_1}\ldots\omega^{i_n j_n}\frac{\partial^n f}{\partial X^{i_1}\ldots\partial X^{i_n}}\frac{\partial^n g}{\partial X^{j_1}\ldots\partial X^{j_n}}$$
$$= f\exp(\tfrac{i\hbar}{2}\overleftrightarrow{\mathcal{P}})g, \quad i_1,\ldots j_1,\cdots = 1,2;$$
$$(X^1, X^2) = (q,p), \quad (\omega^{ij}) = \begin{pmatrix} 0 & 1 \\ -1 & 0 \end{pmatrix} \tag{5}$$

The real parameter $\hbar$ is a *deformation parameter*.

Eq. (4) we call the *master equation* (ME). By a symmetry reduction, using also some representations of the Moyal bracket Lie algebra one can reduce ME to the known heavenly equations, to $su(N)$ SDYM equations or $su(N)$ principal chiral model equations and to many integrable nonlinear equations of mathematical physics [5]. We should point out that $\Theta(\hbar; x,y,\tilde{x},\tilde{y},p,q)$ being a solution of ME is considered to be the formal series with respect to $\hbar$,

$$\Theta = \sum_{n=-N}^\infty \Theta_n \hbar^n, \quad N < \infty, \quad \Theta_n = \Theta_n(x,y,\tilde{x},\tilde{y},p,q) \tag{6}$$

In our recent paper [6] some evidence for the integrability of ME has been provided. It has been shown that ME admits infinite number of nonlocal conservation laws and linear systems (Lax pairs) for ME have been found. Moreover, a twistor construction has been also done.

The aim of the present work is to consider in some details and develop the results of [6]. First, in Section 2, we find infinite number of new nonlocal conservation laws such that the conserved "charges" define hidden symmetries of ME. In Section 3 new linear systems for ME are obtained and the dressing operators leading to solutions of these systems are found. The dressing operators appear to be the solutions of the linear systems presented in [6]. In Section 4 the infinite Lie algebra of the hidden symmetries of ME is given. Finally, Section 5 is devoted to a twistor construction for ME.

2. Conservation laws and hidden symmetries

Let $\eta^{(0)} = \eta^{(0)}(\hbar; x, y, \tilde{x}, \tilde{y}, p, q)$ be some function. Define

$$j_x^{(1)} := \mathcal{D}_{\tilde{x}} \eta^{(0)}, \quad j_y^{(1)} := \mathcal{D}_{\tilde{y}} \eta^{(0)} \tag{1}$$

where

$$\mathcal{D}_{\tilde{x}} := \partial_{\tilde{x}} - \frac{1}{i\hbar} \partial_y \Theta *, \quad \mathcal{D}_{\tilde{y}} := \partial_{\tilde{y}} + \frac{1}{i\hbar} \partial_x \Theta * \tag{2}$$

and $\Theta = \Theta(\hbar; x, y, \tilde{x}, \tilde{y}, p, q)$ is a solution of ME (4). We have

$$\partial_x j_x^{(1)} + \partial_y j_y^{(1)} = (\partial_x \mathcal{D}_{\tilde{x}} + \partial_y \mathcal{D}_{\tilde{y}}) \eta^{(0)} = (\mathcal{D}_{\tilde{x}} \partial_x + \mathcal{D}_{\tilde{y}} \partial_y) \eta^{(0)}$$
$$= \partial_x \partial_{\tilde{x}} \eta^{(0)} + \partial_y \partial_{\tilde{y}} \eta^{(0)} + \frac{1}{i\hbar}(\partial_x \Theta * \partial_y \eta^{(0)} - \partial_y \Theta * \partial_x \eta^{(0)}) \tag{3}$$

Consequently $\partial_x j_x^{(1)} + \partial_y j_y^{(1)} = 0$ iff $\eta^{(0)}$ satisfies the following linear equation

$$\partial_x \partial_{\tilde{x}} \eta^{(0)} + \partial_y \partial_{\tilde{y}} \eta^{(0)} + \frac{1}{i\hbar}(\partial_x \Theta * \partial_y \eta^{(0)} - \partial_y \Theta * \partial_x \eta^{(0)}) = 0 \tag{4}$$

Given a function $\eta^{(0)}$ satisfying Eq.(4), there exists the following function $\eta^{(1)} = \eta^{(1)}(\hbar; x, y, \tilde{x}, \tilde{y}, p, q)$ such that

$$\partial_x \eta^{(1)} = \mathcal{D}_{\tilde{y}} \eta^{(0)}$$
$$-\partial_y \eta^{(1)} = \mathcal{D}_{\tilde{x}} \eta^{(0)} \tag{5}$$

From (4) with (2) one gets

$$(\mathcal{D}_{\tilde{x}} \partial_x + \mathcal{D}_{\tilde{y}} \partial_y) \eta^{(1)} = (\mathcal{D}_{\tilde{x}} \mathcal{D}_{\tilde{y}} - \mathcal{D}_{\tilde{y}} \mathcal{D}_{\tilde{x}}) \eta^{(0)}$$
$$= \frac{1}{i\hbar}(\partial_x \partial_{\tilde{x}} \Theta + \partial_y \partial_{\tilde{y}} \Theta + \{\partial_x \Theta, \partial_y \Theta\}_\mathcal{M}) * \eta^{(0)} \stackrel{\text{by ME}}{=} 0 \tag{6}$$

Therefore, as Θ is a solution of ME $\eta^{(1)}$ satisfies the same equation (4) as $\eta^{(0)}$ does. This enables us to define the current $j^{(2)}$

$$j^{(2)}_x := \mathcal{D}_{\bar{x}}\eta^{(1)} \; , \quad j^{(2)}_y := \mathcal{D}_{\bar{y}}\eta^{(1)} \tag{7}$$

which satisfies the equation $\partial_x j^{(1)}_x + \partial_y j^{(1)}_y = 0$. Hence there exists a function $\eta^{(2)}$ such that

$$\begin{aligned}\partial_x \eta^{(2)} &= \mathcal{D}_{\bar{y}}\eta^{(1)} \\ -\partial_y \eta^{(2)} &= \mathcal{D}_{\bar{x}}\eta^{(1)}\end{aligned} \tag{8}$$

Continuing this procedure we arrive at the series of functions (*conserved charges*) $\eta^{(1)}, \eta^{(2)}, \ldots$ and currents $j^{(1)}, j^{(2)}, \ldots$, defined by the recursion equations

$$\begin{aligned}j^{(n+1)}_x &= -\partial_y \eta^{(n+1)} = \mathcal{D}_{\bar{x}}\eta^{(n)} \\ j^{(n+1)}_y &= \partial_x \eta^{(n+1)} = \mathcal{D}_{\bar{y}}\eta^{(n)} \; , \quad n = 0, 1, \ldots\end{aligned} \tag{9}$$

It is evident that as Θ satisfies ME and $\eta^{(0)}$ satisfies (4) all $\eta^{(n)}$ satisfy the linear equation

$$\partial_x \partial_{\bar{x}} \eta^{(n)} + \partial_y \partial_{\bar{y}} \eta^{(n)} + \frac{1}{i\hbar}(\partial_x\Theta * \partial_y\eta^{(n)} - \partial_y\Theta * \partial_x\eta^{(n)}) = 0 \tag{10}$$

Thus we obtain infinite number of conservation laws

$$\partial_x j^{(n)}_x + \partial_y j^{(n)}_y = 0 \; , \quad n = 1, 2, \ldots \tag{11}$$

and the conserved charges

$$\eta^{(n)} = \int^x dx^{(n)} \mathcal{D}_{\bar{y}} \int^{x^{(n)}} dx^{(n-1)} \mathcal{D}_{\bar{y}} \ldots \int^{x^{(2)}} dx^{(1)} \mathcal{D}_{\bar{y}} \eta^{(0)} \; , \quad n = 1, 2, \ldots \tag{12}$$

In particular, an interesting case is when one puts

$$\eta^{(0)} = 1. \tag{13}$$

Then (taking appropriate boundary conditions) we get from (4) with (13)

$$\eta^{(1)} = \frac{1}{i\hbar}\Theta \tag{14}$$

Observe that Eq. (10) for $\eta^{(1)}$ given by (14) is exactly ME. [The case of $\eta^{(0)} = 1$ has been analyzed in our previous work [6] and in fact the considerations of [6] closely follow E.Brezin et al [7], M.K.Prasad et al [8],

L.L.Chau et al [9] and L.L.Chau [10] where nonlocal conservation laws for some 2-dimensional nonlinear field theories and for SDYM equations have been found.]

Now we are going to look for another collection of conserved charges. Here we follow V.Husain [1] and M.Dunajski and L.J.Mason [11] who obtained infinite number of conservation laws for some heavenly equations in four dimensions.[1]

Assume that $\sigma^{(0)} = \sigma^{(0)}(\hbar; x, y, \tilde{x}, \tilde{y}, p, q)$ is any solution of the *linearized master equation* (LME)

$$\partial_x \partial_{\tilde{x}} \sigma^{(0)} + \partial_y \partial_{\tilde{y}} \sigma^{(0)} + \{\partial_x \Theta, \partial_y \sigma^{(0)}\}_\mathcal{M} - \{\partial_y \Theta, \partial_x \sigma^{(0)}\}_\mathcal{M} = 0 \quad (15)$$

Define

$$J_x^{(1)} := \mathcal{L}_{\tilde{x}} \sigma^{(0)}, \quad J_y^{(1)} := \mathcal{L}_{\tilde{y}} \sigma^{(0)} \quad (16)$$

where

$$\mathcal{L}_{\tilde{x}} := \partial_{\tilde{x}} - \{\partial_y \Theta, \cdot\}_\mathcal{M}, \quad \mathcal{L}_{\tilde{y}} := \partial_{\tilde{y}} + \{\partial_x \Theta, \cdot\}_\mathcal{M}. \quad (17)$$

Then

$$\partial_x J_x^{(1)} + \partial_y J_y^{(1)} = (\partial_x \mathcal{L}_{\tilde{x}} + \partial_y \mathcal{L}_{\tilde{y}})\sigma^{(0)} = (\mathcal{L}_{\tilde{x}} \partial_x + \mathcal{L}_{\tilde{y}} \partial_y)\sigma^{(0)}$$
$$= \partial_x \partial_{\tilde{x}} \sigma^{(0)} + \partial_y \partial_{\tilde{y}} \sigma^{(0)} + \{\partial_x \Theta, \partial_y \sigma^{(0)}\}_\mathcal{M} - \{\partial_y \Theta, \partial_x \sigma^{(0)}\}_\mathcal{M}$$
$$\stackrel{\text{by (15)}}{=} 0. \quad (18)$$

Hence, there exists a function $\sigma^{(1)}$ such that

$$\partial_x \sigma^{(1)} = \mathcal{L}_{\tilde{y}} \sigma^{(0)}, \quad -\partial_y \sigma^{(1)} = \mathcal{L}_{\tilde{x}} \sigma^{(0)}. \quad (19)$$

From (19) we get

$$(\mathcal{L}_{\tilde{x}} \partial_x + \mathcal{L}_{\tilde{y}} \partial_y)\sigma^{(1)} = (\mathcal{L}_{\tilde{x}} \mathcal{L}_{\tilde{y}} - \mathcal{L}_{\tilde{y}} \mathcal{L}_{\tilde{x}})\sigma^{(0)}$$
$$= \{\partial_x \partial_{\tilde{x}} \Theta + \partial_y \partial_{\tilde{y}} \Theta + \{\partial_x \Theta, \partial_y \Theta\}_\mathcal{M}, \sigma^{(0)}\}_\mathcal{M} \stackrel{\text{by ME}}{=} 0. \quad (20)$$

It means that $\sigma^{(1)}$ satisfies LME. Analogously as before we arrive at the series of conserved charges $\sigma^{(1)}, \sigma^{(2)}, \ldots$ and currents $J^{(1)}, J^{(2)}, \ldots$ which are defined by the following recursion equations

$$J_x^{(n+1)} = -\partial_y \sigma^{(n+1)} = \mathcal{L}_{\tilde{x}} \sigma^{(n)}$$
$$J_y^{(n+1)} = \partial_x \sigma^{(n+1)} = \mathcal{L}_{\tilde{y}} \sigma^{(n)}, \quad n = 0, 1, \ldots \quad (21)$$

From the assumption that Θ is a solution of ME and $\sigma^{(0)}$ satisfies LME (15) it follows that all $\sigma^{(n)}$'s satisfy LME

$$\partial_x \partial_{\tilde{x}} \sigma^{(n)} + \partial_y \partial_{\tilde{y}} \sigma^{(n)} + \{\partial_x \Theta, \partial_y \sigma^{(n)}\}_\mathcal{M} - \{\partial_y \Theta, \partial_x \sigma^{(n)}\}_\mathcal{M} = 0, \quad (22)$$

with $n = 0, 1, \ldots$. This means that $\sigma^{(n)}$, $n = 0, 1, \ldots$ are the *hidden symmetries* of ME.
Equation (20) defines the *recursion operator* R by

$$\sigma^{(n+1)} = R\sigma^{(n)} \tag{23}$$

(Compare with [11]).
For example taking

$$\sigma^{(0)} = \tilde{x} \tag{24}$$

we find

$$\begin{aligned}
\sigma^{(1)} &= -y + f^{(1)}(\hbar; \tilde{x}, \tilde{y}, p, q), \\
\sigma^{(2)} &= x\partial_{\tilde{y}} f^{(1)} - y\partial_{\tilde{x}} f^{(1)} + \{\Theta, f^{(1)}\}_{\mathcal{M}} + f^{(2)}(\hbar; \tilde{x}, \tilde{y}, p, q), \\
&\ldots etc
\end{aligned} \tag{25}$$

If one puts

$$\sigma^{(0)} = \tilde{y} \tag{26}$$

then

$$\begin{aligned}
\sigma^{(1)} &= x + f^{(1)}(\hbar; \tilde{x}, \tilde{y}, p, q), \\
\sigma^{(2)} &= x\partial_{\tilde{y}} f^{(1)} - y\partial_{\tilde{x}} f^{(1)} + \{\Theta, f^{(1)}\}_{\mathcal{M}} + f^{(2)}(\hbar; \tilde{x}, \tilde{y}, p, q), \\
&\ldots etc.
\end{aligned} \tag{27}$$

3. Linear systems for ME and dressing operators

We deal here with conserved charges $\eta^{(0)}, \eta^{(1)}, \ldots$, etc., defined by (13), (14) and (12). For this especial choice we replace the $\eta^{(0)}, \eta^{(1)}, \ldots, \eta^{(n)}, \ldots$ by $\psi^{(0)}, \psi^{(1)}, \ldots, \psi^{(n)}, \ldots$ so we have

$$\begin{aligned}
\psi^{(0)} &= 1, \qquad \psi^{(1)} = \frac{1}{i\hbar}\Theta, \\
\psi^{(n)} &= \int^x dx^{(n)} \mathcal{D}_{\tilde{y}} \int^{x^{(n)}} dx^{(n-1)} \mathcal{D}_{\tilde{y}} \cdots \int^{x^{(2)}} dx^{(1)} \mathcal{D}_{\tilde{y}} 1
\end{aligned} \tag{28}$$

Define

$$\Psi(\lambda) := 1 + \sum_{n=1}^{\infty} \lambda^n \psi^{(n)}, \qquad \lambda \in \overline{C} - \{\infty\} \tag{29}$$

One can easily check that $\Psi(\lambda)$ satisfies the following linear system of differential equations

$$\begin{aligned}
\partial_x \Psi(\lambda) &= \lambda \mathcal{D}_{\tilde{y}} \Psi(\lambda) \\
-\partial_y \Psi(\lambda) &= \lambda \mathcal{D}_{\tilde{x}} \Psi(\lambda), \qquad \lambda \in \overline{C} - \{\infty\}
\end{aligned} \tag{30}$$

and, in fact, this system is a Lax pair for ME. Employing the results of [12, 13, 14] one can show that $\Psi(\lambda)$ has the form of

$$\Psi(\lambda) = \exp_*\{\frac{1}{i\hbar}\sum_{n=1}^{\infty}\lambda^n \Lambda^{(n)}\}, \qquad \Lambda^{(1)} = \Theta \qquad (31)$$

$$\partial_x \Lambda^{(n)} = \partial_{\tilde{y}} \Lambda^{(n-1)} + \sum_{l=1}^{n-1}\frac{B_l}{l!}\sum_{\substack{k_1+...+k_l=n-1 \\ k_1,...,k_l \geq 1}}\{\Lambda^{(k_1)}, ..., \{\Lambda^{(k_l)}, \partial_x \Theta\}_{\mathcal{M}}...\}_{\mathcal{M}}$$

$$\partial_y \Lambda^{(n)} = -\partial_{\tilde{x}} \Lambda^{(n-1)} + \sum_{l=1}^{n-1}\frac{B_l}{l!}\sum_{\substack{k_1+...+k_l=n-1 \\ k_1,...,k_l \geq 1}}\{\Lambda^{(k_1)}, ..., \{\Lambda^{(k_l)}, \partial_y \Theta\}_{\mathcal{M}}...\}_{\mathcal{M}},$$

with $n > 1$ and where B_l are the Bernoulli numbers, $\frac{t}{e^t-1} = \sum_{l=0}^{\infty} B_l \frac{t^l}{l!}$. The formula (31) proves that if Θ is analytic in $\hbar$ then all $\Lambda^{(n)}$s are also analytic in $\hbar$ [15]. Then $\Psi_*^{-1}(\lambda)$ defined by

$$\Psi_*^{-1}(\lambda) * \Psi(\lambda) = \Psi(\lambda) * \Psi_*^{-1}(\lambda) = 1 \qquad (32)$$

has the following form

$$\Psi_*^{-1}(\lambda) = \exp_*\{-\frac{1}{i\hbar}\sum_{n=1}^{\infty}\lambda^n \Lambda^{(n)}\} \qquad (33)$$

and it fulfills the following system

$$\begin{aligned}\partial_x \Psi_*^{-1}(\lambda) &= \lambda(\partial_{\tilde{y}}\Psi_*^{-1}(\lambda) - \frac{1}{i\hbar}\Psi_*^{-1}(\lambda) * \partial_x \Theta) \\ -\partial_y \Psi_*^{-1}(\lambda) &= \lambda(\partial_{\tilde{x}}\Psi_*^{-1}(\lambda) + \frac{1}{i\hbar}\Psi_*^{-1}(\lambda) * \partial_y \Theta)\end{aligned} \qquad (34)$$

with $\lambda \epsilon \overline{C} - \{\infty\}$. Analogously, defining

$$\sigma(\lambda) := \sum_{n=0}^{\infty}\lambda^n \sigma^{(n)}, \qquad \lambda \epsilon \overline{C} - \{\infty\} \qquad (35)$$

where $\sigma^{(n)}$, $n = 0, 1, ...$ are the conserved charges introduced in the previous section (see (20)) one arrives at the system

$$\begin{aligned}\partial_x \sigma(\lambda) &= \lambda \mathcal{L}_{\tilde{y}}\sigma(\lambda) \\ -\partial_y \sigma(\lambda) &= \lambda \mathcal{L}_{\tilde{x}}\sigma(\lambda), \qquad \lambda \epsilon \overline{C} - \{\infty\}\end{aligned} \qquad (36)$$

which is also a Lax pair for ME. Let F be a function such that

$$\sigma(\lambda) = \Psi(\lambda) * F * \Psi_*^{-1}(\lambda) \tag{37}$$

It is evident that such a function F always exists and is uniquely defined by $\sigma(\lambda)$. Moreover, from (29), (33) and (35) one quickly finds that F must be of the form

$$F = F(\hbar; \tilde{y} + \lambda x, \tilde{x} - \lambda y, \lambda, p, q) \tag{38}$$

It means that F is a *twistor function*, as the equations

$$\tilde{y} + \lambda x =: w^1 = \text{const}, \quad \tilde{x} - \lambda y =: w^2 = \text{const}, \quad \lambda =: w^3 = \text{const} \tag{39}$$

define a totally null anti–self–dual 2–surface in C^4 (the *twistor surface*). This twistor surface is the integral manifold for the following anti–self–dual 2–form ω

$$\begin{aligned}\omega &= (d\tilde{y} + \lambda dx) \wedge (d\tilde{x} - \lambda dy) \\ &= -d\tilde{x} \wedge d\tilde{y} + \lambda(dx \wedge d\tilde{x} + dy \wedge d\tilde{y}) - \lambda^2 dx \wedge dy \end{aligned} \tag{40}$$

The formula (37) says that $\Psi(\lambda)$ is the *dressing operator* for the linear system (35). For example, taking

$$F := \tilde{x} - \lambda(y + f^{(1)}(\hbar; p, q)) \tag{41}$$

one recovers the solution given by (24), (24) with $f^{(1)} = f^{(1)}(\hbar; p, q)$; taking

$$F := \tilde{y} + \lambda(x + f^{(1)}(\hbar; p, q)) \tag{42}$$

we arrive at (26), (26). (Compare with [15].)

Analogously as in our previous work [6] consider the linear system

$$\begin{aligned}\frac{1}{\lambda}\partial_x \Phi(\frac{1}{\lambda}) &= \mathcal{D}_{\tilde{y}}\Phi(\frac{1}{\lambda}) \\ -\frac{1}{\lambda}\partial_y \Phi(\frac{1}{\lambda}) &= \mathcal{D}_{\tilde{x}}\Phi(\frac{1}{\lambda}), \quad \lambda \in \overline{C} - \{0\} \\ \Phi(\frac{1}{\lambda}) &= \Phi^{(0)} + \sum_{n=1}^{\infty}(\frac{1}{\lambda})^n \Phi^{(n)}. \end{aligned} \tag{43}$$

This is also a linear system for ME. One quickly finds that

$$\begin{aligned}\partial_x \Theta &= i\hbar\, \Phi^{(0)} * \partial_{\tilde{y}}[\Phi_*^{(0)}]^{-1} \\ \partial_y \Theta &= -i\hbar\, \Phi^{(0)} * \partial_{\tilde{x}}[\Phi_*^{(0)}]^{-1} \end{aligned} \tag{44}$$

The solution $\Phi(\frac{1}{\lambda})$ can be written in the form

$$\Phi(\frac{1}{\lambda}) = \exp_*\{\frac{1}{i\hbar}\sum_{n=0}^{\infty}(\frac{1}{\lambda})^n \Omega^{(n)}\}$$

$$\exp_*\{\frac{1}{i\hbar}\Omega^{(0)}\} = \Phi^{(0)}. \tag{45}$$

Then we consider $\tilde{\sigma}(\frac{1}{\lambda}) = \sum_{n=0}^{\infty}(\frac{1}{\lambda})^n \sigma^{(n)}$

$$\tilde{\sigma}(\frac{1}{\lambda}) = \Phi(\frac{1}{\lambda}) * \tilde{F} * \Phi_*^{-1}(\frac{1}{\lambda}). \tag{46}$$

Straightforward calculations show that $\tilde{\sigma}(\frac{1}{\lambda})$ satisfies the following linear system

$$\frac{1}{\lambda}\partial_x \tilde{\sigma}(\frac{1}{\lambda}) = \mathcal{L}_{\tilde{y}}\tilde{\sigma}(\frac{1}{\lambda})$$

$$-\frac{1}{\lambda}\partial_y \tilde{\sigma}(\frac{1}{\lambda}) = \mathcal{L}_{\tilde{x}}\tilde{\sigma}(\frac{1}{\lambda}), \quad \lambda \in \overline{C} - \{0\} \tag{47}$$

iff the function $\tilde{F}$ is of the form

$$\tilde{F} = \tilde{F}(\hbar; x + \frac{1}{\lambda}\tilde{y}, -y + \frac{1}{\lambda}\tilde{x}, \frac{1}{\lambda}, p, q) \tag{48}$$

i.e., $\tilde{F}$ is a twistor function.
Of course the system (46) is also a linear system for ME (a Lax pair for ME) and Eq. (46) expresses the fact that $\Phi(\frac{1}{\lambda})$ is the dressing operator for this system.

4. Infinite algebra of hidden symmetries

From the previous sections one can conclude that the general solution of LME (15) i.e., the general symmetry of ME is given by

$$\delta_{(F\tilde{F})}\Theta = \frac{1}{2\pi i}\oint_\gamma \frac{d\lambda}{\lambda^2}(-\Psi(\lambda) * F * \Psi_*^{-1}(\lambda) + \Phi(\frac{1}{\lambda}) * \tilde{F} * \Phi_*^{-1}(\frac{1}{\lambda}))$$

$$F = F(\hbar; \tilde{y} + \lambda x, \tilde{x} - \lambda y, \lambda, p, q)$$

$$\tilde{F} = \tilde{F}(\hbar; x + \frac{1}{\lambda}\tilde{y}, -y + \frac{1}{\lambda}\tilde{x}, \frac{1}{\lambda}, p, q) \tag{49}$$

where a curve γ does not contain singularities of functions which are integrated.
One can compare (48) with respective formulas given by Q.H.Park [2, 16]. In (48) the first sign $(-)$ and the factor $\frac{1}{\lambda^2}$ are chosen for further convenience.

Now we are looking for the algebra of hidden symmetries of ME. To this end consider the commutator

$$[\delta_{(F_1\tilde{F}_1)}, \delta_{(F_2\tilde{F}_2)}]\Theta = \delta_{(F_1\tilde{F}_1)}(\Theta + \delta_{(F_2\tilde{F}_2)}\Theta) - \delta_{(F_1\tilde{F}_1)}\Theta$$
$$- \delta_{(F_2\tilde{F}_2)}(\Theta + \delta_{(F_1\tilde{F}_1)}\Theta) + \delta_{(F_2\tilde{F}_2)}\Theta. \quad (50)$$

Simple calculations with the use of (48) give

$$[\delta_{(F_1\tilde{F}_1)}, \delta_{(F_2\tilde{F}_2)}]\Theta = \frac{(i\hbar)}{2\pi i}\oint_\gamma \frac{d\lambda}{\lambda^2}[-\{\delta_{(F_2\tilde{F}_2)}\Psi * \Psi_*^{-1}, \Psi * F_1 * \Psi_*^{-1}\}_{\mathcal{M}}$$
$$+ \{\delta_{(F_1\tilde{F}_1)}\Psi * \Psi_*^{-1}, \Psi * F_2 * \Psi_*^{-1}\}_{\mathcal{M}}$$
$$+ \{\delta_{(F_2\tilde{F}_2)}\Phi * \Phi_*^{-1}, \Phi * \tilde{F}_1 * \Phi_*^{-1}\}_{\mathcal{M}}$$
$$- \{\delta_{(F_1\tilde{F}_1)}\Phi * \Phi_*^{-1}, \Phi * \tilde{F}_2 * \Phi_*^{-1}\}_{\mathcal{M}}]. \quad (51)$$

Performing variation of the system (29) and employing (33) one obtains

$$\partial_x(\delta_{(F_1\tilde{F}_1)}\Psi * \Psi_*^{-1}) = \lambda[\mathcal{L}_{\tilde{y}}(\delta_{(F_1\tilde{F}_1)}\Psi * \Psi_*^{-1}) + \frac{1}{i\hbar}\partial_x\delta_{(F_1\tilde{F}_1)}\Theta]$$
$$-\partial_y(\delta_{(F_1\tilde{F}_1)}\Psi * \Psi_*^{-1}) = \lambda[\mathcal{L}_{\tilde{x}}(\delta_{(F_1\tilde{F}_1)}\Psi * \Psi_*^{-1}) - \frac{1}{i\hbar}\partial_y\delta_{(F_1\tilde{F}_1)}\Theta] \quad (52)$$

The solution of (51) can be found and it reads

$$i\hbar\delta_{(F_1\tilde{F}_1)}\Psi(\lambda) * \Psi_*^{-1}(\lambda) = \frac{1}{2\pi i}\sum_{n=1}^\infty \lambda^n \oint_\gamma \frac{d\lambda'}{(\lambda')^{n+1}}$$
$$\times (-\Psi' * F_1' * \Psi_*'^{-1} + \Phi' * \tilde{F}_1' * \Phi_*'^{-1})$$
$$= \frac{1}{2\pi i}\oint_{\gamma'_>} d\lambda' \frac{\lambda}{\lambda'(\lambda' - \lambda)}$$
$$\times (-\Psi' * F_1' * \Psi_*'^{-1} + \Phi' * \tilde{F}_1' * \Phi_*'^{-1}) \quad (53)$$

where $\gamma'_>$ is any curve closing a domain containing the following circle $\overline{K}(0;r) : \gamma \subset K(0;r)$. Therefore, $\Psi' := \Psi(\lambda')$, and consequently $F_1' := F_1(\hbar; \tilde{y} + \lambda'x, \tilde{x} - \lambda'y, \lambda', p, q)$,... etc.

Analogously for $\delta_{(F_1\tilde{F}_1)}\Phi * \Phi_*^{-1}$ one gets the system of equations

$$\frac{1}{\lambda}\partial_x(\delta_{(F_1\tilde{F}_1)}\Phi * \Phi_*^{-1}) = \mathcal{L}_{\tilde{y}}(\delta_{(F_1\tilde{F}_1)}\Phi * \Phi_*^{-1}) + \frac{1}{i\hbar}\partial_x\delta_{(F_1\tilde{F}_1)}\Theta$$
$$-\frac{1}{\lambda}\partial_y(\delta_{(F_1\tilde{F}_1)}\Phi * \Phi_*^{-1}) = \mathcal{L}_{\tilde{x}}(\delta_{(F_1\tilde{F}_1)}\Phi * \Phi_*^{-1}) - \frac{1}{i\hbar}\partial_y\delta_{(F_1\tilde{F}_1)}\Theta \quad (54)$$

The solution of (53) reads

$$i\hbar\delta_{(F_1\tilde{F}_1)}\Phi(\frac{1}{\lambda}) * \Phi_*^{-1}(\frac{1}{\lambda}) = -\frac{1}{2\pi i}\sum_{n=0}^{\infty}\frac{1}{\lambda^n}\oint_\gamma d\lambda'(\lambda')^{n-1}$$
$$\times (-\Psi' * F_1' * \Psi_*'^{-1} + \Phi' * \tilde{F}_1' * \Phi_*'^{-1}) \quad (55)$$

Since for calculating (50) we need $\lambda \in \gamma$ one can write down (54) in the form of

$$i\hbar\delta_{(F_1\tilde{F}_1)}\Phi(\frac{1}{\lambda}) * \Phi_*^{-1}(\frac{1}{\lambda})|_{\lambda\in\gamma} = \frac{1}{2\pi i}\oint_{\gamma'_<} d\lambda' \frac{\lambda}{\lambda'(\lambda'-\lambda)}$$
$$\times (-\Psi' * F_1' * \Psi_*'^{-1} + \Phi' * \tilde{F}_1' * \Phi_*'^{-1}) \quad (56)$$

where now $\gamma'_< \subset K(0;r)$ and $\overline{K}(0;r)$ is a circle belonging to the domain closed by γ. Note that γ, $\gamma'_>$ and $\gamma'_<$ must be chosen so that they close the same singularities of the integrated functions. Substituting (52) and (55) and also analogous formulas for $\delta_{(F_2\tilde{F}_2)}\Psi(\lambda) * \Psi_*^{-1}(\lambda)$ and $\delta_{(F_2\tilde{F}_2)}\Phi(\frac{1}{\lambda}) * \Phi_*^{-1}(\frac{1}{\lambda})$ into (50), performing then integrations and applying the residue theorem one gets

$$[\delta_{(F_1\tilde{F}_1)}, \delta_{(F_2\tilde{F}_2)}]\Theta = \delta_{(\{F_1,F_2\}_\mathcal{M}\{\tilde{F}_1,\tilde{F}_2\}_\mathcal{M})}\Theta \quad (57)$$

Hence, the *hidden symmetries for ME form a closed algebra associated with the Moyal bracket Lie algebra.*
This coincides with the results of [2, 11, 16] where the algebra of hidden symmetries is given for Plebański's heavenly equations and for SDYM equations.
As ME can be reduced to all known heavenly equations, to $su(N)$ SDYM equations, to $su(N)$ chiral equations etc., the algebra (57) contains the hidden symmetry algebras for the reduced equations.

5. Twistor construction.

For completeness we describe here briefly a twistor construction for ME given in our previous work [6]. This construction is similar to the one for SDYM equations [17, 18].
What must be noted, and it has been pointed out by K.Takasaki [15], is that the twistor construction we propose is valid for the case of Θ being analytic in $\hbar$, i.e., $\Theta = \sum_{n=0}^{\infty}\hbar^n\Theta_n$, $\Theta_n = \Theta_n(x,y,\tilde{x},\tilde{y},p,q)$. The case with negative powers of $\hbar$ should be considered separately and as we know from [6] it can be solved by the Fairlie–Leznov method [19] but we still have not succeeded in a twistor image for this case.

We start with a twistor function

$$H = H(\lambda) = H(\hbar; \tilde{y} + \lambda x, \tilde{x} - \lambda y, \lambda, p, q) = \exp_*\{\tfrac{1}{i\hbar}\sum_{n=-\infty}^{\infty}\lambda^n \Delta^{(n)}\}$$
$$\lambda \epsilon (\overline{C} - \{0\}) \cap (\overline{C} - \{\infty\})$$
$$\Delta^{(n)} = \sum_{m=0}^{\infty} \hbar^m \Delta_m^{(n)}(x, y, \tilde{x}, \tilde{y}, p, q) \tag{58}$$

Let $\Psi = \Psi(\lambda)$ be a function of the form

$$\Psi(\lambda) = \exp_*\{\tfrac{1}{i\hbar}\sum_{n=1}^{\infty}\lambda^n \Lambda^{(n)}\}, \quad \lambda \epsilon (\overline{C} - \{\infty\})$$
$$\Lambda^{(n)} = \Lambda^{(n)}(\hbar; x, y, \tilde{x}, \tilde{y}, p, q) = \sum_{m=0}^{\infty} \hbar^m \Lambda_m^{(n)}(x, y, \tilde{x}, \tilde{y}, p, q),$$
$$\Lambda^{(1)} = \Theta \tag{59}$$

and let $\Phi = \Phi(\tfrac{1}{\lambda})$ be a function of the form

$$\Phi(\tfrac{1}{\lambda}) = \exp_*\{\tfrac{1}{i\hbar}\sum_{n=1}^{\infty}(\tfrac{1}{\lambda})^n \Omega^{(n)}\}, \quad \lambda \epsilon (\overline{C} - \{0\})$$
$$\Omega^{(n)} = \Omega^{(n)}(\hbar; x, y, \tilde{x}, \tilde{y}, p, q) = \sum_{m=0}^{\infty} \hbar^m \Omega_m^{(n)}(x, y, \tilde{x}, \tilde{y}, p, q), \tag{60}$$

These functions are chosen so that the following factorization holds

$$H(\lambda) = \Phi_*^{-1}(\tfrac{1}{\lambda}) * \Psi(\lambda), \quad \lambda \epsilon (\overline{C} - \{0\}) \cap (\overline{C} - \{\infty\}). \tag{61}$$

(This is the *Riemann–Hilbert problem* or the *Birkhoff factorization* [18]). One easily finds that from the conditions: $(\lambda \partial_{\tilde{y}} - \partial_x)H(\lambda) = 0$ and $(\lambda \partial_{\tilde{x}} + \partial_y)H(\lambda) = 0$ it follows that

$$[(\lambda \partial_{\tilde{y}} - \partial_x)\Psi(\lambda)] * \Psi_*^{-1}(\lambda) = \lambda[(\partial_{\tilde{y}} - \tfrac{1}{\lambda}\partial_x)\Phi(\tfrac{1}{\lambda})] * \Phi_*^{-1}(\tfrac{1}{\lambda})$$
$$[(\lambda \partial_{\tilde{x}} + \partial_y)\Psi(\lambda)] * \Psi_*^{-1}(\lambda) = \lambda[(\partial_{\tilde{x}} + \tfrac{1}{\lambda}\partial_y)\Phi(\tfrac{1}{\lambda})] * \Phi_*^{-1}(\tfrac{1}{\lambda})$$
$$\lambda \epsilon (\overline{C} - \{0\}) \cap (\overline{C} - \{\infty\}) \tag{62}$$

The left-hand side of Eq. (62) can be analytically extended on $\overline{C}$ and in the gauge (58) we get

$$[(\lambda \partial_{\tilde{y}} - \partial_x)\Psi(\lambda)] * \Psi_*^{-1}(\lambda) = -\lambda \tfrac{1}{i\hbar}\partial_x \Theta$$
$$[(\lambda \partial_{\tilde{x}} + \partial_y)\Psi(\lambda)] * \Psi_*^{-1}(\lambda) = \lambda \tfrac{1}{i\hbar}\partial_y \Theta, \quad \lambda \epsilon \overline{C} \tag{63}$$

Thus we recover the linear system (29) of ME. Substituting (63) into (62) one recovers the linear system (42) of ME as well.

It means that our procedure gives a twistor construction for ME

Acknowledgments

We are grateful to Maciej Dunajski for many discussions on the problems considered in the paper.

This work was partially supported by the CONACyT (México) grant 32427–E and by KBN (Poland) grant Z/370/S.

Notes

1. We are indebted to Maciej Dunajski for pointing out to us the method how to obtain new conservation laws.

References

[1] V. Husain, *Class. Quantum Grav.* **11** (1994) 927.
[2] Q.H. Park, *Int. J. Mod. Phys.* **A7** (1992) 1415.
[3] J.F. Plebański and M. Przanowski, *Phys. Lett.* **A212** (1996) 22.
[4] J.F. Plebański and M. Przanowski, *The universal covering of heavenly equations via Weyl–Wigner–Moyal formalism*, in: *Gravitation, Electromagnetism and Geometrical Structures*, ed. G.Farrarese (Pitagora Editrice, Bologna 1996) pp. 15.
[5] M.Przanowski and S.Formański, *Acta Phys. Pol.* **B30** (1999) 863.
[6] M. Przanowski, J.F. Plebański and S. Formański, *Integrability of SDYM equations for the Moyal bracket Lie algebra*, in: *Exact Solutions and Scalar Fields in Gravity*, A. Mac´initialas, J.L. Cervantes–Cota and C. Lämmerzahl eds. (Kluwer Academic Publishers, USA. 2001) pp. 77.
[7] E. Brezin, C. Itzykson, J. Zinn–Justin andJ.B.Zuber, *Phys. Lett.* **B82** (1979) 442.
[8] M.K. Prasad, A. Sinha and L.L. Chau Wang, *Phys. Lett.* **B87** (1979) 237.
[9] L.L. Chau, M.K.Prasad and A.Sinha, *Phys. Rev.* **D24** (1981) 1574.
[10] L.L. Chau, *Chiral fields, self–dual Yang–Mills fields as integrable systems, and the role of the Kac–Moody algebra*, in: *Nonlinear Phenomena*, ed. K.B.Wolf, Lecture Notes in Physics **189** (Springer, New York, 1983) pp. 110.
[11] M. Dunajski and L.J. Mason, *Commun. Math. Phys.* **213** (2000) 641.
[12] J.F. Plebański, *On the Generators of Unitary and Pseudo–Unitary Groups*, monograph of CINVESTAV (Mexico 1966).
[13] I. Bialynicki–Birula, B.Mielnik and J.F.Plebański, *Ann.Phys.* **51** (1969) 187.
[14] B.Mielnik and J.F.Plebański, *Ann. Inst. Henri Poincaré* **XII** (1970) 215.
[15] K.Takasaki, *J. Geom. Phys.* **14** (1994) 111.
[16] Q.H.Park, *Phys. Lett.* **B238** (1990) 287.
[17] R.S.Ward, *Phys. Lett.* **A61** (1977) 81.
[18] L.J.Mason and N.M.J.Woodhouse, *Integrability, Self–Duality, and Twistor Theory* (Clarendon Press, Oxford 1996).
[19] D.B.Fairlie and A.N.Leznov, *J. Phys. A: Math. Gen.* **33** (2000) 4657.

IV

EXPERIMENTS AND OTHER TOPICS

NONLOCALITY AND SUPERLUMINAL SIGNAL–VELOCITY IN PHOTONIC TUNNELLING

Günter Nimtz and Astrid Haibel
Universität zu Köln, II. Physikalisches Institut,
Zülpicher Str. 77, D-50937 Köln, Germany
g.nimtz@uni-koeln.de

Abstract The time behavior of partial reflection by opaque photonic barriers was measured with microwaves. It was observed that unlike the duration of partial reflection by dielectric sheets, the measured reflection duration of barriers is independent of their length. The experimental results point to a nonlocal behavior of evanescent modes.

Keywords: Nonlocality, Signalvelocity, Superluminal.

1. Introduction

We are used to measuring a reflection time determined by sheet thickness from partial reflection of light by a sheet of glass. The reflection is observed only after a time span corresponding to twice the layer thickness multiplied by the group velocity of light in glass. Three hundred years ago Newton conjectured that light was composed of corpuscles and argued in the case of partial reflection by two or more surfaces: Light striking the first surface sets off a kind of wave or field that travels along with the light and predisposes it to reflect or not reflect off the second surface. He called this process "fits of easy reflection or easy transmission" [1]. As theory and experiments have shown this is not true in the case of dielectric media with a real part of the refractive index.

Amazingly in the case of reflection by the surface of an opaque photonic barrier, where the refractive index is purely imaginary, Newton's conjecture seems to be close to reality: The partial reflection by barriers suffers a short and constant time delay independent of length. For the photonic barrier investigated here we found that the reflection duration

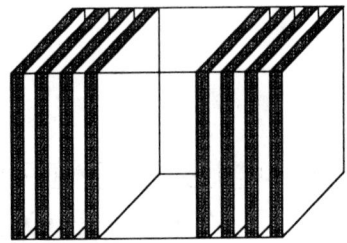

Figure 1. Two one–dimensional periodic quarter wavelength hetero–structures of perspex and air which are separated by a distance of 189 mm forming a resonant cavity (cross–section is 400×400 mm^2).

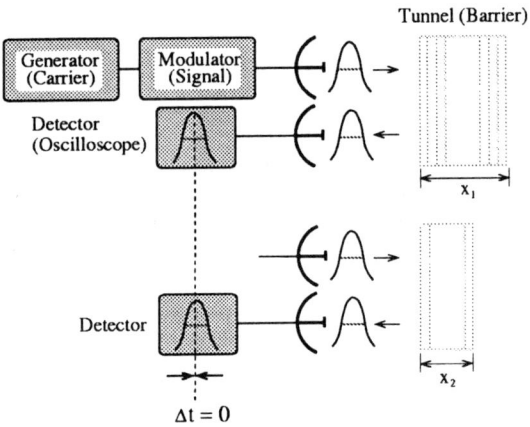

Figure 2. Experimental set–up for the periodic dielectric quarter wavelength heterostructure to measure the group velocity. The diagram shows two resonant photonic barriers with 2×4 and 2×2 perspex sheets separated by the same air gap.

equals the transmission time observed in photonic tunnelling experiments [2] (see Fig. 1).

We are going to explain the experimental set–up and the experiments and discuss the unexpected observation.

2. Experimental set–up

The experimental set–up is displayed in Fig. 2. Pulse–like microwave signals with a half width of 8.5 ns are transmitted from a parabolic antenna. The carrier frequency of the pulse is 9.15 GHz ($\lambda = 32.8$ mm) the frequency-band width is 80 MHz. The reflected signal is received with a second antenna and detected by an HP 54825 oscilloscope. The measurement is performed asymptotically, so that any coupling between generator, detector, and devices under test (a perspex sheet and photonic

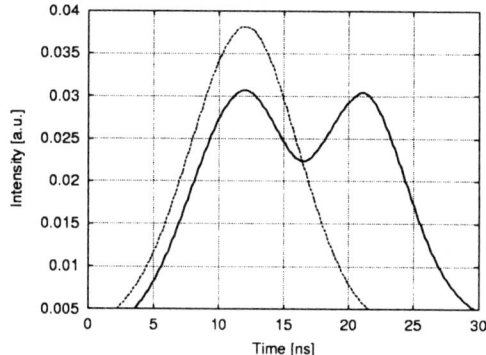

Figure 3. Partial reflection of a microwave pulse by a thick perspex sheet. The dashed pulse is the result of reflection by a metallic front surface only. The layer is 800 mm thick, its refractive index is 1.6. The double peak is due to the superposition of reflection of the signal pulse by front and back surfaces. The delay of the second peak is 8.5 ns in agreement with the propagation time in perspex. In order to enhance reflection the back surface was partially coated with a metal film.

barriers) is avoided by the long optical distances ($\approx$ 3m) and by uniline devices in the microwave circuit. The time resolution of the set-up is ±10 ps.

To check the experimental arrangement we measured the time response of partial reflection by the two surfaces of a perspex sheet of 0.8 m thickness. The second peak of the signal corresponding to the partial reflection by the back surface arrived $\approx$ 8.5 ns later than the first one related to the front surface reflection. For a clear demonstration of the partial reflection by the two surfaces, the intensity reflected by the back surface was adjusted to that by the front side as shown in Fig. 3. This was obtained by a partial metallic coating of the perspex layer's back surface. The measured time delay of the two peaks is in agreement with the calculated propagation time 8.53 ns considering the refractive index n=1.6 of perspex.

3. Partial reflection by photonic barriers

The investigated photonic barrier device is sketched in Fig. 1. It consists of two photonic lattices each 45.5 mm long which are separated by an air gap of 189 mm resulting in a total length of 280 mm. Each lattice consists of four perspex slabs each 5 mm thick in an equidistant distance of 8.5 mm air.

For such a structure in the 'normal dielectric case' we would expect a broadened pulse composed of the partial reflections of all the slabs simi-

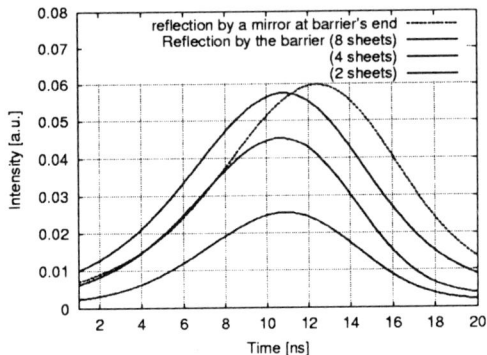

Figure 4. Signals reflected by barriers of different length. The largest one had a total barrier length of 280 mm, the two smaller one were recorded after the barrier was shortened to a total length of 226 mm and 199 mm, respectively. The procedure of shortening is illustrated in Fig. 2. For comparison the reflection time of a mirror (without barrier) at the back surface position of the longest photonic barrier is displayed. The expected travel time between front and end position of 1.87 ns has been in fact measured.

lar to the signal shown in Fig. 3. The last surface reflected signal should be seen ≈ 1.9 ns after the reflected one by the front surface. However, the partial reflection by photonic barriers revealed a strange behavior. If the barrier is shortened to 4 or 2 sheets (see Figs. 2 and 4) the reflection duration keeps constant whereas the amplitude decreases as a result of the increasing transmission. The measured reflection duration is ≈ 100 ps. A back surface reflected signal from opaque barriers, however, has never been detected (see Fig. 4).

Transmission and reflection dispersion relations of the long barrier are displayed in Fig. 5. In Transmission there are five pronounced forbidden bands separated by resonance peaks in the frequency range displayed. The reflection behavior in these forbidden bands was used for the measurements.

4. Tunnelling time and superluminal velocity

Amazingly, the reflection time equals the tunnelling time [10, 11]. An empirical relationship has been proposed which gives the tunnelling time τ as a function of the signal's carrier frequency ν_c or of a wave packet's energy E divided by the Planck constant h [12]:

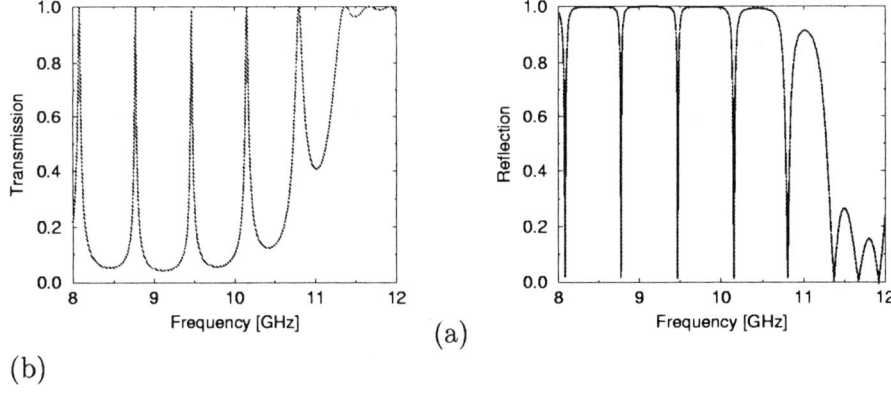

Figure 5. The graphs show the dispersion relations for the resonant heterostructure vs frequency of Fig. 1. (a) The transmission dispersion of the periodic heterostructure displays five forbidden gaps, which correspond to the photonic tunnelling regime, for details see Ref. [4]. (b) shows the reflection dispersion relation.

$$\tau \approx T = \frac{1}{\nu_c}$$
$$\tau \approx \frac{h}{E} \qquad (1)$$

The simple relationship was confirmed by several experiments carried out with different photonic barriers. Recently the tunnelling time of electrons were measured for the first time [13]. As conjectured by Haibel and Nimtz the observed tunnelling time of the electrons are in agreement with the above universal formula [12]. The short tunnelling time may result in superluminal group and signal velocities [14, 15]. However, the principle of causality is not violated in the case of a superluminal signal velocity. The time span between cause and effect becomes shortened compared with local action by the velocity of light. Cause and effect can not be exchanged in consequence of a signal's finite frequency band and time length [14, 16].

5. Conclusions

In measuring the reflection duration by photonic barriers we observed that the partial reflection by the back surface is an instantaneous effect on the amplitude, whereas the reflection duration is not changed. This strange behavior is opposite to the measured partial reflection by a perspex layer (see Fig. 3). The behavior may be explained by a nonlocality of evanescent modes. As a result of our experiments evanescent modes constituted by ensembles of photons behave like a quantum mechanical particle. Nonlocality and causality were investigated in Ref. [6] and quite recently with respect to superluminal photonic tunneling nonlocality was discussed by Perel'man (Ref. [7]).

In our experiments the applied signal pulse had a carrier frequency of 9.15 GHz in the center of a forbidden band gap (see Fig. 5) and a narrow 1 % frequency–band width. Consequently all frequency components of the signal were evanescent. In this case there is no finite phase–time expected nor observed for a signal inside a barrier [8, 9]. Such a behavior seem to explain the experimental data of reflection by opaque barriers.

Obviously the information on photonic barrier length is available at the front surface already. This is a property which Newton suggested erroneously to explain partial reflection of corpuscles by dielectric layers [1]. Evanescent modes appear to be nonlocal at least within a range of some ten wavelengths as experiments have shown in this study [7]. The distance of observing nonlocality effects is limited by the exponential decay of the field intensity of evanescent modes, i.e. of the probability in the wave mechanical tunnelling analogy.

Acknowledgments

We gratefully acknowledge discussions with H. Aichmann, P. Mittelstaedt, A. Stahlhofen, and R.-M. Vetter. We thank M. E. Perel'man for giving us the paper on his investigation prior to publication.

References

[1] R. P. Feynman, QED, The strange Theory of Light and Matter, (Princeton University Press, Princeton NJ 1988) p.22.

[2] A. Haibel and G. Nimtz, *Ann. Phys.* (Leipzig) **9** (2001) 1.

[3] R. P. Feynman, R. B. Leighton, and M. Sands, The Feyman Lectures on Physics, (Addison–Wesley Publishing Company) **II** (1964) 33.

[4] G. Nimtz, A. Enders, and H. Spieker, *J. Phys. I.*, France **4** (1994) 565.

[5] E. Merzbacher, Quantum Mechanics, 2nd ed., (John Wiley & Sons, New York 1970) p.6.

[6] E. Recami, *Time's Arrows, Quantum Measurement and Superluminal Behavior*, ISBN 88 8080 024 8, (2000) pp. 17–35.

[7] M. E. Perel'man, preprint (2001).

[8] Th. Hartman, *J. Appl. Phys.* **33** (1962) 3427.

[9] G. Nimtz and W. Heitmann, *Prog. Quantum Electronics* **21** (1997) 81.

[10] A. A. Stahlhofen, *Phys. Rev.* **A62** (2000) 12112.

[11] A. Haibel, G. Nimtz, and A. A. Stahlhofen, *Phys. Rev.* **E63** (2001) 047601.

[12] A. Haibel and G. Nimtz, *Ann. Phys.* (Leipzig) **10** (2001) 707.

[13] S. K. Sekatskii and V. S. Letokhov, *Phys. Rev.* **B64** (2001) 233311.

[14] G. Nimtz and A. Haibel, *Ann. Phys.* (Leipzig) **11** (2002) 163.

[15] S. Longhi, M. Marano, P. Laporta, and M. Belmonte, *Phys. Rev.* **E64** (2001) 055602.

[16] G. Nimtz, *Eur. Phys. J.* **B7** (1999) 523.

AT THE INTERFACE OF QUANTUM AND GRAVITATIONAL REALMS

D. V. Ahluwalia
Theoretical Physics Group, Facultad de Fisica, UAZ, A. P. C-600, Zacatecas, ZAC 98062, Mexico
ahluwalia@phases.reduaz.mx; http://phases.reduaz.mx

Abstract In this talk I review a series of recent conceptual developments at the interface of the quantum and gravitational realms. Wherever possible, I comment on the possibility to probe the interface experimentally. It is concluded that the underlying spacetime for a quantum theory of gravity must be non-commutative, that wave-particle duality suffers significant modification at the Planck scale, and that the latter forbids probing spacetime below Planck length. Furthermore, study of quantum test particles in classical and quantum sources of gravity puts forward theoretical challenges and new experimental possibilities. It is suggested that existing technology may allow to probe gravitationally–modified wave particle duality in the laboratory,

Keywords: Non–commutative spacetime, Gravitationally–induced phases, Cosmological matter–antimatter asymmetry.

1. Introduction

The purpose of this written version of the talk given at the "Mexican Meeting on Mathematical and Experimental Physics (El Colegio Nacional, 10-14 September 2001)" is to briefly review some of the conceptual developments at the interface of the quantum and gravitational realms. The version presented here primarily confines to the contributions I have made, and is by no means intended as a review of this developing field. I am also happy, when possible, to point reader's attention to the relevant experimental literature.

2. Non–commutative nature of spacetime and gravitationally–modified wave particle duality

Setting the Stage: Quantum Measurement with Gravitational Effects Ignored.

The experimental foundations of quantum mechanics reside in the *in principle* lower limit on the extent to which the unavoidable disturbance that the position and momentum measurements carry can be reduced.[1] This circumstance arises from the experimental implication of the photo–electric effect. It tells us that the energy carried by a light beam is not a continuous variable. Its intensity can only be changed in discrete units of $\hbar\omega$, where ω is the angular frequency of the probing light beam. This is encoded in fundamental commutators, such as:

$$[x, p_x] = i\hbar, \quad [x, y] = 0. \tag{1}$$

In configuration space, a solution of the above commutators is:

$$p_x = \frac{\hbar}{i} \frac{\partial}{\partial x}. \tag{2}$$

The operator p_x carries with it non–renormalizable eigenfunctions of the form

$$\psi(x) \sim \exp\left(\frac{i p_x x}{\hbar}\right). \tag{3}$$

However, integrating over different–momenta eigenfunctions one may obtain well–defined and normalizable wave packets that describe space–localizable particles. Now let λ be the spatial periodicity of $\psi(x)$. That is, $x \to x \pm \lambda$, advances the phase of $\psi(x)$ by $\pm 2\pi$. Then, we find that a particle of momentum p_x carries with a de Broglie wave length:

$$\lambda_{dB} = \frac{h}{|p_x|}. \tag{4}$$

In a commutative spacetime this lies at the heart of the quantum-mechanical wave particle duality.

Quantum measurement with gravitational effects incorporated.

The usual measurement process in quantum mechanics ignores any gravitational effects that may be inherent in it. Such effects become important in quantum gravity because, in general, there is an *unavoidable* change in the energy–momentum tensor associated with the collapse of a wave function.[2] Consequently, quantum measurements of spatial–temporal locations of events cannot ignore inherent gravitational effects.

These effects, we hasten to add are intrinsic to the events under consideration. In the first approximation, they do not directly refer to the background curvature [1, 2, 3, 4, 5, 6, 7, 8, 9, 10, 11, 12].

When gravitational effects in a quantum measurement process are incorporated it is found that there is an *in-principle and unavoidable* non–commutativity of the position and temporal measurements. Furthermore, the fundamental commutator undergoes a change of the form [12]:

$$[x, p_x] = i\hbar \left[1 + \epsilon \frac{\lambda_P^2 p_x^2}{\hbar^2}\right], \tag{5}$$

where

$$\lambda_P = \sqrt{\frac{\hbar G}{c^3}}, \tag{6}$$

and ϵ is a number of the order of unity (to be set to unity, now onwards). Given Eq. (6), the introductory discussion contained in "*Setting the Stage: Quantum Measurement with Gravitational Effects Ignored*" immediately implies that wave particle duality must suffer a fundamental change at the Planck scale. For one–dimensional motion, the change in wave–particle duality takes the form [13, 12]:

$$\lambda = \frac{\overline{\lambda}_P}{\tan^{-1}(\overline{\lambda}_P/\lambda_{dB})} \tag{7}$$

In the above equation we have defined, $\overline{\lambda}_P = 2\pi\lambda_P$. It is readily seen that in the low–energy limit the gravitationally modified wave length λ reduces to the well–known de Broglie wave length. In the Planck regime, however, something surprising, and welcome, happens. The gravitationally modified wave length λ saturates to (4 times, with $\epsilon = 1$) the Planck length. A similar result was later obtained by Bruno, Amelino–Camelia, and Kowalski–Glikman [14] (Also, see important observations of Padmanabhan in Refs. [15, 16]). Theoretically, this result may be interpreted as that no particle wavelengths are available to probe spacetime below the Planck length distances (λ_P^2 areas, and λ_P^3 volumes). Moreover, this saturation also suggests that in some sense the relativity, special and general, must suffer changes so that length contractions below λ_P do not occur.

Theoretically, one is, therefore, called upon to develop a relativity that carries not only an inertial–observer independent velocity, i.e. c, but also a similarly independent length scale. The latter may be identified with $\epsilon\lambda_P$. Amelino–Camelia has already undertaken the task of building such a modification to the relativity theory [17].

Experimentally, the derived saturation of λ, implies freezing of neutrino oscillations at the Planck energies, and carries several phenomenological implications [13]. However, the phenomenological implications may not be confined to early universe alone, as we now argue.

Superconducting quantum interference devices (SQIDs), when cooled sufficiently below the critical temperature, may carry temperature-tunable superconducting currents with total superconducting mass

$$m_s \sim f(T)\, N_a\, m_c, \tag{8}$$

behaving as one quantum object (under certain circumstances). In Eq. (8), $N_a \approx 6 \times 10^{23}$ mole^{-1}, $m_c \approx 2 \times 0.9 \times 10^{-27}$ gm, and $f(T)$ encodes fraction of the available electrons that are in a superconducting Cooper state at temperature, T. Sufficiently below the critical temprature, $f(T)$ may approach unity. The temperature–tunable, m_s, can easily compete with Planck mass,

$$m_P = \sqrt{\frac{\hbar c}{G}} \approx 2.2 \times 10^{-5}\ \text{gm}. \tag{9}$$

Thus, SQUIDs carry significant potential to probe wave–particle duality near the Planck scale. The theoretical and experimental problem that remains to be attended is to devise an experiment that invokes m_s, and not m_c.

3. Quantum test particles in classical sources of gravity

Principle of equivalence with classical test particles in classical source of gravity has been verified to a remarkable accuracy, see, e.g., [18]. In this section we devote our attention to quantum test particles when the source of gravity is treated as a classical background.

The quantum behavior of a mass eigenstate in the classical source of gravity is well studied in the pioneering experiments on neutron interferometry [19, 20]. In recent such experiments an apparent violation of the principle of equivalence seems evident at roughly a part in one thousand [20]. If this result is not due to a yet unknown systematic error, then quantum mechanical motion of neutron in classical source of gravity poses serious theoretical challenge for its understanding [21, 22].

In atomic interferometry [23] the principle of equivalence is confirmed to a few parts in 10^9.

To study the possibilities that quantum test particles offer to study gravitational field let's us first note that the local gravitational potential in the solar system carries two sources: (a) Solar-system sources, such as Earth, (b) Cosmological sources, such as the local super–cluster of galaxies. The former, on the surface of Earth, when measured in dimensionless units is, -7×10^{-10}, and varies as $R_\oplus/(z + R_\oplus)$ — where z is the vertical distance from the surface of the Earth, and $R_\oplus$ is Earth's radius. While the latter can be estimated to be roughly 3×10^{-5} — see, Ref. [21]. It is roughly constant over the solar system.

Potentials of the type "a" induce not only gravitational forces, but they also are also responsible for observable gravitationally–induced phases and the accompanying quantum interference effects. The type–b potentials are essentially force free, and they have the net effect of red–shifting local clocks. However, an experiment that seeks to study a possible violation of equivalence principle must treat such potentials with due care. In particular, given a violation of equivalence principle, the type–b potentials acquire a *local* observability. Quantum system — modelled after flavor–oscillation clocks [24, 25, 26, 27, 28, 29, 30] — appear to be most sensitive experimental probes for type–b potentials. There is now a significant and growing literature on the subject of flavor–oscillation clocks and I refer to the just cited list of references for the involved details.

4. Quantum test particles in quantum sources of gravity

The next level of theoretical and experimental sophistication is called upon when one treats both the sources and the test particles as quantum objects. The example of the former is provided by a SQUID in a linear superposition of two counter–propagating super–currents [31]. An example of the latter is once again a system of flavor–oscillations clocks.

In the absence of a complete theory of quantum gravity, it is a non-trivial theoretical task to model the gravitational field of a quantum source of gravity. Yet, in the weak field, non-relativistic, regime the gravitational field may simply be taken as a quantum linear superposition of configurations with classical counterparts. To our knowledge, not even a preliminary analysis exists on the subject. However, as is apparent, such a theoretical undertaking is likely to prove a fruitful playing ground on the interface of gravitational and quantum realms.

5. Spatial and temporal fluctuations in spacetime foam

Amelino–Camelia has argued that *spatial* fluctuations of the space–time foam can be experimentally probed in gravity wave interferometers.

This is a new and unexpected observation and may provide direct evidence for quantum–gravity induced effects.

Complementing Amelino–Camelia's work, Kirchbach and I have put forward a thesis that the observed matter–antimatter asymmetry may arise from asymmetric space–time fluctuations and their interplay with the Stückelberg–Feynman interpretation of antimatter [32]. The thesis also argues that the effect of spacetime fluctuations is to diminish the fine structure constant, $\alpha = e^2/\hbar c$, in the past. Recent studies of the QSO absorption lines provide a 4.1 standard deviation support for this prediction [33, 34, 35]. It is entirely possible that the empirical data on the fine structure constant has already detected first signatures of the quantum–gravity induced spacetime fluctuations.

6. Concluding remarks

Towards building a quantum theory of gravity, the interface of the gravitational and quantum realms is a rich conceptual and experimental arena. Here, theorists and experimentalists alike may play with much profit. In this talk I have outlined mostly my personal contributions. It is apparent that the underlying spacetime for a quantum theory of gravity must be non-commutative, that wave–particle duality suffers significant modification at the Planck scale, and that the latter forbids probing spacetime below Planck length. Furthermore, study of quantum test particles in classical and quantum sources of gravity puts forward theoretical challenges and new experimental possibilities.

Acknowledgments

It is my pleasure to thank A. Macias for an invitation to present these results, and for arranging a well–attended conference in the stimulating environment of El Colegio Nacional. I also thank T. Padmanabhan and D. Sudarsky for several stimulating discussions on the subject, and extend my apologies for not being able to track down several of his relevant publications under the tight submission deadline for this manuscript.

This work is being supported by CONACyT project No 32067-E.

Notes

1. This remark applies equally well to any set of canonically conjugate variables.

2. Such a collapse may be associated with a position measurement, or position measurements of different components of a position vector.

References

REFERENCES

[1] D. V. Ahluwalia, *Phys. Lett.* **B339** (1994) 301.
[2] J. Madore, gr-qc/9709002.
[3] G. Veneziano, *Europhys. Lett.* **2** (1986) 199.
[4] S. de Haro, *Class. Quantum Grav.* **15** (1998) 519.
[5] R. J. Adler, D. I. Santiago, *Mod. Phys. Lett.* **A14** (1999) 1371.
[6] G. Amelino-Camelia, *Mod. Phys. Lett.* **A12** (1997) 1387.
[7] J. Y. Ng, gr-qc/0201022.
[8] M. Maggiore, *Phys. Lett.* **B304** (1993) 65.
[9] N. Sasakura, *Prog. Theor. Phys.* **102** (1999) 169.
[10] F. Scardigli *Phys. Lett.* **B452** (1999) 39.
[11] S. Capozziello, G. Lambiase, G. Scarpetta, *Int. J. Theor. Phys.* **39** (2000) 15.
[12] A. Kempf, G. Mangano, R. B. Mann, *Phys. Rev.* **D52** (1995) 1108.
[13] D. V. Ahluwalia, *Phys. Lett.* **A275** (2000) 31.
[14] N. R. Bruno, G. Amelino-Camelia, J. Kowalski-Glikman *Phys. Lett.* **B522** (2001) 133.
[15] T. Padmanabhan *Phys. Rev. Lett.* **78** (1997) 1854.
[16] T. Padmanabhan, *Class. Quantum Grav.* **4** (1987) L107.
[17] G. Amelino-Camelia, *Int. J. Mod. Phys.* **D11** (2002) 35.
[18] G. L. Smith *et al.*, *Phys. Rev.* **D61** (2000) 022001.
[19] R. Colella, A. W. Overhauser, S. A. Werner, *Phys. Rev. Lett.* **34** (1975) 1472.
[20] K. C. Littrel, B. E. Allman, S. A. Werner, *Phys. Rev. A* **A56** (1997) 1767.
[21] G. Z. Adunas, E. Rodriguez-Milla, D. V. Ahluwalia, *Phys. Lett. B* **B485** (2000) 215.
[22] G. Z. Adunas, E. Rodriguez-Milla, D. V. Ahluwalia, *Gen. Rel. Grav.* **33** (2001) 183.
[23] A. Peters, K. Y. Chung, S. Chu, *Nature* **400** (1999) 849.
[24] D. V. Ahluwalia, C. Burgard, *Gen. Rel. Grav.* **28** (1996) 1161. Erratum **29** (1997) 681.
[25] D. V. Ahluwalia, *Gen. Rel. Grav.* **29** (1997) 1491.
[26] D. V. Ahluwalia, C. Burgard, *Phys. Rev.* **D57** (1998) 4724.
[27] K. Konno, M. Kasai, *Prog. Theor. Phys.* **100** (1998) 1145.
[28] J. Wudka, *Phys. Rev.* **D64** (2001) 065009.
[29] S. Capozziello, G. Lambiase, *Mod. Phys. Lett.* **A14** (1999) 2193.
[30] A. Camacho, *Mod. Phys. Lett.* **A14** (1999) 2545.
[31] A. Cho, in: News of the Week, *Science* **287** (2000) 2395.
[32] D. V. Ahluwalia, M. Kirchbach, *Int. J. Mod. Phys.* **D10** (2001) 811.
[33] J. K. Webb *et al.*, *Phys. Rev. Lett.* **82** (1999) 884.
[34] J. K. Webb *et al.*, *Phys. Rev. Lett.* **87** (2001) 091301.
[35] M. T. Murphy *et al. Mon. Not. Roy. Astron. Soc.* **327** (2001) 1223.

NON–NEWTONIAN GRAVITY AND COHERENCE PROPERTIES OF LIGHT

A. Camacho
Department of Physics,
Instituto Nacional de Investigaciones Nucleares
Apartado Postal 18-1027, México, D. F., México.
acamacho@nuclear.inin.mx

Abstract In the present work we will consider a Young–type experiment with light from two atoms, and evaluate the effects of a non–Newtonian gravitational term on the resulting first–order correlation function. Also the consequences upon the Hanbury–Brown–Twiss effect of these kind of terms will be considered.

Keywords: Hanbury–Brown–Twiss effect, non–Newtonian gravity.

1. Introduction

Optical interferometry has played a fundamental role in some experimental aspects of gravitational physics [1], for instance, we may mention that there are gravity–waves detectors which are built following Michelson interferometer [2], or that Sagnac ring interferometer [3] constitutes the bedrock for the so–called ring laser gyroscopic device, the one could be used to test the different metric theories of gravity in the weak–field and slow motion limit [4].

It is also needless to say that general relativity (GR) is one of the milestones of modern physics, and that nowadays many of its predictions have been already confronted against some experiments [5]. Nevertheless, we may find several theoretical attempts to construct a theory of elementary particles, which naturally predict the existence of new forces whose effects extend over macroscopic distances [6]. A crucial characteristic of these forces is that they are not described by an inverse–square law, and even more, they, generally, violate the Weak Equivalence Principle (WEP) [6].

After more than a decade of experiments [7], there is no compelling evidence for any kind of deviations from the predictions of Newtonian gravity. Nevertheless, Gibbons and Whiting (GW) phenomenological analysis of gravity data [8] has proved that the very precise agreement between the Newtonian gravity and the observation of planetary motion does not preclude the existence of large non–Newtonian effects over smaller distance scales. GW conclusions allowed them to affirm that the current experimental constraints over possible deviations did not test Newtonian gravity over the 10–1000m distance scale, usually denoted as the "geophysical window".

Looking at the experimental efforts that have been done in order to test the inverse–square law we will find that they can be separated into two large classes: (i) those experiments which involve the direct measurement of the magnitude of G [9]; and (ii) the direct measurement of the magnitude of $G(r)$ with r [10]. At this point it is noteworthy to comment that recently some new proposals have been considered [11], which do not fall in these aforementioned two cases.

In the present work we will analyze the possibility of detecting a Yukawa–type contribution to the gravitational potential (which is one of the possibilities in this direction [12]) considering the effects of this fifth force upon the first and second–order coherence properties of light. In other words, we will consider a Young–type experiment with light from two atoms [13], and evaluate the effects of a non–Newtonian gravitational term on the resulting first–order correlation function. Also the consequences upon the Hanbury–Brown–Twiss effect [14] of these kind of terms will be considered. Some possible experimental scenarios will also be shown.

2. Young's experiment and non–Newtonian gravity

Let us consider two identical atoms (located at P and P'), where each one of them has two levels, and a single photon, such that only one of these atoms will be excited. The initial state of our system reads

$$a(|0,1'> +|1,0'>)|\tilde{0}> +b|0,0'>|\phi>. \qquad (1)$$

Here $|0>, |1>, |0'>, |1'>$ denote the ground and excited states of the two atoms, while $|\tilde{0}>$ is the vacuum of the electromagnetic field, and $|\phi>$ designates the photon. After a time larger than the mean decay time, t_m, the system decays to

$$|\alpha> = \frac{1}{\sqrt{2}}|0,0'> [|\gamma> +|\gamma'>]. \qquad (2)$$

In this last expression $|\gamma>$ and $|\gamma'>$ denote the photon states emitted from sites P and P', respectively.

Let us now assume that the the gravitational interaction contains a Yukawa–type term [12]

$$V(r) = -\frac{G_\infty mM}{r}\left[1 + \alpha \exp(-r/\lambda)\right]. \tag{3}$$

G_∞ describes the interaction between M and m in the limit case $r \to \infty$, i.e., $G_N = G_\infty(1 + \alpha)$, where G_N is the Newtonian constant [7]. Hence the gravitational potential generated by M reads

$$U(r) = -\frac{G_\infty M}{r}\left[1 + \alpha \exp(-r/\lambda)\right]. \tag{4}$$

The interference experiment will detect at point S the light that results from the decay of the system. But here we will take into account the redshift in the frequency that appears as a consequence of the fact that the electromagnetic field *climbs* in a region where a non–vanishing gravitational field is present. In other words, if the frequency at the emission point is ν, and the radiation is detected at a point, which respect to the emission point has a difference ΔU in the gravitational potential, then the frequency at the detection point reads [5]

$$\tilde{\nu} = \frac{\nu}{1 + \Delta U/c^2}. \tag{5}$$

As is already known the electromagnetic field operator can be separated into two parts, namely, with positive and negative frequency parts [15]. Nevertheless, in the case of an experiment which employs absorptive detectors the measurements are destructive, and in consequence only that part of the field operator containing annihilation operators, $\mathbf{E}^{(+)}(\mathbf{r}, t)$, has to be considered. In order to simplify the model we will assume that the field is linearly polarized, and that the radiation emitted from P (or P') is monochromatic.

One of the ideas behind this proposal is to consider the possibility of performing this kind of experiment near the Earth's surface, hence we will assume that

$$r = R + z, \tag{6}$$

where $R \gg |z|$.

Under these conditions the field operator containing the annihilation operator reads, approximately

$$E^{(+)}(\mathbf{r}, t) = \Xi \hat{a} \exp\left\{-i\nu\left[1 - \frac{g_0}{c^2}h\frac{1 + \alpha e^{(-R/\lambda)}}{1 + \alpha}\right]\left[t - \hat{k}\cdot\mathbf{r}\right]\right\}. \tag{7}$$

Here h is the *climbed* distance, $\hat{k}$ denotes the unitary vector in the direction of propagation, Ξ is a constant with dimensions of electric field,

$\hat{a}$ is the corresponding annihilation operator, and $g_0 = g_\infty(1+\alpha)$ is the effective acceleration of gravity at laboratory distances.

The first–order correlation function is given by [13]

$$G^{(1)}(\mathbf{r},\mathbf{r};t,t) = <\alpha|E^{(-)}(\mathbf{r},t)E^{(+)}(\mathbf{r},t)|\alpha>. \tag{8}$$

The radiation stemming from P (and also from P') has to be described, according to the rules of quantum theory, as a superposition of plane wave states [13], neverwithstanding, we may suppose, without introducing unphysical assumptions, that the state vector of $|\gamma>$ is given by a plane wave [15]. Hence we have that

$$G^{(1)}(\mathbf{r},\mathbf{r};t,t) = |\Xi|^2 \left\{1 + \cos\left([\mathbf{k}-\mathbf{k}']\cdot\mathbf{r} + \tilde{g}[h\mathbf{k}-h'\mathbf{k}']\cdot\mathbf{r} + \tilde{g}\nu t\Delta h\right)\right\}. \tag{9}$$

At this last expression we have introduced the following definition

$$\tilde{g} = \frac{g_0}{c^2}\frac{1+\alpha e^{(-R/\lambda)}}{1+\alpha}. \tag{10}$$

Where we have that $\Delta h = h' - h$. It is also readily seen that if $g_0 = 0$, then we recover the usual Young's interference pattern [15].

3. Interference patterns

At this point we must mention that a remarkable difference of expression (9) with respect to the case in which gravity is absent concerns the explicit time dependence of the interference pattern, a fact that can be understood noting that in the case without gravitational field time disappears from the corresponding expression because both waves do have the same frequency [15], a fact that in our case does not happen, indeed the difference in the *climbed* distance renders different frequencies.

3.1. Time independent interference pattern

Concerning expression (9) a possibility comprises the case in which $h = h'$, i.e., $\Delta h = 0$. In this case expression (9) may be rewritten as

$$G^{(1)}(\mathbf{r},\mathbf{r};t,t) = |\Xi|^2 \left\{1 + \cos\left(A\left[1+\tilde{g}(h+h')\right]\right)\right\}. \tag{11}$$

A is a factor present in the case in which there is no gravitational field, and it depends upon the geometry of the inteferometer, and also on the wavelength of the emitted radiation [15].

If we try to detect the effects of a fifth force inside the so–called "geophysical window" [8], we may consider the following values $\alpha = 1/3$ and $\lambda = 10$m [7]. Then, if $\tilde{g}[h+h'] \sim 10^{-8}$

$$(h+h')/R^2 \sim 10^{-4} m. \tag{12}$$

3.2. Time dependent interference pattern

From expression (9) we may see that there are certain time values, t_n ($n\,\text{inter}\,N$), such that

$$\tilde{g}\Delta h \nu t_n = 2\pi n. \tag{13}$$

Hence the interval between t_{n+1} and t_n is

$$\Delta t_n \equiv t_{n+1} - t_n = \frac{2\pi}{\tilde{g}\nu\Delta h}. \tag{14}$$

If we consider this last expression for the purely Newtonian case

$$\Delta t_n^{(N)} = \frac{2\pi c^2}{g_0 \nu \Delta h}, \tag{15}$$

and compare it against the non–Newtonian situation

$$\Delta t_n^{(NN)} = \frac{2\pi c^2}{g_0 \nu \Delta h} \frac{1+\alpha}{1+\alpha e^{-R/\lambda}}, \tag{16}$$

we deduce that

$$\Delta t_n^{(NN)}/\Delta t_n^{(N)} = \frac{1+\alpha}{1+\alpha e^{-R/\lambda}}. \tag{17}$$

Employing the aforementioned values for α and λ we have, approximately, that

$$\Delta t_n^{(NN)}/\Delta t_n^{(N)} = 4/3. \tag{18}$$

This last expression seems to be very promising, indeed, if we were able to detect $\Delta t_n^{(N)}$, then the detection $\Delta t_n^{(NN)}$ should be not a difficult task.

In order to have a realistic experimental situation we must take into account the fact that the emitted radiation is really a pulse. The lifetime of this pulse, assuming that the radiative energy loss (of the dipole oscillator that acts as emitter) is very slow compared with a period of atomic dipole oscillation, has an order of magnitude of $\tau \sim 0.1\mu s$ [16]. The possibility of detecting time intervals of 50fs, based on the interference of two–photon probability amplitudes in two–photon detection [17], implies that the aforementioned differences ($\sim 0.1\mu s$) could represent no technological difficulty.

As was mentioned before, one possibility is to consider the radiation within the optical spectrum, hence we may introduce the following wavelength, for the emitted field, $\lambda^{(r)} \sim 400\text{nm}$. Therefore, if $\Delta t_n^{(NN)} \sim 0.01\mu s$, then we obtain, from (16), a constraint upon Δh, as function of R

$$\Delta h/R^2 \sim 10^{-4} m^{-1}. \tag{19}$$

If the experiment were performed near the Earth's surface ($R \sim 10^6 m$), then $\Delta h \sim 10^4 m$.

4. Hanbury–Brown–Twiss effect

Let us now consider the consequences of a fifth force of Yukawa–Type upon the so called Hanbury–Brown–Twiss effect (HBT) [14], i.e., we must now analyze the second–order coherence properties of light.

Once again we have two atoms, located at points P and P', but now there are two detection points, S_1 and S_2. Initially the atoms are excited, but there is no electromagnetic field, hence the initial state vector reads

$$|\alpha(t=0)>= |1,1'>|\tilde{0}> . \qquad (20)$$

After an interval much larger than the atomic decay time, t_m, the system becomes

$$|\alpha(t>>t_m)>= |0,0'>|\gamma,\gamma'> . \qquad (21)$$

Resorting to the definition of second–order correlation function [14], we find, assuming once again the plane wave approximation for the emitted radiation, that the interference term is given by

$$\cos\left\{ \left[\mathbf{k}-\mathbf{k}'\right]\cdot\left[\mathbf{r_2}-\mathbf{r_1}\right] + \tilde{g}[h'_2\mathbf{k}'-h_2\mathbf{k}]\cdot\mathbf{r_2} - \tilde{g}[h'_1\mathbf{k}'-h_1\mathbf{k}]\cdot\mathbf{r_1} \right.$$
$$\left. +\ \tilde{g}\nu t\left[\Delta h - \Delta h'\right] \right\} . \qquad (22)$$

Here we have that h_1 and h_2 are the *climbed* distances for the radiation emitted by P and detected at S_1 and S_2, respectively (we have the same argument if the light is emitted in P'). Additionally, $\Delta h = h_2 - h_1$, $\Delta h' = h'_2 - h'_1$. Imposing the condition $g_0 = 0$, we recover the usual HBT situation [14].

Acknowledgments

The author would like to thank A.A. Cuevas–Sosa for his help. This work was partially supported by CONACYT (México) Grant No. I35612–E.

References

[1] W. Schleich and M. O. Scully, in: *Modern Trends in Atomic and Molecular Physics, Proceedings of Les Houches Summer School, Session XXXVIII*, eds. R. Stora and G. Grynberg, Amsterdam: North–Holland 1984.

[2] K. S. Thorne, *Rev. Mod. Phys.* **52** (1980) 299.

[3] G. Sagnac, *R. Acad. Sci.* **157** (1913) 708.

[4] M. O. Scully, M. S. Zubairy, and M. P. Haugan, *Phys. Rev.* **A24** (1981) 2009.

[5] I. Ciufolini and J. A. Wheeler, *Gravitation and Inertia*, Princeton, New Jersey, (Princeton University Press, 1995).

REFERENCES

[6] E. Fishbach, G. T. Gillies, D. E: Krause, J. G. Schwan, and C. L. Talmadge, *Metrologia* **29** (1992) 213.

[7] E. Fishbach and C. L. Talmadge, *The Search for Non–Newtonian Gravity*, New York, (Springer–Verlag, 1999).

[8] G. W. Gibbons and B. F. Whiting, *Nature*, **291** (1981) 636.

[9] M. V. Moody and H. J. Paik, *Phys. Rev. Lett.* **70** (1991) 1195.

[10] M. A. Zumberge et al., *Phys. Rev. Lett.* **67** (1991)) 3051.

[11] O. Bertolami, *Mod. Phys. Lett.* **A1** (1986) 383.

[12] F. Fujii, *Nature* **234** (1971) 5.

[13] M. O. Scully and K. Drühl, *Phys. Rev.* **A25** (1982) 2208.

[14] H. Hanbury–Brown and R. Q. Twiss, *Nature* **178** (1956) 1046.

[15] M. O. Scully and M. S. Zubairy, *Quantum Optic*, Cambridge (Cambridge University Press, 1997).

[16] L. Allen and J. H. Eberly, *Optical Resonance and Two–Level Atom*, New York (Dover Publications, Inc., 1987).

[17] C. K. Hong, Z. Y. Ou, and L. Mandel, *Phys. Rev. Lett.*, **59** (1987) 2044.

ON A POSSIBLE NEW TYPE OF A T ODD SKEWON FIELD LINKED TO ELECTROMAGNETISM

Friedrich W. Hehl
Institute for Theoretical Physics, University of Cologne
50923 Köln, Germany
hehl@thp.uni-koeln.de

Yuri N. Obukhov*
Institute for Theoretical Physics, University of Cologne
50923 Köln, Germany
yo@thp.uni-koeln.de

Guillermo F. Rubilar
Institute for Theoretical Physics, University of Cologne
50923 Köln, Germany
gr@thp.uni-koeln.de

Abstract In the framework of generally covariant (pre–metric) electrodynamics ("charge & flux electrodynamics"), the Maxwell equations can be formulated in terms of the electromagnetic excitation $H = (\mathcal{D}, \mathcal{H})$ and the field strength $F = (E, B)$. If the spacetime relation linking H and F is assumed to be *linear*, the electromagnetic properties of (vacuum) spacetime are encoded into 36 components of the vacuum constitutive tensor density χ. We study the propagation of electromagnetic waves and find that the metric of spacetime emerges eventually from the principal part $^{(1)}\chi$ of χ (20 independent components). In this article, we concentrate on the remaining skewon part $^{(2)}\chi$ (15 components) and the axion part $^{(3)}\chi$ (1 component). The skewon part, as we'll show for the first time, can be represented by a 2nd rank traceless tensor $\not{S}_i{}^j$. By means of the Fresnel equation, we discuss how this tensor disturbs

---

*On leave from: Department of Theoretical Physics, Moscow State University, 117234 Moscow, Russia.

the light cones. Accordingly, this is a mechanism for violating Lorentz invariance and time symmetry. In contrast, the (Abelian) axion part $^{(3)}\chi$ does *not* interfere with the light cones.

Keywords: Electrodynamics, light cone, metric, skewon, abelian axion.

1. The constitutive tensor density χ of vacuum spacetime and its irreducible decomposition

In pre–metric electrodynamics [2] the axioms of electric charge and of magnetic flux conservation manifest themselves in the Maxwell equations for the excitation $H = (\mathcal{D}, \mathcal{H})$ and the field strength $F = (E, B)$:

$$dH = J, \qquad dF = 0. \tag{1}$$

If local coordinates x^i are given, with $i, j, \ldots = 0, 1, 2, 3$, we can decompose the excitation and field strength 2–forms into their components according to

$$H = \frac{1}{2} H_{ij}\, dx^i \wedge dx^j, \qquad F = \frac{1}{2} F_{ij}\, dx^i \wedge dx^j. \tag{2}$$

Then a *linear* spacetime relation, see [10], reads

$$H_{ij} = \frac{1}{2} \kappa_{ij}{}^{kl} F_{kl} = \frac{1}{4} \hat{\epsilon}_{ijkl} \chi^{klmn} F_{mn}. \tag{3}$$

The constitutive tensor density χ has 36 independent components. If we decompose it with respect to the 4–dimensional linear group into irreducible pieces, then we find

$$\chi = {}^{(1)}\chi + {}^{(2)}\chi + {}^{(3)}\chi, \qquad \text{with} \quad 36 = 20 \oplus 15 \oplus 1 \tag{4}$$

independent components, respectively. In components, we have the following definitions for the irreducible pieces of χ:

$${}^{(2)}\chi^{ijkl} := \frac{1}{2}\left(\chi^{ijkl} - \chi^{klij}\right), \quad {}^{(3)}\chi^{ijkl} := \chi^{[ijkl]}, \tag{5}$$

$${}^{(1)}\chi^{ijkl} := \chi^{ijkl} - {}^{(2)}\chi^{ijkl} - {}^{(3)}\chi^{ijkl}. \tag{6}$$

Then, by explicit substitution of the definitions, it is straightforward to show that the irreducible piece a of the irreducible piece b, for $a, b = 1, 2, 3$, with the Kronecker δ^{ab}, behaves as follows:

$${}^{(a)}\left[{}^{(b)}\chi^{ijkl}\right] = \delta^{ab}\, {}^{(a)}\chi^{ijkl}, \qquad \text{no sum over } a. \tag{7}$$

More explicitly, we have, e.g.,

$${}^{(1)}\chi^{ijkl} = {}^{(1)}\chi^{klij}, \qquad {}^{(2)}\chi^{ijkl} = -{}^{(2)}\chi^{klij}. \tag{8}$$

A simple way to see the correctness of this irreducible decomposition is to recall that χ^{ijkl} is antisymmetric in the first and in the last pair of indices; then it is possible to map it to a 6×6 matrix. And this matrix can be decomposed in its symmetric tracefree piece corresponding to $^{(1)}\chi^{ijkl}$, the antisymmetric piece $^{(2)}\chi^{ijkl}$, and its trace $^{(3)}\chi^{ijkl}$, see (4).

1.1. Principal piece $^{(1)}\chi$ and light cone structure

Let us try to get a rough picture of the physical meaning of these different irreducible pieces. From $^{(1)}\chi^{ijkl}$ alone, if *electric/magnetic reciprocity* is assumed additionally to hold for (3), then, up to an arbitrary conformal factor, the Lorentzian metric of spacetime can be derived [9, 3, 8, 1, 13]. One can think of this reduction in the way that electric/magnetic reciprocity cuts the 20 components of $^{(1)}\chi^{ijkl}$ into half, that is, only 10 components for the metric are left. Modulo the undetermined function, we have then 9 remaining components.

These 9 components determine the *light cone* at each point of spacetime. Accordingly, in $^{(1)}\chi^{ijkl}$ the light cone of spacetime is hidden and thereby conventional Maxwell–Lorentzian vacuum electrodynamics as well. To put it more geometrically, the first irreducible piece of χ, via electric/magnetic reciprocity, yields the *conformal* structure of spacetime. In this sense, there is no doubt that $^{(1)}\chi^{ijkl}$ is the principal part of the constitutive tensor density χ of the vacuum.

1.2. Abelian axion α

Thus one could be inclined to believe that it is best to require the vanishing of the second and the third irreducible piece of χ. But this would appear to be premature before one has inquired into the possible meanings of these pieces. In fact, the *abelian axion* $\alpha(x)$, introduced by Ni [4, 5] in 1973, is represented by $^{(3)}\chi^{ijkl}$. This field is P odd (P stands for parity), i.e., it is a pseudo– or axial–scalar in conventional language. Experimentally, this field hasn't been found so far. Nevertheless, as we shall see below, the axion does not interfere with the light cone structure of spacetime at all. Therefore, this chapter is not yet closed, the abelian axion remains a serious option for a particle search in experimental high energy physics.

1.3. Skewon piece $\not{S}_i{}^j$ and dissipation

If two irreducible pieces of a quantity don't vanish possibly, it is not too far-fetched to reflect on the remaining piece $^{(2)}\chi^{ijkl}$ and on what its existence may mean. The conventional argument for discarding $^{(2)}\chi$ runs as follows, see Post [10]. Suppose a Lagrangian 4–form L exists for the electromagnetic field. In general $H \sim \partial L/\partial F$. If H is assumed to be linear in F, as is done in (3), then L reads

$$L \sim H \wedge F = \chi \cdot F \wedge F = {}^{(1)}\chi \cdot F \wedge F + {}^{(3)}\chi \cdot F \wedge F. \tag{9}$$

The term with ${}^{(1)}\chi$ eventually becomes the Maxwell Lagrangian, the term with ${}^{(3)}\chi$ part of the axion Lagrangian. The piece with ${}^{(2)}\chi$ drops out of the Lagrangian L because of the antisymmetry of ${}^{(2)}\chi$ according to ${}^{(2)}\chi^{ijkl} = -{}^{(2)}\chi^{klij}$. Since we conventionally assume that all information of a physical system is collected in its Lagrangian, we reject ${}^{(2)}\chi^{ijkl} \neq 0$ as being unphysical. This presents the state of the art.

However, we would like to point out that the argument with the Lagrangian does *not* forbid the existence of a non–vanishing ${}^{(2)}\chi \neq 0$. It only implies that L is 'insensitive' to ${}^{(2)}\chi$. In other words, if ${}^{(2)}\chi \neq 0$, then not all information about the system is contained in the Lagrangian.

Remember that pre–metric electrodynamics is based on the conservation laws of electric charge and magnetic flux and on an axiom about the (kinematic) electromagnetic energy–momentum current ${}^k\Sigma_\alpha$. No Lagrangian is needed nor assumed. But, of course, the proto–Lagrangian $\Lambda := -H \wedge F/2$ exists anyway and has indeed been introduced in the context of the discussion of ${}^k\Sigma_\alpha$. Accordingly, in pre–metric electrodynamics, even when linearity is introduced according to (3), Λ has no decisive meaning — and that Λ does not depend on ${}^{(2)}\chi$ is interesting to note but no reason for a headache.

This reminds us of a complementary property of the axion α or of ${}^{(3)}\chi$. It features in Λ, see (9), but it drops out of ${}^k\Sigma_\alpha$, as we will see below. Should we be alarmed that the axion doesn't contribute to the electromagnetic energy–momentum current? No, not really. As has been shown by Ni, in spite of this 'insensitivity' of ${}^k\Sigma_\alpha$ against α, one can set up a reasonable theory of the axion.

Consequently, in linear pre–metric electrodynamics, it is not alarming that ${}^{(2)}\chi$ drops out from the proto–Lagrangian Λ, and in future we will take *the possible existence of* ${}^{(2)}\chi^{ijkl}$ *for granted*.

What is then the possible physical meaning of ${}^{(2)}\chi$? In pre–metric electrodynamics [2], the energy–momentum current reads

$$^k\Sigma_\alpha := \frac{1}{2}\left[F \wedge (e_\alpha \rfloor H) - H \wedge (e_\alpha \rfloor F)\right]. \tag{10}$$

We can specify a certain vector field $\xi = \xi^\alpha e_\alpha$, with the basis e_α of the tangent vector space at each point of spacetime. Then we can transvect the energy–momentum current with ξ^α:

$$\mathcal{Q} := \xi^\alpha {}^k\Sigma_\alpha = \frac{1}{2}\left[F \wedge (\xi \rfloor H) - H \wedge (\xi \rfloor F)\right]. \tag{11}$$

The scalar–valued 3–form $\mathcal{Q}$ is expected to be related to conserved quantities provided we can find suitable (Killing type) vector fields ξ. Therefore we determine its exterior derivative and find after some algebra, see Rubilar [12],

$$d\mathcal{Q} = (\xi \rfloor F) \wedge J + \frac{1}{2}(F \wedge \mathcal{L}_\xi H - H \wedge \mathcal{L}_\xi F), \quad (12)$$

or, in holonomic components, with $\mathcal{Q}^i := \epsilon^{ijkl} Q_{jkl}/6$, $\mathcal{J}^i := \epsilon^{ijkl} J_{jkl}/6$, and $\mathcal{H}^{ij} := \epsilon^{ijkl} H_{kl}/2$,

$$\partial_i \mathcal{Q}^i = \xi^k F_{kl} \mathcal{J}^l + \frac{1}{4}\left(F_{kl} \mathcal{L}_\xi \mathcal{H}^{kl} - \mathcal{H}^{kl} \mathcal{L}_\xi F_{kl}\right). \quad (13)$$

Here $\mathcal{L}_\xi$ denotes the Lie derivative along ξ. Now we substitute the *linear* relation (3), or $\mathcal{H}^{kl} = \chi^{klmn} F_{mn}/2$, and find

$$\partial_i \mathcal{Q}^i = \xi^k F_{kl} \mathcal{J}^l + \frac{1}{8}\left[F_{kl} \mathcal{L}_\xi(\chi^{klmn} F_{mn}) - \chi^{klmn} F_{mn} \mathcal{L}_\xi F_{kl}\right]. \quad (14)$$

We apply the Leibniz rule of the Lie derivative and rearrange a bit:

$$\partial_i \mathcal{Q}^i = \xi^k F_{kl} \mathcal{J}^l + \frac{1}{8}\left[(\mathcal{L}_\xi \chi^{ijkl}) F_{ij} F_{kl} + (\chi^{ijkl} - \chi^{klij}) F_{ij} \mathcal{L}_\xi F_{kl}\right]. \quad (15)$$

We substitute the irreducible pieces of χ^{ijkl}. Then we have

$$\partial_i \mathcal{Q}^i = \xi^k F_{kl} \mathcal{J}^l + \frac{1}{8} \mathcal{L}_\xi \left(^{(1)}\chi^{ijkl} + {}^{(3)}\chi^{ijkl}\right) F_{ij} F_{kl}$$
$$+ \frac{1}{4} {}^{(2)}\chi^{ijkl} F_{ij} \mathcal{L}_\xi F_{kl}. \quad (16)$$

If $^{(1)}\chi$ and $^{(3)}\chi$ carry a reasonable symmetry, namely $\mathcal{L}_\xi {}^{(1)}\chi = \mathcal{L}_\xi {}^{(2)}\chi = 0$, and are thus well–behaved, then, in vacuum, i.e., for $\mathcal{J}^i = 0$, we have non–conservation of energy, for example, because of the offending term $^{(2)}\chi F\dot{F}$. Here the dot symbolizes the 'time' derivative along ξ.

In any case we see that $^{(2)}\chi$ induces a *dissipative* term with first 'time' derivative. This is what we might have expected since dissipative phenomena in general cannot be described in a Lagrangian framework.

It is then our hypothesis that $^{(2)}\chi$ can represent fields which are *odd under T transformations*. Of course, we must investigate how these skewons, as we may call them in a preliminary way, disturb the light cone and whether there is perhaps a viable subclass of the skewons.

2. The skewon field $\mathcal{S}_i{}^j$

The *skewon* piece of the constitutive tensor density χ^{ijkl} is defined in $(5)_1$. Therefrom we can read off the algebraic symmetries

$$^{(2)}\chi^{ijkl} = -^{(2)}\chi^{klij}, \qquad ^{(2)}\chi^{[ijkl]} = 0. \quad (17)$$

From χ^{ijkl}, the skewon piece inherits the antisymmetry in the first and the second pair of indices:

$$^{(2)}\chi^{(ij)kl} = 0, \qquad ^{(2)}\chi^{ij(kl)} = 0. \tag{18}$$

Thus $^{(2)}\chi^{ijkl}$ can be mapped to an antisymmetric (or *skew*symmetric) 6×6 matrix. For this reason we called it the skewon piece of χ^{ijkl}. Since this matrix has 15 independent components, we expect that it is equivalent to a 2nd rank tensor in 4 dimensions (16 components) with vanishing trace (1 component). Accordingly, we define the skewon field by

$$S_i{}^j := \frac{1}{4} \hat{\epsilon}_{iklm} {}^{(2)}\chi^{klmj}. \tag{19}$$

Because of (17)$_2$, its trace vanishes, indeed,

$$S_n{}^n = \frac{1}{4} \hat{\epsilon}_{nklm} {}^{(2)}\chi^{[klmn]} = 0. \tag{20}$$

If we define the tracefree part of $S_i{}^j$ by

$$\not{S}_i{}^j := S_i{}^j - \frac{1}{4} S_k{}^k \delta_i^j, \tag{21}$$

then for our $S_i{}^j$, we have $S_i{}^j = \not{S}_i{}^j$.

Let us invert (19). We multiply by ϵ^{inpq} and find

$$\epsilon^{inpq} S_i{}^j := \frac{1}{4} \epsilon^{inpq} \hat{\epsilon}_{iklm} {}^{(2)}\chi^{klmj} = \frac{1}{4} \delta^{npq}_{klm} {}^{(2)}\chi^{[klm]j} \tag{22}$$

or

$$^{(2)}\chi^{[ijk]l} = -\frac{2}{3} \epsilon^{ijkm} S_m{}^l. \tag{23}$$

We expand the bracket:

$$^{(2)}\chi^{ijkl} + {}^{(2)}\chi^{jkil} + {}^{(2)}\chi^{kijl} = -2 \epsilon^{ijkm} S_m{}^l. \tag{24}$$

The second term on the left hand side of this equation, by means of the symmetries (17)$_1$ and (18)$_1$, can be rewritten as $^{(2)}\chi^{jkil} = -{}^{(2)}\chi^{iljk} = {}^{(2)}\chi^{lijk}$. Thus,

$$^{(2)}\chi^{ijkl} + 2{}^{(2)}\chi^{(k|ij|l)} = -2 \epsilon^{ijkm} S_m{}^l \tag{25}$$

or, because of (18)$_2$,

$$^{(2)}\chi^{ijkl} = 2 \epsilon^{ijm[k} S_m{}^{l]}. \tag{26}$$

In order to make the symmetry (17)$_1$ manifest, we rename the indices

$$^{(2)}\chi^{klij} = 2 \epsilon^{klm[i} S_m{}^{j]} \tag{27}$$

and subtract (27) from (26). This yields the final result

$$^{(2)}\chi^{ijkl} = \epsilon^{ijm[k} S_m{}^{l]} - \epsilon^{klm[i} S_m{}^{j]}. \tag{28}$$

For (28), all the symmetries (17) and (18) can be verified straightforwardly. In (19), we chose the conventional factor as $1/4$ in order to find in (28) a formula free of inconvenient factors.

3. Decomposing the "dual" constitutive tensor density $\kappa_{ij}{}^{kl}$ and recovering the skewon field

Let us recall that the starting point for the discussion of the constitutive (spacetime) relation is the κ–map. Namely, we have the tensor density $\kappa_{ij}{}^{kl}$ with 36 components, see [2] Eq.(D.1.11). One can decompose this object into its irreducible pieces. Obviously, contraction is the only tool for such a decomposition. Following Post [11], we can define the contacted tensor of type (1,1),

$$\kappa_i{}^k := \kappa_{il}{}^{kl}, \tag{29}$$

with 16 independent components. The second contraction yields the pseudo–scalar function

$$\kappa := \kappa_k{}^k = \kappa_{kl}{}^{kl}. \tag{30}$$

The traceless piece

$$\not{\kappa}_i{}^k := \kappa_i{}^k - \frac{1}{4}\kappa\,\delta_i^k \tag{31}$$

has 15 independent components. These pieces can now be subtracted from the original constitutive tensor. Then,

$$\kappa_{ij}{}^{kl} = {}^{(1)}\kappa_{ij}{}^{kl} + {}^{(2)}\kappa_{ij}{}^{kl} + {}^{(3)}\kappa_{ij}{}^{kl} \tag{32}$$

$$= {}^{(1)}\kappa_{ij}{}^{kl} + 2\,\not{\kappa}_{[i}{}^{[k}\delta_{j]}^{l]} + \frac{1}{6}\kappa\,\delta_{[i}^k\delta_{j]}^l. \tag{33}$$

By construction, $^{(1)}\kappa_{ij}{}^{kl}$ is the totally traceless part of the constitutive map:

$$^{(1)}\kappa_{il}{}^{kl} = 0. \tag{34}$$

Thus, we split κ according to $36 = 20 + 15 + 1$, and the (2,2) tensor $^{(1)}\kappa_{ij}{}^{kl}$ is subject to the 16 constraints (34) and carries $20 = 36 - 16$ components.

Now we are prepared to proceed with the analysis of the χ–picture. By definition, we have

$$\chi^{ijkl} := \frac{1}{2}\epsilon^{ijmn}\kappa_{mn}{}^{kl}. \tag{35}$$

Substituting here the decomposition (33), we find

$$\chi^{ijkl} = {}^{(1)}\chi^{ijkl} + {}^{(2)}\chi^{ijkl} + {}^{(3)}\chi^{ijkl}. \tag{36}$$

In correspondence with (33), we have the irreducible pieces:

$$^{(1)}\chi^{ijkl} = \frac{1}{2}\epsilon^{ijmn}\,{}^{(1)}\kappa_{mn}{}^{kl}, \tag{37}$$

$$^{(2)}\chi^{ijkl} = \frac{1}{2}\epsilon^{ijmn}\,{}^{(2)}\kappa_{mn}{}^{kl} = -\epsilon^{ijm[k}\rlap{/}{\kappa}_m{}^{l]}, \tag{38}$$

$$^{(3)}\chi^{ijkl} = \frac{1}{2}\epsilon^{ijmn}\,{}^{(3)}\kappa_{mn}{}^{kl} = \frac{1}{12}\epsilon^{ijkl}\kappa. \tag{39}$$

Let us identify the skewon and the axion fields by

$$S_i{}^j = -\frac{1}{2}\rlap{/}{\kappa}_i{}^j, \qquad \alpha = \frac{1}{12}\kappa. \tag{40}$$

Using the S–identity (A.4), we have

$$^{(2)}\chi^{ijkl} = 2\epsilon^{ijm[k}S_m{}^{l]} = -2\epsilon^{klm[i}S_m{}^{j]} \tag{41}$$

or

$$^{(2)}\chi^{ijkl} = \epsilon^{ijm[k}S_m{}^{l]} - \epsilon^{klm[i}S_m{}^{j]}. \tag{42}$$

Thus, the S-identity (A.4) guarantees the *skew*–symmetry of $^{(2)}\chi$ under exchange of the first with the second index pair:

$$^{(2)}\chi^{ijkl} = -\,{}^{(2)}\chi^{klij}. \tag{43}$$

On the other hand, the K–identity (A.11) provides the *symmetry* of the $^{(1)}\chi$:

$$^{(1)}\chi^{ijkl} = {}^{(1)}\chi^{klij}. \tag{44}$$

This holds true because of the tracelessness property (34).

It is thus very satisfactory to find the one–to–one correspondence of the irreducible decomposition (33) of $\kappa_{ij}{}^{kl}$ [based on the trace extraction] and the irreducible decomposition (36) of χ^{ijkl} [based on the separation into symmetric and skew–symmetric parts].

4. The skewon field as 6 × 6 matrix

It is convenient to put the $S_i{}^j$ also into the conventional 6 × 6 matrix since this provides the interpretation of the spacetime relation in terms of the 3–dimensional electromagnetic field $\mathcal{D}, \mathcal{H}, B, E$. Therefore we compute the 3 × 3 matrices with the help of (28) in a fairly messy but straightforward way, see [12]:

$$^{(2)}\mathcal{A}^{ab} := {}^{(2)}\chi^{0a0b} = \epsilon^{abc} S_c{}^0, \tag{45}$$

$$^{(2)}\mathcal{B}_{ab} := \frac{1}{4} \epsilon_{acd} \epsilon_{bef} {}^{(2)}\chi^{cdef} = -\epsilon_{abc} S_0{}^c, \tag{46}$$

$$^{(2)}\mathcal{C}_a{}^b := \frac{1}{2} \epsilon_{acd} {}^{(2)}\chi^{cd0b} = -S_a{}^b + \delta_a^b S_c{}^c, \tag{47}$$

$$^{(2)}\mathcal{D}^a{}_b := \frac{1}{2} \epsilon_{bcd} {}^{(2)}\chi^{0acd} = S_b{}^a - \delta_b^a S_c{}^c. \tag{48}$$

Quite generally, we have

$$\chi^{IJ} = \begin{pmatrix} \mathcal{A}^{ab} & \mathcal{D}^a{}_b \\ \mathcal{C}_a{}^b & \mathcal{B}_{ab} \end{pmatrix}. \tag{49}$$

For the skewon piece[1], we have specifically

$$^{(2)}\chi^{IJ} = \begin{pmatrix} \mathcal{A}^{[ab]} & \frac{1}{2}(\mathcal{D}^a{}_b - \mathcal{C}_b{}^a) \\ \frac{1}{2}(\mathcal{C}_a{}^b - \mathcal{D}^b{}_a) & \mathcal{B}_{[ab]} \end{pmatrix} = \begin{pmatrix} {}^{(2)}\mathcal{A}^{ab} & {}^{(2)}\mathcal{D}^a{}_b \\ {}^{(2)}\mathcal{C}_a{}^b & {}^{(2)}\mathcal{B}_{ab} \end{pmatrix}$$

$$= \begin{pmatrix} \epsilon^{abc} S_c{}^0 & +S_b{}^a - \delta_b^a S_c{}^c \\ -S_a{}^b + \delta_a^b S_c{}^c & -\epsilon_{abc} S_0{}^c \end{pmatrix}. \tag{50}$$

The rest is done easily. We substitute (50) into the spacetime relation

$$\begin{pmatrix} \mathcal{H}_a \\ \mathcal{D}^a \end{pmatrix} = \begin{pmatrix} C^b{}_a & B_{ab} \\ A^{ab} & D_b{}^a \end{pmatrix} \begin{pmatrix} -E_b \\ B^b \end{pmatrix} \tag{51}$$

and find quite generally for the skewon contributions of the 3D excitations,

$$^{(2)}\mathcal{D}^a = -\epsilon^{abc} S_c{}^0 E_b + (-\delta_b^a S_c{}^c + S_b{}^a) B^b, \tag{52}$$

$$^{(2)}\mathcal{H}_a = (-\delta_a^b S_c{}^c + S_a{}^b) E_b - \epsilon_{abc} S_0{}^c B^b. \tag{53}$$

The diagonal terms of the 3D tensor $(-\delta_a^b S_c{}^c + S_a{}^b)$ are of the type as those postulated by Nieves and Pal [7] for describing a "...third electromagnetic constant of an isotropic medium". However, our "medium" is spacetime, i.e., the vacuum itself.

We stress that for the derivation of (52) and (53) we neither specialized the skewon field $S_i{}^j$ nor did we apply any metric distilled from ${}^{(1)}\chi^{ijkl}$. Therefore the 1+3 decompositions in (52) and (53) are generally valid for *any* linear spacetime relation.

5. Spatially isotropic skewon field and the ansatz of Nieves and Pal

If we specialize first to 3-dimensional *isotropy*, then we have, with the 3D pseudo-scalar function $s = s(x)$,

$$S_a{}^b = \frac{s}{2}\delta_a^b, \qquad S_0{}^a = 0, \qquad S_b{}^0 = 0. \tag{54}$$

It was possible to formulate isotropy for the skewon piece *without* taking recourse to a metric tensor since $S_i{}^j$ is a mixed variant tensor of 2nd rank. Thus,

$$S_i{}^j = \frac{s}{2}\begin{pmatrix} -3 & 0 & 0 & 0 \\ 0 & 1 & 0 & 0 \\ 0 & 0 & 1 & 0 \\ 0 & 0 & 0 & 1 \end{pmatrix} \tag{55}$$

and

$$-\delta_a^b S_c{}^c + S_a{}^b = -s\,\delta_a^b. \tag{56}$$

Consequently, equations (52) and (53) become

$$^{(2)}\mathcal{D}^a = -s\,B^a, \qquad ^{(2)}\mathcal{H}_a = -s\,E_a, \tag{57}$$

exactly what Nieves and Pal had postulated and discussed subsequently [7]. Accordingly, the spacetime relations (52) and (53) are *anisotropic* generalizations of the Nieves and Pal ansatz. The off diagonal terms with $S_0{}^a$ and $S_b{}^0$ lead, respectively, to magnetic and electric Faraday type of effects of the spacetime under consideration, i.e., these terms rotate the polarization of a wave propagating in such a spacetime.

6. On the four electromagnetic constants for vacuum spacetime with spatial isotropy

Since the axion field also contributes to the spacetime relation of the type (57), we are going to determine it. We have

$$^{(3)}\chi^{ijkl} := \alpha\,\epsilon^{ijkl}, \qquad \alpha = \frac{1}{4!}\hat{\epsilon}_{ijkl}\,^{(3)}\chi^{ijkl}, \tag{58}$$

$$^{(3)}\chi^{IJ} = \alpha\begin{pmatrix} 0_3 & 1_3 \\ 1_3 & 0_3 \end{pmatrix} = \alpha\,\epsilon^{IJ}. \tag{59}$$

Because of (51), we find

$$^{(3)}\mathcal{D}^a = +\alpha\,B^a, \tag{60}$$

$$^{(3)}\mathcal{H}_a = -\alpha\,E_a. \tag{61}$$

If we take (77) from below in Cartesian coordinates, i.e., we have a Lorentz metric $o_{ij} \stackrel{*}{=} (c^2, -1, -1, -1)$ with $o_{ab} \stackrel{*}{=} -\delta_{ab}$, then the spacetime relation becomes

$$\mathcal{D}^a = \varepsilon_0 \, \delta^{ab} E_b + (-s + \alpha) \, B^a, \qquad (62)$$
$$\mathcal{H}_a = (-s - \alpha) \, E_a + \mu_0^{-1} \, \delta_{ab} B^b. \qquad (63)$$

In the special case when skewon and axion become *constant* fields, we can say that we found 4 electromagnetic constants for a spacetime (vacuum) with spatial isotropy: The electric constant ε_0, the magnetic constant μ_0, the isotropic part s of the skewon $S_a{}^b$, and the axion α.

7. How does the skewon field affect light propagation [13, 12]?

For any linear spacetime relation, the Fresnel equation can be written as

$$\mathcal{G}^{ijkl} q_i q_j q_k q_l = 0, \qquad (64)$$

where q_i is the wave covector. The fourth order tensor density of weight $+1$, the *Fresnel tensor*, as we may call it, is defined by

$$\mathcal{G}^{ijkl} := \frac{1}{4!} \hat{\epsilon}_{mnpq} \hat{\epsilon}_{rstu} \chi^{mnr(i} \chi^{j|ps|k} \chi^{l)qtu}. \qquad (65)$$

First, we recall [13] that the Fresnel equation is independent of the axion piece $^{(3)}\chi$ of the constitutive tensor:

$$\mathcal{G}^{ijkl}(\chi) = \mathcal{G}^{ijkl}(^{(1)}\chi + {}^{(2)}\chi). \qquad (66)$$

Thus, for arbitrary $^{(1)}\chi$ and $^{(2)}\chi$, we have

$$\mathcal{G}^{ijkl}(^{(3)}\chi) = 0. \qquad (67)$$

Furthermore, due to the skewsymmetry (17)$_1$ of $^{(2)}\chi$, we have

$$\mathcal{G}^{ijkl}(^{(2)}\chi) = 0. \qquad (68)$$

By explicit calculations, we used computer algebra for it, we find

$$\mathcal{G}^{ijkl}(^{(2)}\chi + {}^{(3)}\chi) = 0. \qquad (69)$$

This identity is non–trivial since $\mathcal{G}$ depends cubically on the constitutive tensor χ. The identity (69) shows that the symmetric piece $^{(1)}\chi$ is indispensable in order to obtain well behaved wave properties: If $^{(1)}\chi = 0$, the Fresnel equation is trivially satisfied and thus no light cone structure could be induced.

Furthermore, since in general

$$\mathcal{G}^{ijkl}(^{(1)}\chi + {}^{(2)}\chi) \neq \mathcal{G}^{ijkl}(^{(1)}\chi), \tag{70}$$

the skewon field *does* influences the Fresnel equation, and therefore, eventually the light cone structure. An example of this general result can be found in the asymmetric constitutive tensor studied by Nieves and Pal [6, 7]. Actually, after some algebra one finds

$$\begin{aligned}\mathcal{G}^{ijkl}(^{(1)}\chi + {}^{(2)}\chi) &= \mathcal{G}^{ijkl}(^{(1)}\chi) \\ &+ \frac{2}{4!}\hat{\epsilon}_{mnpq}\hat{\epsilon}_{rstu}\,{}^{(1)}\chi^{mnr(i}\,{}^{(2)}\chi^{j|ps|k}\,{}^{(2)}\chi^{l)qtu} \\ &+ \frac{1}{4!}\hat{\epsilon}_{mnpq}\hat{\epsilon}_{rstu}\,{}^{(2)}\chi^{mnr(i}\,{}^{(1)}\chi^{j|ps|k}\,{}^{(2)}\chi^{l)qtu}\end{aligned} \tag{71}$$

or, in a (more of less) obvious notation, see the definition (65),

$$\begin{aligned}\mathcal{G}^{ijkl}(\chi,\chi,\chi) &= \mathcal{G}^{ijkl}(^{(1)}\chi,{}^{(1)}\chi,{}^{(1)}\chi) + 2\,\mathcal{G}^{ijkl}(^{(1)}\chi,{}^{(2)}\chi,{}^{(2)}\chi) \\ &+ \mathcal{G}^{ijkl}(^{(2)}\chi,{}^{(1)}\chi,{}^{(2)}\chi).\end{aligned} \tag{72}$$

The other terms vanish due to the symmetry properties of each irreducible piece.

Take now (70) and substitute the parametrization of $^{(2)}\chi$ in terms of $S_i{}^j$, see (19). After some lengthy but straightforward algebra, one finds that the two last contributions to the right hand side of (70) are actually equal, namely

$$\mathcal{G}^{ijkl}(^{(1)}\chi,{}^{(2)}\chi,{}^{(2)}\chi) = \mathcal{G}^{ijkl}(^{(2)}\chi,{}^{(1)}\chi,{}^{(2)}\chi) \tag{73}$$

$$= \frac{1}{3}\,{}^{(1)}\chi^{m(i|n|j}S_m{}^k S_n{}^{l)}. \tag{74}$$

Therefore, the final result reads

$$\mathcal{G}^{ijkl}(\chi) = \mathcal{G}^{ijkl}(^{(1)}\chi + {}^{(2)}\chi) = \mathcal{G}^{ijkl}(^{(1)}\chi) + {}^{(1)}\chi^{m(i|n|j}S_m{}^k S_n{}^{l)}, \tag{75}$$

a very simple expression, indeed.

For the particular case of Nieves and Pal, we will use the ansatz

$$S_i{}^j = \mathcal{S}_i{}^j = \omega_i v^j - \frac{1}{4}\omega_k v^k \delta_i^j \tag{76}$$

and assume additionally the existence of a metric g_{ij} (resulting from $^{(1)}\chi$!) for raising and lowering indices: $\omega_i = g_{il}v^l$. This determines $^{(2)}\chi$ via (28). Furthermore, we assume for the principal part the usual metric dependent expression for the vacuum in a Riemannian spacetime, namely

$$^{(1)}\chi^{ijkl} = 2\sqrt{\frac{\varepsilon_0}{\mu_0}}\sqrt{-g}\,g^{i[k}g^{l]j}. \tag{77}$$

Because of (66), the axion piece is not required. Accordingly, we substitute $^{(1)}\chi$ and $^{(2)}\chi$ into (65). Then,

$$\mathcal{G}^{ijkl}q_i q_j q_k q_l = -\sqrt{\frac{\varepsilon_0}{\mu_0}}\sqrt{-g}\times$$
$$\left[\frac{\varepsilon_0}{\mu_0}(q\cdot q)^2 - (q\cdot q)(v\cdot v)(v\cdot q)^2 + (v\cdot q)^4\right] = 0. \tag{78}$$

We now use $q_i \stackrel{*}{=} (\omega, -\vec{k})$, $v^i \stackrel{*}{=} (v,0,0,0)$, $g_{ij} = o_{ij} \stackrel{*}{=} (c^2, -1, -1, -1)$, and $c^2 = \frac{1}{\varepsilon_0\mu_0}$. Thus,

$$-\frac{1}{\varepsilon_0^3}\mathcal{G}^{ijkl}q_i q_j q_k q_l \stackrel{*}{=} \left[\omega^2 - (c\vec{k})^2\right]^2 + \left[\frac{cv^2}{\varepsilon_0}\,\omega\,c\vec{k}\right]^2 = 0. \tag{79}$$

This equation describes how the skewon piece, via v, affects light propagation. In general, for $v \neq 0$, the Fresnel equation will have complex solutions. This is again a manifestation of the dispersive properties described by the skewon piece $^{(2)}\chi$ of the constitutive tensor. Modulo different conventions, our result (79) agrees with that of Nieves and Pal [7] Eq.(5.7).

8. Discussion

The skewon and the axion part of χ are explicitly known:

$$^{(2)}\chi^{ijkl} = \epsilon^{ijm[k}S_m{}^{l]} - \epsilon^{klm[i}S_m{}^{j]}, \tag{80}$$
$$^{(3)}\chi^{ijkl} = \alpha\,\epsilon^{ijkl}. \tag{81}$$

Accordingly, we found a traceless 2nd rank tensor field $S_i{}^j = \mathcal{S}_i{}^j$ and a pseudo–scalar α. For the principal part $^{(1)}\chi$ with its 20 independent components things are more difficult.

We can tentatively assume

$$^{(1)}\chi^{ijkl} \sim g^{[i|[k}h^{l]|j]} + g^{[k|[i}h^{j]|l]} + \epsilon^{ijm[k}a_m{}^{l]} + \epsilon^{klm[i}a_m{}^{j]}, \tag{82}$$

with two symmetric $g^{ij} = g^{ji}$, $h^{ij} = h^{ji}$ and a traceless tensor $a_i{}^j$, with $a_k{}^k = 0$. However, a second look convinces us that in (82) the two tast terms (with a) should be deleted. The reason is the S–identity, see

Appendix. In view of (A.4), a traceless (1,1) tensor can only contribute to the $^{(2)}\chi$, not to $^{(1)}\chi$. So, the structure of $^{(1)}\chi$ is most probably determined only by the two symmetric tensors g^{ij} and h^{ij} with $10 + 10$ independent components.

Furthermore, it is clear that the two terms on the r.h.s. of (82) are equal: $g^{[i|[k} h^{l]|j]} = g^{[k|[i} h^{j]|l]}$. Indeed:

$$4g^{[i|[k} h^{l]|j]} = g^{ik}h^{lj} - g^{il}h^{kj} - g^{jk}h^{li} + g^{jl}h^{ki} = 4g^{[k|[i} h^{j]|l]} \quad (83)$$
$$= g^{ki}h^{jl} - g^{kj}h^{il} - g^{li}h^{jk} + g^{lj}h^{ik}. \quad (84)$$

Since both tensors are symmetric, $g^{ij} = g^{ji}$ and $h^{ij} = h^{ji}$, the two expressions are equivalent. Thus only one term is left over:

$$^{(1)}\chi^{ijkl} \sim g^{[i|[k} h^{l]|j]} \quad (85)$$

By the way, if we turn to the κ-representation, we have:

$$^{(1)}\kappa_{ij}{}^{kl} \sim \epsilon_{ijmn} g^{m[k} h^{l]n}. \quad (86)$$

For the symmetric $g^{ij} = g^{ji}$ and $h^{ij} = h^{ji}$, a contraction is automatically zero, $^{(1)}\kappa_{il}{}^{kl} \equiv 0$. This means that such a term belongs indeed to the first irreducible part, in accordance with Sec.3. Such a structure looks very much as the most general parametrization of the first irreducible part, but a proof is still not available.

Summarizing: it seems that the general structure of the first irreducible part reads

$$^{(1)}\chi^{ijkl} \sim g^{[i|[k} h^{l]|j]}. \quad (87)$$

It would be desirable to find a corresponding proof.

Acknowledgments

G.F.R. would like to thank the German Academic Exchange Service (DAAD) for financial support. We are grateful to Alfredo Macias, UAM–Iztapalapa, for inviting us to contribute to the 1st Mexican Meeting on Mathematical and Experimental Physics. This project has been partly supported by CONACyT Grants: 28339E, 32138E, by a FOMES Grant: P/FOMES 98-35-15, and by the joint German–Mexican project CONACyT — DFG: E130–655 — 444 MEX 100.

Appendix: Two identities

We can rederive our results (19) and (28) in an alternative way in order to get more insight in the relevant structure. In 4D, any object with five completely antisymmetrized indices is zero, $Z^{[ijmkl]} \equiv 0$. When 4 of these 5 indices belong to the Levi–Civita symbol, we have the identity:

$$\epsilon^{ijmk} Z^l_{...} \equiv \epsilon^{ljmk} Z^i_{...} + \epsilon^{ilmk} Z^j_{...} + \epsilon^{ijlk} Z^m_{...} + \epsilon^{ijml} Z^k_{...}. \quad (A.1)$$

APPENDIX A: Two identities

Applying this for the case when $Z = S_m{}^l$, we find the identity

$$\epsilon^{ijmk} S_m{}^l \equiv \epsilon^{ljmk} S_m{}^i + \epsilon^{ilmk} S_m{}^j + \epsilon^{ijlk} S_m{}^m + \epsilon^{ijml} S_m{}^k. \quad (A.2)$$

Suppose that $S_i{}^j$ is a *traceless* tensor, i.e. $S_m{}^m = 0$. Then a simple rearrangement of the terms in the above identity yields:

$$\epsilon^{ijmk} S_m{}^l - \epsilon^{ijml} S_m{}^k \equiv -\epsilon^{klmi} S_m{}^j + \epsilon^{klmj} S_m{}^i. \quad (A.3)$$

Summarizing, we proved the **S-identity**: *Every (1,1) tensor $S_i{}^j$ which is traceless, $S_k{}^k = 0$, has the property*

$$\epsilon^{ijm[k} S_m{}^{l]} \equiv -\epsilon^{klm[i} S_m{}^{j]}. \quad (A.4)$$

This identity always holds true in four dimensions, just because of the properties of the Levi–Civita symbol. Eq.(A.4) can also be found by adding (26) and (27). The S–identity underlies the possibility to express $^{(2)}\chi$ in terms of the skewon tensor.

Let us now demonstrate another identity which holds for a (2,2) tensor $K_{ij}{}^{kl}$. Consider the contraction

$$\epsilon^{ijmn} K_{mn}{}^{[kl]} = \frac{1}{2}\left(\epsilon^{ijmn} K_{mn}{}^{kl} - \epsilon^{ijmn} K_{mn}{}^{lk}\right). \quad (A.5)$$

Apply the identity (A.1) to the indices $[ijmnk]$ in the first term on the r.h.s. and to the indices $[ijmnl]$ in the second term:

$$\epsilon^{ijmn} K_{mn}{}^{kl} = \epsilon^{kjmn} K_{mn}{}^{il} + \epsilon^{ikmn} K_{mn}{}^{jl} + \epsilon^{ijkn} K_{mn}{}^{ml} + \epsilon^{ijmk} K_{mn}{}^{nl}, \quad (A.6)$$

$$\epsilon^{ijmn} K_{mn}{}^{lk} = \epsilon^{ljmn} K_{mn}{}^{ik} + \epsilon^{ilmn} K_{mn}{}^{jk} + \epsilon^{ijln} K_{mn}{}^{mk} + \epsilon^{ijml} K_{mn}{}^{nk}. \quad (A.7)$$

Suppose that the tensor K is *traceless*: $K_{mn}{}^{mk} = 0$. Then using (A.6)-(A.7) in (A.5), we find:

$$\epsilon^{ijmn} K_{mn}{}^{[kl]} = \epsilon^{jmn[k} K_{mn}{}^{l]i} - \epsilon^{imn[k} K_{mn}{}^{l]j}. \quad (A.8)$$

Now we once again apply the identity (A.1):

$$\epsilon^{jmnk} K_{mn}{}^{li} = \epsilon^{jmnl} K_{mn}{}^{ki} + \epsilon^{jmlk} K_{mn}{}^{ni} + \epsilon^{jlnk} K_{mn}{}^{mi} + \epsilon^{lmnk} K_{mn}{}^{ji}. \quad (A.9)$$

Taking into account the tracelesness, $K_{mn}{}^{mk} = 0$, we can rearrange the terms (move the first term from the r.h.s. to the l.h.s.) and find

$$\epsilon^{jmn[k} K_{mn}{}^{l]i} = \frac{1}{2}\epsilon^{klmn} K_{mn}{}^{ij}. \quad (A.10)$$

Finally, using (A.10) in (A.8), we prove the **K-identity**: *Every (2,2) tensor $K_{ij}{}^{kl}$ that is traceless $K_{ki}{}^{kj} = 0$ has the property*

$$\epsilon^{ijmn} K_{mn}{}^{[kl]} = \epsilon^{klmn} K_{mn}{}^{[ij]}. \quad (A.11)$$

[Incidentally, this property applies in particular to the Weyl curvature tensor $C_{ij}{}^{kl}$. In this case, we recover from (A.11) the well known anti–self double–duality of the Weyl tensor: $C_{ij}{}^{kl} = -\frac{1}{4}\epsilon^{klmn}\epsilon_{ijpq}C_{mn}{}^{pq}$].

Notes

1. For future reference we display here also the principal and the axion pieces as 6×6 matrices, respectively, namely

$$^{(1)}\chi^{IJ} = \begin{pmatrix} \mathcal{A}^{(ab)} & \frac{1}{2}(\mathcal{D}^a{}_b + \mathcal{C}_b{}^a) \\ \frac{1}{2}(\mathcal{C}_a{}^b + \mathcal{D}^b{}_a) & \mathcal{B}_{(ab)} \end{pmatrix} = \begin{pmatrix} ^{(1)}\mathcal{A}^{ab} & ^{(1)}\mathcal{D}^a{}_b \\ ^{(1)}\mathcal{C}_a{}^b & ^{(1)}\mathcal{B}_{ab} \end{pmatrix},$$

here we used the notation $\not{M}_a{}^b := M_a{}^b - M_c{}^c \delta_a^b/3$, and

$$^{(3)}\chi^{IJ} = \frac{1}{6}(\mathcal{C}_c{}^c + \mathcal{D}^c{}_c) \begin{pmatrix} 0_3 & 1_3 \\ 1_3 & 0_3 \end{pmatrix} = \frac{1}{6}(\mathcal{C}_c{}^c + \mathcal{D}^c{}_c) \epsilon^{IJ}. \tag{A.12}$$

References

[1] A. Gross and G.F. Rubilar, *Phys. Lett.* **A285** (2001) 267. gr-qc/0103016.

[2] F.W. Hehl, Yu.N. Obukhov, *Foundations of Classical Electrodynamics*. Birkhäuser, Boston, MA (2002) to be published.

[3] F.W. Hehl, Yu.N. Obukhov, and G.F. Rubilar, *Spacetime metric from linear electrodynamics II*, talk given at Internat. European Conf. on Gravitation "Journées Relativistes 99", Weimar, Germany, 12-17 Sep 1999. Ann. d. Phys. (Leipzig) **9** (2000) Special issue, SI-71. Guest editors: G. Neugebauer and R. Collier, Jena. gr-qc/9911096.

[4] W.-T. Ni, *A non-metric theory of gravity*, Dept. Physics, Montana State University, Bozeman. Preprint December 1973. The paper is available via http://gravity5.phys.nthu.edu.tw/.

[5] W.-T. Ni, *Phys. Rev. Lett.* **38** (1977) 301.

[6] J.F. Nieves and P.B. Pal, *Phys. Rev.* **D39** (1989) 652.

[7] J.F. Nieves and P.B. Pal, *Am. J. Phys.* **62** (1994) 207.

[8] Yu.N. Obukhov, T. Fukui, and G.F. Rubilar, *Phys. Rev.* **D62** (2000) 044050 (5 pages). gr-qc/0005018

[9] Yu.N. Obukhov and F.W. Hehl, *Phys. Lett.* **B458** (1999) 466. gr-qc/9904067

[10] E.J. Post, *Formal Structure of Electromagnetics – General Covariance and Electromagnetics* (North Holland: Amsterdam, 1962, and Dover: Mineola, New York, 1997).

[11] E.J. Post, *Ann. Phys. (NY)* **71** (1972) 497.

[12] G.F. Rubilar, *Thesis*, University of Cologne (March 2002).

[13] G.F. Rubliar, Yu.N. Obukhov, F.W. Hehl, *Generally covariant Fresnel equation and the emergence of the light cone structure in linear pre-metric electrodynamics*. Preprint University of Cologne (Jan:2002).

A PROPOSAL FOR TESTING THE WEAK EQUIVALENCE PRINCIPLE FOR CHARGED PARTICLES IN SPACE

Hansjoerg Dittus
ZARM, University of Bremen, 28359 Bremen, Germany
dittus@zarm.uni-bremen.de

Claus Laemmerzahl
Institute for Experimental Physics, Heinrich-Heine-University Düsseldorf,
40225 Düsseldorf, Germany
claus.laemmerzahl@uni-duesseldorf.de

Abstract A short review of the experiment of Witteborn and Fairbank aimed at testing the Weak Equivalence Principle for electrons is given. The various gravity induced additional electromagnetic fields are discussed. Since these additional gravity induced fields are not very well under control a spaceborne version of this experiment will reduce these disturbances considerably. The corresponding estimates for these kind of tests in space are presented. As a result, gravity–induced stray field can be indeed reduced considerably, but other effects like patch–effects remain a big problem which has to be solved by better maschining or by accurately measuring out the corresponding stray fields.

Keywords: Equivalence principle, charged particles, antimatter, Schiff–Barnhill effect.

1. Introduction

The Einstein Equivalence Principle is the basis for establishing the gravitational interaction as a metric theory [1]. It consists of the Weak Equivalence Principle (WEP), Local Lorentz Invariance (LLI), and Local Position Invariance (LPI). The WEP states that in a gravitational field all structureless pointlike particles follow the same path. The WEP has been confirmed for neutral bulk matter with an accuracy of 10^{-12} [2] and for quantum matter with an accuracy of 10^{-9} [3]. At least for neutral bulk matter it is planned to increase this accuracy by 6 orders of magnitude with the Satellite Test of Equivalence Principle (STEP) [4, 5]. The main consequence of the WEP is that the gravitational interaction can be geometrized.

LLI states that locally Special Relativity should be valid which has as consequence that at each space–time point one can define a Lorentzian metric. Within a widely used kinematical framework [6, 7], see [8] for a review, the validity of Special Relativity is connected with the isotropy of light propagation, the independence of the velocity of light from the state of motion of the laboratory, and with a specific outcome for Doppler shift experiments. Today, the best tests confirm Special Relativity in terms of modified Mansouri–Sexl-parameters [8] with an accuracy of $|\beta + \delta - \frac{1}{2}| \leq 2 \cdot 10^{-9}$ [9], $|\alpha - \beta - 1| \leq 7 \cdot 10^{-5}$ [10], and $|\alpha| \leq 8 \cdot 10^{-7}$ [11], respectively.

LPI describes that all kinds of matter and non–gravitational interactions couple to gravity in the same way. This can be recast in the term of the universality of the gravitational red shift: Each clock, irrespective of the physical interaction and matter they are built of, is influenced by gravity in the same way. Today, this is confirmed to an accuracy of 10^{-2} to 10^{-4} [12, 13, 10] depending on the type of clocks used in the experiment. At last, LPI is connected with the constancy of physical 'constants' [14, 15]. Very recently, a variation of the fine structure constant has been inferred from the analysis of astrophysical data [16].

Here we are concerned with the WEP for charged particles. If all electromagnetic fields are shielded, then charged particles which contain no inner structure should fall along the same path as neutral particles. Until now, there is only one single experiment which was dedicated to a test of the WEP for charged matter, the Witteborn–Fairbank experiment [17]. The reported accuracy was 0.1 which, compared with neutral matter, is poor. The reason for that was the appearance of stray electric fields and of gravity–induced electric fields of which the Schiff–Barnhill effect [18] is the most popular one. For a thorough survey of experimental constraints on free fall experiments with charged matter, see [19].

There are at least two reasons for considering charged particles: First, tests of the WEP for charged particles search for anomalous couplings between charge and gravity. Second, tests of the WEP for charged particles are a first step in testing the WEP for antimatter.

There are of course more theoretical reasons for a violation of the WEP for antimatter than just for charged particles. There are theories that predict anomalous behavior of antimatter. Models of quantum gravity usually lead to additional gravitational spin–0 and spin–1 fields which lead to additional long–range gravitational forces if they have small masses. While for ordinary matter in the gravitational field of the Earth the vector gravitational field of the Earth is attractive, it is repulsive for anti–particles [20]. Thus, anti–matter behaves differently than matter in generalized theories of gravity.

There are mainly three arguments against violations of the WEP for antimatter, see e.g. Nieto and Goldman [22]: (i) The CPT–theorem predicts that the mass of antiparticles is the same as that for particles, (ii) the $K - \overline{K}$-system shows that the anti–K, the $\overline{K}$, behaves in the same way as the K, and (iii) the quantum field theoretical argument that, due to vacuum fluctuations, inside bulk matter there are so many antiparticles present which contribute to the total bulk mass so that a verification of the WEP for bulk mass immediately also applies to antimatter (for a different way of indirect reasoning, see [21]).

These arguments are quite strong but not totally conclusive. In order to be true, in each case one always has to make assumptions: (i) the CPT theorem has been shown to be true in flat space only. In principle it is an open question whether the CPT invariance is valid in the presence of gravitation. Furthermore, it only implies that the inertial masses of particles and antiparticles are the same. It makes no statement about the equality of the inertial and gravitational mass of the antiparticle. (ii) in order to describe the $K - \overline{K}$-system one has to assume some theoretical model. Also for (iii) one has to assume the validity of the whole quantum field theoretical formalism.

Since there are still loopholes in arguments against a violation of the WEP for antimatter, this should be tested by experiments in any case. And a first step in testing the WEP for antimatter is to perform tests for charged matter.

For completeness we like to mention that there is an additional theoretical problem concerning the WEP for charged matter: Accelerated charges are subject to a back reaction of the field of the accelerated charge to its motion. This additional acceleration is $\sim \frac{2}{3}(e^2/c^3)\dddot{x}(t)$. The question is whether a charged particle attached to a fixed position on the surface of the Earth, or a particle freely falling down radiates. This problem has been controversially discussed, see e.g. [23, 24, 25]. In any case, since this effect of relativistic order, it is too small to be of experimental importance and can be safely neglected.

In the following we want to review the only experiment devoted to test the WEP for charged matter, describe its functionality and its problems and discuss whether it makes sense to do this kind of experiment in space. Our result is, that while gravity–induced errors will indeed considerably be reduced there are still some not yet understood errors due to patch effects which limit the accuracy.

2. An Eötvös coefficient for charged particles

Though it will not influence the description of the experiment, we want to define a simple frame and introduce notions by means of which it should be possible to describe tests of the WEP for charged particles. The anticipated result is that the Eötvös coefficient should split into a charge–independent part describing a violation of the WEP for neutral particles or due to the mass only, and a part which scales with the charge-to-mass ratio. Any potential charged–induced violation of the WEP should be larger the bigger the charge-to-mass ratio is.

Since a charged particle possesses an electromagnetic field, the total energy of a charged particle consists of its bare rest mass and the energy of the electromagnetic field. In general, this electromagnetic energy may contribute to a violation of the WEP in a different way than the bare mass. Therefore, we split the inertial mass of a particle into a bare part and a part depending on the electric charge:

$$m_i = m_i^0 \left(1 + \kappa_i \frac{q}{m_i^0}\right). \tag{1}$$

The same will be done with the gravitational mass since in general it might be possible that the electromagnetic energy reacts to the gravitational attraction in a different way than the 'bare' gravitational mass

$$m_g = m_g^0 \left(1 + \kappa_g \frac{q}{m_g^0}\right). \tag{2}$$

In a special case, this point of view is supported by the $TH\epsilon\mu$-formalism. In this formalism it has been shown [1, 26] that the acceleration of a bound system of charged particles in general violates the WEP. The reason for that is that the electromagnetic binding energy depends on the position and on the state of motion. This violation can be related to an Eötvos–coefficient which, of course, is proportional to the charge of the particles.

If we take these modified inertial and gravitational masses, then the equation of motion for a charged particle in a gravitational field (all electromagnetic fields are shielded) is

$$m_i \ddot{x} = m_g \nabla U \tag{3}$$

where U is the Newtonian gravitational potential. The Eötvös coefficient describing the difference in the acceleration of two particles then turns out to be

$$\eta = 2\frac{\ddot{x}_2 - \ddot{x}_1}{\ddot{x}_1 + \ddot{x}_2} \approx \eta^0 + (\kappa_{g2} - \kappa_{i2})\frac{q_2}{m_2^0} - (\kappa_{g1} - \kappa_{i1})\frac{q_1}{m_1^0} \tag{4}$$

where we inserted (3) and the charge–dependent ansatz for m_i and m_g, used the usual Eötvös–coefficient

$$\eta^0 = 2\frac{(m_{g2}^0/m_{i2}^0) - (m_{g1}^0/m_{i1}^0)}{(m_{g2}^0/m_{i2}^0) + (m_{g1}^0/m_{i1}^0)} \tag{5}$$

for the 'bare' masses, made an expansion to first order in the κ's, and neglected combined violations due to the 'bare' masses and due to charges. Most important is, of course, the case that one of the two particles is neutral. Then we end up with the modified Eötvös coefficient

$$\eta = \eta^0 + \kappa \frac{q_2}{m_2^0}, \tag{6}$$

where $\delta\kappa = \kappa_{g2} - \kappa_{i2}$. A charge–induced violation of the WEP is encoded in the parameter $\delta\kappa$. Only if the electromagnetic energy of the charged particles contributes in the same way to the inertial and the gravitational mass (which means $\kappa_g = \kappa_i$), then there will be no charge–induced violation of the WEP. To first order of approximation, charge–induced violations of WEP are independent of any 'bare' violation of the WEP described by η^0.

Consequently, a comparison of a charged with a neutral or a second charged particle gives an estimate on the total η. Only if we can change the charge of the body and thus the charge–to–mass ratio then we can also make statements about the two contributions η^0 and $\kappa_g - \kappa_i$. For a single charged particle like the electron it is possible that η^0 and $\delta\kappa\frac{q}{m}$ compensate. If we use atoms and ionize them to different degrees, then we can get information about both parts of η.

It is also obvious that for smaller charge–to–mass ratios any accuracy for $\delta\kappa$ will become worse. In the case of electrons, which have been used in the Witteborn–Fairbank experiment, we have $e/m = 1.7 \cdot 10^{11}$ C/kg. Then $|\eta^0 + \delta\kappa \cdot 1.7 \cdot 10^{11} \text{ C/kg}| \leq 0.1$. Assuming no cancellation of the two terms, we can conclude $|\eta^0| \leq 0.1$ and $|\delta\kappa| \leq 6 \cdot 10^{-13}$ kg/C. If, instead, we take protons with a mass approximately 2,000 times larger than the mass of electrons, then the estimate on $\delta\kappa$ is 2,000 times worse.

3. The Witteborn–Fairbank setup

To our knowledge, only one experiment has been carried out in the past to test the WEP for charged matter so far. In 1967, Witteborn and Fairbank [17] measured the net force on electrons freely falling in a copper tube. The experimental set–up consisted of a vacuum tank cooled down to liquid helium temperature of 4.2 K as well as a vertical copper tube (drift tube) inside the Dewar vessel to shield stray electrical

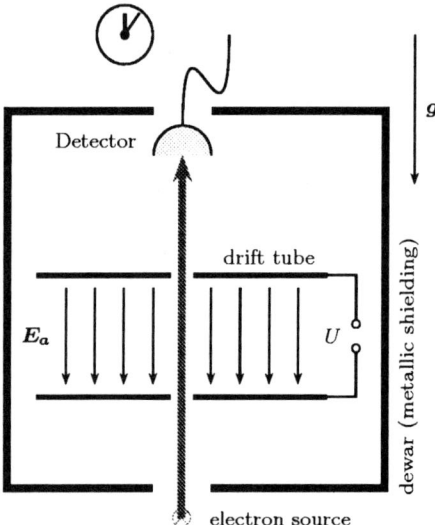

Fig. 1. Schematic view of the Witteborn–Fairbank set-up. Electrons move against gravity inside a metallic shielding. An additional tunable electric field in a drift tube allows to determine the maximum time of flight.

fields. The copper tube had a length of about 1 m and a diameter of 5 ± 0.0003 cm. Electrons moved along the drift tube's symmetry axis and had been forced to do so by a magnetic field of a coaxial solenoid. A cathode at the bottom emitted the electrons and accelerated them upwards. Electrons having passed the drift tube have been detected by an electron multiplier detector (Fig.1). The electrons are emitted as short bursts, so that a mean time–of–flight between the emission from the cathode and the detection can be measured.

First we analyze this experiment in the case that only gravity would act. Furthermore, our test particle is an arbitrary particle with inertial and gravitational masses m_i and m_g and charge q. Then we have the condition of energy conservation $\mathcal{E} = \frac{1}{2}m_i v^2 - m_g g x = const.$ and the equation of motion $m_g \ddot{x} = m_g g$ which has the general solution $x = \frac{1}{2}\frac{m_g}{m_i} g t^2 + v_0 t + x_0$ where v_0 and x_0 are the initial velocity and position, respectively.

There is a maximum time of flight t_{max} which is given by the condition that the electrons arrive the detector with zero velocity. This is characterized by $\mathcal{E} = \frac{1}{2}m_i v_0^2$ and $x - x_0 = h$ for $t = t_{max}$ where h is the height of the drift tube. Solving these equations for t_{max} yields

$$t_{max} = \sqrt{\frac{m_i}{m_g}\frac{2h}{g}}. \tag{7}$$

For known h and measured t_{max} we can determine $\frac{m_g}{m_i} g$. Here g is the gravitational acceleration of the test particle as seen in the frame of the metallic shield (what in the present case is the same as the rest frame of the Earth). A comparison of the same quantity with another kind of particle or with neutral matter gives the corresponding Eötvös–coefficients.

However, the purely gravitational case is not realized in the case of the Witteborn–Fairbank set–up. There are at least three additional electric fields present which are connected with the existing gravitational environment and with the measurement process:

- The Schiff–Barnhill field which is unavoidable on Earth. Schiff and Barnhill [18] calculated (see below) that electrons bound in the metallic shield of the drift tube create an electric field $E_{SB} = m_{eg}g/e \sim 5 \cdot 10^{-11}$V/m which, for electrons, balances the gravitational field exactly. Here the g is the gravitational acceleration experienced by the electrons in the metallic shield.

- The so–called DMRT field, introduced by and named after Dessler, Michel, Rorschach and Trammel [27], which comes from the differential compression of the metallic shielding. If the surfaces of this shielding are aligned to the gravitational acceleration, also this field is proportional to the gravitational acceleration: $E_{DMRT} = \gamma \frac{m_g^{atom}}{e} g$, where m_g^{atom} is the gravitational atomic mass of the material the drift tube is made from. There is still some debate about the value of γ for various substances. Witteborn and Fairbank who carried through their experiment at 4.2° K, experienced an only later recognized vanishing of this extra field. Later experiments showed that the DMRT field at least in that particular set–up vanished for temperatures below 4.5° K [28].

- Another external uniform electrical field $E_a < 2.5 \cdot 10^{-10}$V/m directed parallel to the drift tube's symmetry axis is applied in order to decelerate the electrons. With this additional field, the electrons experience an additional acceleration (or deceleration) which leads to a variation of the time of flight of the electrons. This is necessary in order to be able to determine the maximum time of flight t_{max}.

With these additional fields Eq.(7) must be modified: The observed flight times of charged particles of mass m and charge q now is

$$t_{max} = \sqrt{\frac{2h}{\frac{m_g}{m_i}g - \frac{q}{m_i}(E_{SB} + E_{DMRT} - E_a)}}. \qquad (8)$$

By varying E_a and measuring t_{max} one can determine the difference $\frac{m_g}{m_i}g - \frac{q}{m_i}(E_{SB} + E_{DMRT})$ for electrons.

If we insert the general value of the Schiff–Barnhill field $E_{\text{SB}} = m_{eg}g/e$ and of the DMRT–field, $E_{\text{DMRT}} = \gamma \frac{m_g^{\text{atom}}}{e} g$, then we get for the maximum time of flight

$$t_{\max} = \sqrt{\frac{2h}{\frac{m_g - \frac{q}{e}(m_{eg} + \gamma m_g^{\text{atom}})}{m_i} g + \frac{q}{m_i} E_a}}. \qquad (9)$$

By varying the external electric field E_a, the quantity

$$\frac{m_g - \frac{q}{e}(m_{eg} + \gamma m_g^{\text{atom}})}{m_i} g \qquad (10)$$

can be measured which contains information about a modified ratio of the gravitational to inertial mass, only.

For elementary particles with the charge of the electron, $q = \pm e$ we get

$$t_{\max} = \sqrt{\frac{2h}{\frac{m_g \mp (m_{eg} + \gamma m_g^{\text{atom}})}{m_i} g + \frac{q}{m_i} E_a}}. \qquad (11)$$

If the charge of the particle has the same (opposite) sign than the electron charge, like protons, positron, or antiprotons, only the difference (sum) of the gravitational mass of the particle and the electron can be tested. For positrons the term in question is $2 \frac{m_{eg} + m_{\bar{e}g} + \gamma m_g^{\text{atom}}}{m_{\bar{e}i}} g$.

For electrons we have $q = e$ and $m = m_e$ so that the first part, and thus the gravitational mass of the electron, completely disappears. Therefore, for electrons it is not possible to make statements about a relation between the inertial and gravitational mass. This is only possible for particles different from electrons.

However, the statement is valid only if the Schiff–Barnhill field indeed is of the form $E_{\text{SB}} = m_{eg}g/e$, that is, if the gravitational acceleration of the electrons inside the metallic tube g_{bulk} is the same as for free electrons g_{free}. If we distinguish between these two accelerations by setting $g_{\text{bulk}} \neq g_{\text{free}}$, then we get from (8)

$$t_{\max} = \sqrt{\frac{2h}{\frac{m_{eg}}{m_{ei}}(g_{\text{free}} - g_{\text{bulk}}) + \frac{e}{m_{ei}}(E_{\text{DMRT}} + E_a)}}. \qquad (12)$$

In the Witteborn–Fairbank experiment there happen to be a fortuitous and temperature depended absence of the DMRT–effect, resulting in $\gamma \approx 0$. By varying E_a and measuring $t_{\max}$ the Witteborn–Fairbank experiment aimed to measure the possible difference $\frac{m_{eg}}{m_{ei}}(g_{\text{free}} - g_{\text{bulk}})$ for

Fig. 2. The Schiff–Barnhill effect. An external gravitational field acting on the electrons in the metal changes the boundary condition thus inducing an electric field inside the metal.

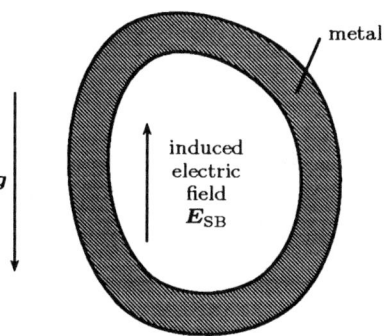

electrons. This value, the subtraction of two large numbers, is extremely small and has been observed to be:

$$\frac{m_{eg}}{m_{ei}}(g_{\text{free}} - g_{\text{bulk}}) \leq (0.13 \pm 0.47) \cdot 10^{-11} \text{eV/m}, \quad (13)$$

setting an upper limit for the Eötvös coefficient for charged matter in the sense of $\eta = 2(g_{\text{free}} - g_{\text{bulk}})/(g_{\text{free}} + g_{\text{bulk}})$ to

$$\eta < 0.1, \quad (14)$$

where one has to insert the calculated value for the Schiff–Barnhill field.

Our analysis shows that Witteborn and Fairbank did not really measure the WEP for charged particles but instead the equality of accelerations of bulk and free electrons so that the significance of the result is rather limited. Tests of the WEP with this device can only be performed for particles *different* from electrons.

4. Side effects in WEP experiments for charged particles

We classify the errors affecting the Witteborn-Fairbank experiment: (1) gravity induced errors which scale with the acceleration g of the apparatus, and (2) all other errors.

4.1. Gravity induced side–effects

The Schiff–Barnhill effect. The Schiff–Barnhill effect [18] was the first unwanted side–effect calculated in order to analyze the Witteborn–Fairbank set–up. It can be understood very easily in terms of solving the Poisson equation with metallic boundary condition in an external homogeneous gravitational field.

A metal is characterized by the mobility of electrons. If there is an external gravitational field, then the electrons try to react and move downwards. However, the excess of electrons at the down–side of the metal creates a repulsive force with the result that both forces in equilibrium should cancel,

$$0 = -e\boldsymbol{E} + m_{eg}\boldsymbol{g}. \tag{15}$$

This relation holds inside the whole metal. Consequently, since $\boldsymbol{E} = -\boldsymbol{\nabla}\phi$, the electrostatic potential ϕ inside the metal is given by $\phi = -\frac{m_{eg}}{e}\boldsymbol{g}\cdot\boldsymbol{x}$. This is also the potential at the (inner and outer) boundary of the metal.

The electric potential inside a hole in the metal, see Fig.2, is given by the Dirichlet problem $\Delta\phi = 0$ with $\phi|_{\text{boundary}} = -\frac{m_{eg}}{e}\boldsymbol{g}\cdot\boldsymbol{x}$. The unique solution to this problem is

$$\phi = -\frac{m_{eg}}{e}\boldsymbol{g}\cdot\boldsymbol{x}, \tag{16}$$

and the corresponding electric field is

$$\boldsymbol{E}_{\text{SB}} = -\boldsymbol{\nabla}\phi = \frac{m_{eg}}{e}\boldsymbol{g} \tag{17}$$

whith $E_{\text{SB}} \sim 6\cdot 10^{-11}$ V/m on Earth. The induced electric field outside the shield can be calculated for a spherical shape of radius R, as the dipole moment $\boldsymbol{d} = \frac{m_{eg}}{e}\boldsymbol{g}R^3$.

Consequently, a charged particle with charge q and mass m_g experiences the total force

$$\boldsymbol{F} = \boldsymbol{F}_{\text{grav}} + \boldsymbol{F}_{\text{SB}} = q\left(\frac{m_g}{q} + \frac{m_{eg}}{e}\right)\boldsymbol{g} = m_{eg}\left(\frac{m_g}{m_{eg}} + \frac{q}{e}\right)\boldsymbol{g}. \tag{18}$$

If this particle is an electron, $q = -e$, then the resulting force vanishes – the electron stays within the metallic hole without falling down, even if the WEP is not valid, that is, even for $m_{ei} \neq m_{eg}$.

The DMRT–field. The DMRT–field [27] is a dominating error source. It results from the differential compression of the crystal layers of the metal of the drift tube in the gravitational field. In the lower part of the metal the crystal layers have smaller distances than in the upper parts which leads to more positive charges (the atomic cores) per unit volume in the lower than in the upper part.

The value of an electric field induced by lattice deformation can be calculated from $\boldsymbol{E} = \frac{1}{e}\boldsymbol{\nabla}W$ where W is the work function of the metal. One gets as result [19]

$$\boldsymbol{E} \approx \frac{2n_0}{9K}\epsilon_F \frac{m_g^{\text{atom}}}{e}\boldsymbol{g}, \tag{19}$$

where m_g^{atom} is the (gravitational) atomic mass of the metallic solid, n_0 the density of the atoms in the shield without gravitational deformation,

K the bulk modulus of compressibility, and ϵ_F the Fermi energy of the free electron gas in the metal. Eq.(19) is the so-called DMRT–field [27].

As expected, the DMRT–field scales with the gravitational acceleration g. Therefore, in principle, one can take the parameters of the metallic shield, insert them into this equation and add this electric field in the equation of motion. However, the value for the quantity $\gamma = \frac{2n_0}{9K}\epsilon_F$ is still in debate. It is expected that under normal conditions this field is up to $2 \cdot 10^4$ larger than the Schiff–Barnhill field. Some other estimates give values which are one order smaller, and other even give an opposite sign.

One way out of this problem may be to measure the DMRT–field for varying accelerations, e.g. on a turntable with varying angular velocities or in free fall (in a drop tower), and use this in order to determine the coefficient γ.

In the experiment carried through by Lockhart, Witteborn and Fairbank [28] they observed a vanishing DMRT field below 4.5 K. There seems to exist up to now no explanation for this behavior of the DMRT–field at low temperatures.

4.2. Other side effects and errors

Beside the disturbances by gravity induced electric fields there are many other influences which have to be discussed in detail for a careful evaluation of the Witteborn–Fairbank experiment. Some of the possible error sources have been already discussed by the experimenters themselves [17], but the most comprehensive analyses can be found in [19]. Systematic error and random noise sources are listed in the following.

Patch effects. Serious effects are caused by inhomogeneities of the surface of the shielding drift tube. Patch effects are different electric potentials at metallic surfaces due to crystal facets or surface contaminations and can be divided into two parts [19]: (i) spatially random electric fields along the drift tube axis required to be not greater than 0.01 nV for electrons and 0.01μV for protons and (ii) axial potential variations which can be modelled by averaging the variations around the tube and convoluting the average with a smoothing function. The analysis shows that, in particular, the latter fields should go up to about 0.3μV for the Witteborn–Fairbank experiment for a typical characteristic length scale for the patches of about 1μm and the tube radius of 5 cm. These patch effects should be much too big for the result claimed by Witteborn and Fairbank. Darling et al. [19] have given two explanations why these disturbances might not have played a dominant role in the experiment: First, exposure to air would have produced an amorphous surface layer masking underlying crystal structures in the copper surface, and second, the shielding appearing at cryogenic temperatures observed by Lockart et al. [28] could also have caused a reduction of the patch effects. Although these explanations might be unsatisfying, one should consider that up–to–date machining techniques, able to produce very clean and absolutely flat surfaces, should reduce the patch effects down to a negligible level. Nevertheless, patch effects are also related to misalignment of the tube walls.

Electric field gradients. Electric field gradients are only affecting polarizable systems like ions and are therefore negligible for electrons. Also, the error calculated for ions is only $F_{\mathrm{grad}}/F_{\mathrm{grav}} \leq 3 \cdot 10^{-9}$ [19].

Thermoelectric fields. Additional disturbing electric fields can also be produced by temperature gradients in the shielding drift tube. The thermoelectric field is given by [19]

$$\boldsymbol{E}_T = \left(\frac{1}{e}\frac{\partial W}{\partial T} + S\right)\boldsymbol{\nabla} T, \qquad (20)$$

where W is the work function of the metal. S is the thermoelectric coefficient which is for copper at liquid helium temperature not greater than 0.05 μV/K. The experimental requirements for $\boldsymbol{\nabla} T \ll 10^{-4}$K/m for electrons and 10^{-1}K/m for protons should be satisfied within the Dewar vessel despite the thermal isolation of the vacuum inside the drift tube. It is interesting that residual helium gas producing a monolayer on the copper drift tube could produce a strong temperature dependence near helium boiling temperature and could explain the unexpected temperature dependent shielding observed by Lockart et al. [28] due to helium desorption above helium boiling temperature [19].

Magnetic fields and field gradients. In the Witteborn–Fairbank experiment guide magnetic fields were needed in order to keep the electrons on the way to the detector. It has been shown that the guide solenoid can be aligned precisely enough with the symmetry axis of the drift tube. A measure of this estimate might be the ratio between the maximum horizontal displacement of the free falling particle from the axis and the length of the drift tube. This ratio has to be at least as small as 1 part in 10,000 for electrons and requires a homogeneity ratio of the horizontal disturbing magnetic field to the aligned vertical field of the solenoid at least one order of magnitude smaller. The analysis showed that this requirement has been fulfilled for the Witteborn–Fairbank experiment.

The spin of electrons or protons couples to magnetic fields and causes deviations of the vertical path by gyromagnetic forces. The deviation from the vertical axis F_z/F_{grav}, is to first order proportional to the magnetic field gradient $\Gamma_z = \partial B_z/\partial z$ ($F_{\text{grav}} = m_g g$ is the gravitational force.). The ratio can be written:

$$F_z/F_{\text{grav}} = \frac{\mu}{B_z}\Gamma_z, \qquad (21)$$

where μ is the effective magnetic moment of the test particle depending on the particle mass, its charge, and its quantum state. In the Witteborn–Fairbank experiment μ is a very small value; because of experimental reasons [17] all freely falling electrons are in the ground state, and the influence by magnetic field gradients can be neglected [19]. By carrying out the same experiment with protons or antiprotons μ has to be averaged over all states for a given temperature. Therefore, eq.(21) must be modified [19]:

$$F_z/F_{\text{grav}} = \frac{kT}{m_g g}\frac{1}{B_z}\Gamma_z, \qquad (22)$$

which gives reason to cool down the protons to at least 10 K.

Radiation pressure. The influence of radiation pressure on the free fall behavior of the test particle can be estimated by a photon scattering analysis. It can be shown that even for room temperature and for electrons (with a much larger cross section than protons), radiation pressure is a negligible effect.

Gas scattering in the tube. Residual gas in the drift tube could cause a serious systematic error due to scattering on helium atoms. Darling et al. [19] calculated the influence for varying residual gas pressure in the tube. The sufficient vacuum conditions inside the tube is determined by the time between two scatterings (τ) which should be greater than the maximum flight time $t_{\max}$ of Eq. (11):

$$\tau = \frac{kT}{p\sigma(v_{\text{test particle}})}v_{\text{helium}} < t_{\max}, \qquad (23)$$

where p is the effective pressure at temperature T, k is the Boltzmann constant, and σ is the cross section of the test particle. This value is small enough at a pressure of about 10^{-10} Pa (attained in the Witteborn–Fairbank experiment) for electrons with drift velocities $v_{\text{test particle}}$ of the order of m/s compared to the drift velocities of helium atoms $v_{\text{helium}} = 140$ m/s. To carry out the experiment with protons or antiprotons an ultrahigh vacuum of at least 10^{-12} Pa has to be attained.

5. Experiments in space
5.1. The idealized case

As we have seen above, the main error in the Witteborn–Fairbank experiment is the DMRT–effect (here we neglect all non–gravity–induced errors). The strength of this effect, unfortunately, cannot be predicted accurately enough by theory. However, since this effect, and also the Schiff–Barnhill effect, scales with the external gravitational field, a first idea is to perform such experiments in space under microgravity conditions. This idea, however, needs some more analysis of the Witteborn–Fairbank experiment.

The various accelerations appearing in $t_{\max}$, Eq.(8), and in the Schiff–Barnhill– and DMRT–fields have to be carefully distinguished, though they are the same on Earth. A complicating circumstance is that we have three kinds of particles involved, the charged test particle, the ions of the metallic shield and the electrons inside this shield, and for each pair of these particles we may have a violation of the WEP. In addition, the Schiff–Barnhill field and the DMRT field are not given absolutely but depend on relative motions of these various particles.

To be more precise: The accelerations $\frac{m_g}{m_i}g$, $\frac{m_{eg}}{m_{ei}}g$, and $\frac{m_g^{\text{atom}}}{m_i^{\text{atom}}}g$ are the accelerations of the charged test particle, of the electrons, and of the atomic ions of the shield, respectively, as described in the rest frame of the gravitating body (g is the gradient of the Newtonian potential $U = GM/r$). In order to calculate the Schiff–Barnhill effect in space which may result from a violation of the WEP between electrons and atoms of the metallic shield, we consider the forces and the equations of motion of the electrons and the atoms of the metallic shield. The forces on the atoms and electrons in the reference system of the Earth are given by

$$F_{\text{atom}} = m_i^{\text{atom}} \ddot{x}_{\text{atom}} = m_g^{\text{atom}} g \qquad (24)$$
$$F_e = m_{ei}\ddot{x}_e = m_{eg}g. \qquad (25)$$

The question now is: what is the force $m_{ei}\ddot{x}'$ of the electron in the rest system of the atoms? We get this force if we perform a coordinate transformation to the frame which accelerates with $\ddot{x}_{\text{atom}}$ with respect to the Earth's system. That is, we look for for new coordinates $x' = f(x)$ so that $\ddot{x}'_{\text{atom}} = 0$.

Such a transformation is given by $x' = x - \frac{1}{2}\ddot{x}_{\text{atom}}t^2$ (the acceleration, of course, is constant). Then we get for the force on the electron

$$\begin{aligned} m_{ei}\ddot{x}'_e &= m_{ei}\frac{d^2}{dt^2}\left(x_e - \frac{1}{2}\ddot{x}_{\text{atom}}t^2\right) \\ &= m_{ei}\ddot{x}_e - m_{ei}\ddot{x}_{\text{atom}} \\ &= \left(m_{eg} - m_{ei}\frac{m_g^{\text{atom}}}{m_i^{\text{atom}}}\right)g\,. \end{aligned} \qquad (26)$$

This force on the electron with respect to the electric shield has to be balanced by an electric field $F'_e = -eE'_{\text{SB}}$ in the rest frame of the atoms. This electric field is then given by

$$eE'_{\text{SB}} = -\left(m_{eg} - m_{ei}\frac{m_g^{\text{atom}}}{m_i^{\text{atom}}}\right)g \qquad (27)$$

This Schiff–Barnhill field has to be used in Eq (8) (in the rest frame of the atoms):

$$t_{\max} = \sqrt{\frac{m_i h}{m_i \ddot{x}' - q\left(E'_{\text{SB}} - E_a\right)}}, \qquad (28)$$

where m_i and m_g are inertial and gravitational mass of the test particle, $m_g \ddot{x}'$ is the force on the test body and $\ddot{x}'$ its acceleration in the rest frame of the atom, that is, with respect to the atoms of the electric shield. We insert the Schiff–Barnhill field and get

$$t_{\max} = \sqrt{\frac{m_i h}{m_i \ddot{x}' - q\left(-\frac{1}{e}\left(m_{eg} - m_{ei}\frac{m_g^{\text{atom}}}{m_i^{\text{atom}}}\right)g - E_a\right)}}. \qquad (29)$$

We express x' with respect to the Earth system, $\ddot{x}' = \ddot{x} - \ddot{x}_{\text{atom}}$ and get

$$t_{\max} = \sqrt{\frac{h}{\left(\frac{m_g}{m_i} - \frac{m_g^{\text{atom}}}{m_i^{\text{atom}}} + \frac{q}{e}\frac{m_{ei}}{m_i}\left(\frac{m_{eg}}{m_{ei}} - \frac{m_g^{\text{atom}}}{m_i^{\text{atom}}}\right)\right)g + \frac{q}{m_i}E_a}}, \qquad (30)$$

where we also used (24) and (25). Therefore, the measured quantity

$$a_{\text{eff}} = \left(\frac{m_g}{m_i} - \frac{m_g^{\text{atom}}}{m_i^{\text{atom}}} + \frac{q}{e}\frac{m_{ei}}{m_i}\left(\frac{m_{eg}}{m_{ei}} - \frac{m_g^{\text{atom}}}{m_i^{\text{atom}}}\right)\right)g \qquad (31)$$

consists of the difference of the ratios of gravitational and inertial masses of the test particles and the atoms (which is an Eötvös coefficient we are looking for) with an additional term consisting of the difference of the gravitational and inertial masses of the electrons and the atoms weighted with the ratio of their charge-to-mass ratios.

In the case that we take electrons as test particles, then $q = -e$ and $m = m_e$, so that we have

$$t_{\max} = \sqrt{\frac{h}{\left(\frac{m_{eg}}{m_{ei}} - \frac{m_g^{\text{atom}}}{m_i^{\text{atom}}} - \left(\frac{m_{eg}}{m_{ei}} - \frac{m_g^{\text{atom}}}{m_i^{\text{atom}}}\right)\right)g + \frac{q}{m_i}E_a}} = \sqrt{\frac{h}{\frac{q}{m_i}E_a}}. \quad (32)$$

As in the corresponding experiment on Earth, no WEP violation can be observed for electrons, even if electrons violate the WEP.

The problem now is how to get, in the general case, that is, for arbitrary test charged bodies, an Eötvös coefficient from expression (31). In principle one may think of two possibilities: (i) to compare the accelerations of two different test particles, and (ii) to compare the test particle with the the electric shield. In the first case we get from (31) for two test particles

$$a_{2\text{eff}} - a_{1\text{eff}} = \left(\frac{m_g^{(2)}}{m_i^{(2)}} - \frac{m_g^{(1)}}{m_i^{(1)}}\right)$$

$$+ \frac{m_{eg}}{e}\left(\frac{q^{(1)}}{m_i^{(1)}} - \frac{q^{(2)}}{m_i^{(2)}}\right)\left(\frac{m_g^{\text{atom}}}{m_i^{\text{atom}}} - \frac{m_{eg}}{m_{ei}}\right)\right) g$$

$$= \left(\eta_{2-1} + \frac{m_{eg}}{e}\left(\frac{q^{(1)}}{m_i^{(1)}} - \frac{q^{(2)}}{m_i^{(2)}}\right)\eta_{\text{atom}-e}\right) g, \quad (33)$$

where we used the Eötvös coefficient $\eta_{2-1} \approx \frac{m_g^{(2)}}{m_i^{(2)}} - \frac{m_g^{(1)}}{m_i^{(1)}}$ and, similarly, for atoms and electrons. That means that the Eötvös coefficient η_{2-1} we are looking for not only depends on the measured difference of accelerations but also on an additional term which will vanish only if there is no WEP violation with respect to electrons and the atoms of the metallic shield or if the charge-to-mass ratios of the two charged test particles are equal. However, if one measures a difference of the two effective accelerations, then there *must* be a violation of the WEP, either between the two test masses or between the electrons and the metallic atoms.

For the second case we have to try to extract from the single equation (31) information about $\delta - \delta_{\text{atom}}$ where $\frac{m_g}{m_i} = 1 + \delta$. We get

$$a_{\text{eff}} = \left(\delta - \delta_{\text{atom}} - \frac{q}{e}\frac{m_{eg}}{m_i}(\delta_{\text{atom}} - \delta_e)\right) g. \quad (34)$$

Though this again is no relation from which one can relate uniquely the searched quantity $\delta - \delta_{\text{atom}}$ with observed quantities, we can, within a small error, conclude that $a_{\text{eff}} \approx (\delta - \delta_{\text{atom}}) g$. This is possible because the last term is very small: the last term consists, as the first term, of a WEP violating term, but for protons as test particle this WEP violating term is multiplied with a small factor of the order $0.5 \cdot 10^{-3}$. Therefore we can neglect the second term. Consequently, in a first approximation, by measuring a_{eff} for one kind of charged particle (not an electron) one can test the validity of the WEP for this particle in comparison to the behavior of the metallic shield.

To sum up: An experiment in space with the Witteborn–Fairbank set–up is well suited to test the WEP for charged matter. The measured quantity directly gives to first order the Eötvös coefficient for the test particle and the electromagnetic shielding.

5.2. Errors from microgravity conditions

Of course, even in space there are disturbances, e.g. small accelerations δa, around. On the ISS, for example, these disturbances might be large as $\delta a \leq 10^{-5} g$ and on a satellite we may have at best $\delta a \leq 10^{-13}$. These residual accelerations will lead to "residual" Schiff–Barnhill– and DMRT–fields. We then get a "microgravity-modified" Eqn.(28)

$$t_{\max} = \sqrt{\frac{2h}{\frac{m_g}{m_i}\ddot{x}' - \frac{q}{m_i}(\delta E'_{\text{SB}} + \delta E'_{\text{DMRT}} - E_a)}} = \qquad (35)$$

$$\sqrt{\frac{2h}{\left(\frac{m_g}{m_i} - \frac{m_g^{\text{atom}}}{m_i^{\text{atom}}} + \frac{q}{e}\frac{m_{ei}}{m_i}\left(\frac{m_{eg}}{m_{ei}} - \frac{m_g^{\text{atom}}}{m_i^{\text{atom}}}\right)\right)g - \frac{q}{e}\frac{m_{eg} + \gamma m_g^{\text{atom}}}{m_i}\delta a + \frac{q}{m_i}E_a}}.$$

where δa is the microgravity acceleration. The quantity $\frac{q}{e}\frac{m_{eg}+\gamma m_g^{\text{atom}}}{m_i}\delta a$ is the microgravity–induced error in the first term of the denominator one aims to measure. Thus,

$$\delta\eta = \frac{q}{e}\frac{m_{eg} + \gamma m_g^{\text{atom}}}{m_i}\frac{\delta a}{g} \qquad (36)$$

is the microgravity induced error in the Eötvös–coefficient. For electrons, positrons, protons, and anti–protons $|q/e| = 1$. If we assume $\delta a \sim 10^{-5} g$, then we have as error for these particles

$$|\delta\eta| = \frac{m_{eg} + \gamma m_g^{\text{atom}}}{m_i} 10^{-5} \qquad (37)$$

where m_i is the inertial mass of the test particle. For a larger m_i the estimate on $\delta\eta$ will become better, but for $\delta\kappa$ it will be worse. Assuming

$\gamma \sim 0.5 \cdot 10^{-6}$, $m_g^{\text{atom}} \sim 10^2 \, m_{pg} \sim 2 \cdot 10^5 \, m_{eg}$ this gives

$$|\delta\eta| = \frac{1.1 \, m_{eg}}{m_i} 10^{-5} \tag{38}$$

(For $\delta a = g$ (as on Earth) and for electrons this gives $\delta\eta \sim 0.1$, as in the Witteborn–Fairbank experiment.) Therefore we get for electron or positrons $|\delta\eta| \sim 10^{-5}$, and for protons or antiprotons $|\delta\eta| \sim 0.5 \cdot 10^{-9}$, — as far as the acceleration–induced errors are concerned.

In order to transcribe this error on η to an error on κ, we use $\delta\eta = \delta\eta^0 + \delta\kappa \frac{q}{m}$. If we assume, for simplicity, $\delta\eta^0 = 0$, then $\delta\kappa = \delta\eta \frac{m}{q}$. The the above estimates on η for electrons/positrons and protons/antiprotons translate to the estimates $|\delta\kappa| \sim 0.5 \cdot 10^{-16}$ kg/C and $|\delta\kappa| \sim 10^{-13}$ kg/C.

5.3. Total estimate of errors in space–born Witteborn–Fairbank experiment

While the gravity–induced errors seem to be reduced considerably when going to space, the other errors play the same role as they do when performing such experiments on Earth. While some errors are really negligible and may not contribute to the gravity induced errors, the patch effect is still not really under control. Even if we assume that due to still unknown mechanisms, the patch effects (as the DMRT–field) can be depressed by cooling the system to cryogenic temperatures, the patch effects will be the most limiting factor in increasing the accuracy for such kind of WEP tests for charged matter. Therefore, going to space just probably will solve one problem, namely the errors due to the DMRT–field, but leaves the other big problem, the patch effects.

6. Outlook

Since in all conceivable set–ups for tests of the WEP for charged matter one needs an electromagnetic shield, the same analysis as above applies to all these tests. Such experiments are ion interferometry or charged particles in a trap. In each case, the induced errors which are mainly due to the physics of the electromagnetic shielding add up to the same error for each of these experiments in the measured Eötvös coefficient.

However, even if it is too difficult to gain knew knowledge about the patch effects, there is one way out: to improve the measurements of such systems in order to have enough data which allows an elimination of these effects. Another way out of this problem might be to use a further particle with a different charge–to–mass ratio in order to eliminate stray fields.

Acknowledgments

C.L. wishes to thank Alfredo Macias and Stephan Schiller for fruitful discussions and furthermore A. Macias for his hospitality at the UAM–Iztapalapa. This research was supported by the German–Mexican DLR–Conacyt project MEX 98/010.

References

[1] C.M. Will: *Theory and Experiment in Gravitational Physics (Revised Edition)*, (Cambridge University Press, Cambridge 1993).

[2] Y. Su, B.R. Heckel, E.G. Adelberger, J.H. Gundlach, M. Harris, G.L. Smith, and H.E. Swanson, *Phys. Rev.* **D50** (1994) 3614.

[3] A. Peters, K.Y. Chung, and S. Chu, *Nature* **400** (1999) 849.

[4] N. Lockerbie, J.C. Mester, R. Torii, S. Vitale, and P.W. Worden: STEP: A Status Report, in: C. Lämmerzahl, C.W.F. Everitt, and F.W. Hehl (Eds.): *Gyros, Clocks, Interferometers ...: Testing Relativistic Gravity in Space*, (Springer-Verlag, Berlin 2001) p. 213.

[5] C. Lämmerzahl and H.-J. Dittus, *Ann. Physik* (Leipzig) **11** (2002) 95.

[6] H.P. Robertson, *Rev. Mod. Phys.* **21** (1949) 378.

[7] R. Mansouri, R.U. and Sexl, *Gen. Rel. Grav.* **8** (1977) 497; *Gen. Rel. Grav.* **8** (1977) 515; *Gen. Rel. Grav.* **8** (1977) 809.

[8] C. Lämmerzahl, C. Braxmaier, H.-J. Dittus, H. Müller, A. Peters, and S. Schiller, *Int. J. Mod. Phys.* (2002) to appear.

[9] H. Müller, S. Herrmann, C. Braxmaier, S. Schiller and A. Peters: New test of the isotropy of space using cryogenic optical resonators, submitted to the Quantum Electronics and Laser Svience Conference, Long Beach 2001.

[10] C. Braxmaier, H. Müller, O. Pradl, J. Mlynek, A. Peters, and S. Schiller, *Phys. Rev. Lett.* **87** (2001).

[11] R. Grieser, R. Klein, G. Huber, S. Dickopf, I. Klaft, P. Knobloch, P. Merz, F. Albrecht, M. Grieser, D. Habs, D. Schwalm, and T. Kühl, *Appl. Phys.* **B 59** (1994) 127.

[12] J.P Turneaure, C.M. Will, B.F. Farrel, E.M. Mattison, and R.F.C. Vessot, *Phys. Rev.* **27** (1983) 1705.

[13] A. Godone, C. Novero, and P. Tavella, *Phys. Rev.* **D 51** (1995) 319.

[14] T. Damour: Equivalence Principle and Clocks, in J. Trân Tanh Vân, J. Dumarchez, S. Reynaud, C. Salomon, S. Thorsett, S. and J.Y. Vinet (Eds.): *Gravitational Waves and Experimental Gravity*, (World Publishers, Hanoi 2000) p. 357.

[15] T. Damour, *C. R. Acad. Sci.* (2001).

[16] J.K. Webb, M.T. Murphy, V.V. Flambaum, V.A. Dzuba, J.D. Barrow, C.W. Churchill, J.X. Prochaska, and A.M. Wolfe, *Phys. Rev. Lett.* **87** (2001) 091301.

[17] F.C. Witteborn and W.M. Fairbank, *Phys. Rev. Lett.*, **19** (1967) 1049.

[18] L.I. Schiff and M.V. Barnhill, *Phys. Rev.* **151** (1966) 1067.

[19] T.W. Darling, F. Rossi, G.I. Opat, and G.F. Moorhead, G.F., *Rev. Mod. Phys.* **64** (1992) 237.

[20] T. Goldman, R.J. Hughes, and M.M. Nieto, *Phys. Lett.* **B 171** (1986) 217.

[21] E.G. Adelberger, B.R. Heckel, C.W. Srtubbs, and Y. Su, *Phys. Rev. Lett.* **66** (1991) 850.

[22] M.M. Nieto and T. Goldman, *Phys. Rep.* **205** (1991) 222.

[23] A.A. Logonov, M.A. Mestvirishvili, and Yu.V. Chugreev, Physics–Uspekhi **39** (1996) 73.

[24] P. Candelas and D. Sciama, *Phys. Rev.* **D 27** (1983) 1715.

[25] V.L. Ginzburg and V.P. Frolov, *Sov. Phys. Usp.* **30** (1987) 1073.

[26] M.P. Haugan, *Ann. Phys.* **118** (1979) 156.

[27] A.J. Dessler, F.C. Michel, H.E. Rorschach, G.T. Trammel, *Phys. Rev.* **168** (1968) 737.

[28] J.M. Lockhart, F.C. Witteborn and W.M. Fairbank, *Phys. Rev. Lett.* **67** (1977) 283.

[29] P.T. Greenland, *Contemp. Phys.* **38** (1997) 181.

[30] F.M. Huber, R.A. Lewis, E.W. Messerschmidt, and G.A. Smith, *Adv. Space Res.* **25** (2000) 1245.

[31] F.M. Huber, E.W. Messerschmidt, and G.A. Smith, *Class. Quantum Grav.* **18** (2001) 2457.

MASSIVE (1/2,1/2) BOSONS

D. V. Ahluwalia
Facultad de Fisica, UAZ, A. P. C-600, Zacatecas, ZAC 98062, Mexico
ahluwalia@phases.reduaz.mx; http://phases.reduaz.mx

M . Kirchbach
Facultad de Fisica, UAZ, A. P. C-600, Zacatecas, ZAC 98062, Mexico
kirchbach@chiral.reduaz.mx; http://chiral.reduaz.mx

Abstract We present an *ab initio* construction of the massive $(1/2, 1/2)$ representation space and obtain a new wave equation that is different from Proca's equation. We demonstrate that the massive $(1/2,1/2)$ does not, in general, lend itself to a pure spin–1 interpretation. It is rather an object containing spin–1 and spin–0 mixed up in a covariantly inseparable manner.

Keywords: Massive Gauge Bosons, Stückelberg propagator.

1. Introduction

Finite–mass gauge fields, such as the electroweak gauge bosons, transform according to the $(1/2, 1/2)$ representation space of the Lorentz group. In that regard, its space-time properties are of current interest. In the present talk we review the *ab initio* investigation of the massive $(1/2, 1/2)$ performed in [1] and show that for massive particles, the $(1/2, 1/2)$ representation space of the Lorentz group does not, in general, lend itself to a single–spin interpretation. We find this multiplet to naturally bifurcate into a triplet and a singlet of opposite relative intrinsic parities. The text–book separation into spin one and spin zero states occurs only for certain limited kinematic settings. We construct a wave equation for the $(1/2, 1/2)$ multiplet that is different from Proca's equation.

The presentation is so composed that to each physical element of the Dirac $(1/2,0) \oplus (0,1/2)$ representation space there is a counterpart in the $(1/2,1/2) = (1/2,0) \otimes (0,1/2)$ representation space.

2. Constructing $(1/2,1/2)$ representation space

The construction of the $(1/2,1/2)$ representation space [1] begins with the observation that it is a direct product of the $(1/2,0)$, $(0,1/2)$ representation spaces, i.e., $(1/2,1/2) = (1/2,0) \otimes (0,1/2)$. Under the Lorentz boost the right–handed $(1/2,0)$ spinors, $\phi_R(\vec{p})$, transform as [2]

$$\begin{aligned}\phi_R(\vec{p}) &= \exp\left(+\frac{\vec{\sigma}}{2} \cdot \vec{\varphi}\right) \phi_R(\vec{0}) \\ &= \frac{1}{\sqrt{2m(E+m)}} \left[(E+m)I_2 + \vec{\sigma} \cdot \vec{p}\right] \phi_R(\vec{0}),\end{aligned} \qquad (1)$$

where $\vec{\varphi}$ is the boost parameter [2], and I_2 is a 2×2 identity matrix. The left–handed $(0,1/2)$ spinors, $\phi_L(\vec{p})$, transform according to

$$\begin{aligned}\phi_L(\vec{p}) &= \exp\left(-\frac{\vec{\sigma}}{2} \cdot \vec{\varphi}\right) \phi_L(\vec{0}) \\ &= \frac{1}{\sqrt{2m(E+m)}} \left[(E+m)I_2 - \vec{\sigma} \cdot \vec{p}\right] \phi_L(\vec{0}).\end{aligned} \qquad (2)$$

Now to describe the physical states inhabiting the $(1/2,1/2)$ representation space, one needs four independent *rest* states, and the *boost*, $\kappa(\vec{p})$, given by

$$\kappa(\vec{p}) = \frac{1}{2m(E+m)} \left[(E+m)I_2 + \vec{\sigma} \cdot \vec{p}\right] \otimes \left[(E+m)I_2 - \vec{\sigma} \cdot \vec{p}\right]. \qquad (3)$$

Associated with $\kappa(\vec{p})$ are the boost generators

$$K_x = \frac{1}{2}\begin{bmatrix} 0 & i & -i & 0 \\ i & 0 & 0 & -i \\ -i & 0 & 0 & i \\ 0 & -i & i & 0 \end{bmatrix}, \quad K_y = \frac{1}{2}\begin{bmatrix} 0 & 1 & -1 & 0 \\ -1 & 0 & 0 & -1 \\ 1 & 0 & 0 & 1 \\ 0 & 1 & -1 & 0 \end{bmatrix}, \qquad (4)$$

and $K_z = \text{diag}(0, -i, i, 0)$. Similarly, the generators of the rotations for the $(1/2,1/2)$ representation space are:

$$J_x = \frac{1}{2}\begin{bmatrix} 0 & 1 & 1 & 0 \\ 1 & 0 & 0 & 1 \\ 1 & 0 & 0 & 1 \\ 0 & 1 & 1 & 0 \end{bmatrix}, \quad J_y = \frac{1}{2}\begin{bmatrix} 0 & -i & -i & 0 \\ i & 0 & 0 & -i \\ i & 0 & 0 & -i \\ 0 & i & i & 0 \end{bmatrix}, \qquad (5)$$

and $J_z=\text{diag}(1,0,0,-1)$. In the *rest* frame, the $(1/2,1/2)$ representation space decomposes into eigenstates of $\vec{J}^2$ and J_z. The spin–1 sector carries three independent rest–frame states:

$$w_{1,+1}(\vec{0}) = \begin{bmatrix} 1 \\ 0 \\ 0 \\ 0 \end{bmatrix}, \quad w_{1,0}(\vec{0}) = \frac{1}{\sqrt{2}}\begin{bmatrix} 0 \\ 1 \\ 1 \\ 0 \end{bmatrix}, \quad w_{1,-1}(\vec{0}) = \begin{bmatrix} 0 \\ 0 \\ 0 \\ 1 \end{bmatrix}, \quad (6)$$

while the spin–0 sector carries one state only

$$w_{0,0}(\vec{0}) = \frac{1}{\sqrt{2}}\begin{bmatrix} 0 \\ 1 \\ -1 \\ 0 \end{bmatrix}. \quad (7)$$

By the Wigner argument, $w_\zeta = \kappa(\vec{p})\,w_{j,m}(\vec{0})$, where $\zeta = 1,2,3,4$ in turn corresponds to $\{j,m\} = \{1,+1\},\{1,0\},\{1,-1\}$, and $\{0,0\}$, can be identified with the states corresponding to momentum $\vec{p}$. The explicit expressions for these are:

$$w_1(\vec{p}) = N\begin{bmatrix} E_- E_+ \\ -p_+ E_+ \\ p_+ E_- \\ -p_+^2 \end{bmatrix}, \quad w_2(\vec{p})\frac{N}{\sqrt{2}}\begin{bmatrix} -2p_- p_z \\ E_+^2 - p_y^2 - p_x^2 \\ E_-^2 - p_y^2 - p_x^2 \\ 2p_+ p_z \end{bmatrix},$$

$$w_3(\vec{p}) = N\begin{bmatrix} -p_-^2 \\ p_- E_+ \\ -p_- E_- \\ E_- E_+ \end{bmatrix}, \quad w_4(\vec{p}) = \frac{N}{\sqrt{2}}\begin{bmatrix} -2p_-(E+m) \\ E_+^2 + p_y^2 + p_x^2 \\ -(E_-^2 + p_y^2 + p_x^2) \\ 2p_+(E+m) \end{bmatrix},$$

where we used the short hand notations $E_\pm = E+m\pm p_z$ and $p_\pm = p_x \pm ip_y$, and the factor takes the value, $N := [2m(E+m)]^{-1}$. It is directly verified that the states $w_{1,3}(\vec{p})$ are eigenstates of $\vec{J}^2$ with eigenvalue 2, while $w_{2,4}(\vec{p})$ are not. However, all $w_\zeta(\vec{p})$'s are eigenstates of the second dragged Casimir operator of the Poincaré group, $\widetilde{C}_2$. The latter is defined as the dragged squared Pauli–Lubánski pseudo–vector $\widetilde{\mathcal{W}}^\mu$ for the $(1/2,1/2)$ space, i.e. $\widetilde{C}_2 = \widetilde{\mathcal{W}}^\mu \widetilde{\mathcal{W}}_\mu$, and $\widetilde{\mathcal{W}}^\mu = \frac{1}{2}\epsilon^{\mu\nu\rho\sigma}I_{\nu\rho}P_\sigma$. Here, P^μ are the operators of translations, while $I_{0i} = K_i$, $I_{ij} = \epsilon_{ijk}J_k$, and each Latin index takes space-like values only. Notice, that $\widetilde{C}_2$ and $\widetilde{\mathcal{W}}_\mu$ for any finite dimensional representations are based upon operators of translations that commute with the Lorentz generators [3]. One still finds

$$\widetilde{C}_2 w_\zeta(\vec{p}) = -m^2 \lambda_\zeta w_\zeta(\vec{p}), \quad \lambda_\zeta = 2 \quad \text{for} \quad \zeta = 1,2,3, \quad \text{and} \quad \lambda_4 = 0,$$

though the commutator $\left[\widetilde{C}_2, \vec{J}^2\right] = -4E\vec{J} \cdot \vec{p}$ does not vanish. Because of that, $w_\zeta(\vec{p})$ are in general *not* single–spin valued states. Stated differently, though $\widetilde{C}_2$ still has frame independent eigenvalues with respect to the basis vectors w_ζ, it does not behave as a genuine Casimir invariant. In addition, the generators of rotations commute with the boost generators only in case the rotation plane is perpendicular to the direction of motion, so that again, the decomposition of the (1/2,1/2) representation space into spin 1 and spin 0 sectors is not possible in all boosted frames. The orthonormality relations for $w_\zeta(\vec{p})$ are

$$\overline{w}_\zeta(\vec{p}) w_\zeta(\vec{p}) = \begin{cases} -1 \text{ for } \zeta = 1,2,3 \\ +1 \text{ for } \zeta = 4, \end{cases}$$

while the completeness relation reads

$$w_4(\vec{p}) \overline{w}_4(\vec{p}) - \sum_{\zeta=1}^{3} w_\zeta(\vec{p}) \overline{w}_\zeta(\vec{p}) = I_4, \qquad (8)$$

where I_4 equals 4×4 identity matrix. In the above expressions we have defined

$$\overline{w}_\zeta(\vec{p}) := w_\zeta(\vec{p})^\dagger \lambda_{00}, \quad \text{where} \quad \lambda_{00} = \begin{bmatrix} -1 & 0 & 0 & 0 \\ 0 & 0 & -1 & 0 \\ 0 & -1 & 0 & 0 \\ 0 & 0 & 0 & -1 \end{bmatrix}. \qquad (9)$$

It will be seen below that λ_{00} is a part of a larger set of covariant matrices. For the moment it suffices to note that $S\lambda_{00}S^{-1}$, with

$$S = \frac{1}{\sqrt{2}} \begin{bmatrix} 0 & i & -i & 0 \\ -i & 0 & 0 & i \\ 1 & 0 & 0 & 1 \\ 0 & i & i & 0 \end{bmatrix}, \qquad (10)$$

is the flat space-time metric with the diagonal $\{1, -1, -1, -1\}$. We note that λ_{00} also serves as the *Parity* operator:

$$\lambda_{00} w_\zeta(-\vec{p}) = \begin{cases} -w_\zeta(\vec{p}) \text{ for } \zeta = 1,2,3 \\ +w_\zeta(\vec{p}) \text{ for } \zeta = 4. \end{cases}$$

So, while $w_\zeta(\vec{p})$ may not carry a definite spin they are endowed with a definite relative intrinsic parity. The relevant projectors onto sub–spaces of opposite relative intrinsic parities are:

$$\pi_+ = w_4(\vec{p}) \overline{w}_4(\vec{p}), \text{ and } \pi_- = -\sum_{\zeta=1}^{3} w_\zeta(\vec{p}) \overline{w}_\zeta(\vec{p}). \qquad (11)$$

Furthermore, we note that $Sw_\zeta(\vec{p})$ transforms as a Lorentz four vector and carries a Lorentz index that can be lowered and raised with the flat space–time metric pointed out above. Thus, we introduce: $A^\mu_\zeta(\vec{p}) := (Sw_\zeta(\vec{p}))^\mu$. In addition, we introduce the following S-transformed projectors and Λ matrices as $\Pi_\pm := S\pi_\pm S^{-1}$, and $\Lambda_{\mu\nu} := S\lambda_{\mu\nu}S^{-1}$. In a way reminiscent of the Dirac case, the wave equation for $A^\mu_\zeta(\vec{p})$ can be read off from the projectors and is seen to be:

$$\left(\Lambda_{\mu\nu}p^\mu p^\nu - \epsilon m^2 I_4\right) A_\zeta(\vec{p}) = 0, \tag{12}$$

where ϵ equals $+1$ for $\zeta = 4$ and is -1 for $\zeta = 1,2,3$. The $\Lambda_{\mu\nu}$ matrices can be read off from $\Lambda_{\mu\nu}p^\mu p^\nu = m^2(\Pi_+ - \Pi_-)$. We find, $\Lambda_{00} = \mathrm{diag}(1,-1,-1,-1)$, $\Lambda_{11} = \mathrm{diag}(1,-1,1,1)$, $\Lambda_{22} = \mathrm{diag}(1,1,-1,1)$, $\Lambda_{33} = \mathrm{diag}(1,1,1,-1)$, $\Lambda_{03} = \mathrm{\not{d}iag}(-1,0,0,1)$, $\Lambda_{12} = \mathrm{\not{d}iag}(0,-1,-1,0)$, where diag stands for principle diagonal, while $\not{d}$iag denotes the cross diagonal. Furthermore,

$$\Lambda_{01} = \begin{bmatrix} 0 & -1 & 0 & 0 \\ 1 & 0 & 0 & 0 \\ 0 & 0 & 0 & 0 \\ 0 & 0 & 0 & 0 \end{bmatrix}, \quad \Lambda_{02} = \begin{bmatrix} 0 & 0 & -1 & 0 \\ 0 & 0 & 0 & 0 \\ 1 & 0 & 0 & 0 \\ 0 & 0 & 0 & 0 \end{bmatrix},$$

$$\Lambda_{13} = \begin{bmatrix} 0 & 0 & 0 & 0 \\ 0 & 0 & 0 & -1 \\ 0 & 0 & 0 & 0 \\ 0 & -1 & 0 & 0 \end{bmatrix}, \quad \Lambda_{23} = \begin{bmatrix} 0 & 0 & 0 & 0 \\ 0 & 0 & 0 & 0 \\ 0 & 0 & 0 & -1 \\ 0 & 0 & -1 & 0 \end{bmatrix}. \tag{13}$$

The remaining $\Lambda_{\mu\nu}$ are obtained from the above expressions by using the symmetry relation $\Lambda_{\mu\nu} = \Lambda_{\nu\mu}$. The matrix Λ_{00} numerically coincides with the flat space-time metric η. [To avoid confusion, note that in the usual notation $\eta_{\mu\nu}$, the indices μ and ν enumerate rows and columns of η. In $\Lambda_{\mu\nu}$, the indices μ and ν parallel the Lorentz index on Dirac's γ_μ, while the row– and column indices are suppressed.]

By studying $\det\left(\Lambda_{\mu\nu}p^\mu p^\nu - \epsilon m^2 I_4\right)$, we find that: (a) For $\epsilon = -1$, the above equation carries three "positive–energy–" and three "negative–energy–" solutions with the correct dispersion relation, $E^2 = \vec{p}^2 + m^2$, while (b) For $\epsilon = +1$, there is one "positive–energy–"and one "negative–energy–" solution. A complete CPT analysis shows these to be particle–antiparticle solutions. Thus the $(1/2, 1/2)$ representation space manifestly carries the particle and antiparticle states with equal masses. This result is not contained in the usual considerations based on the Proca equation. It is further apparent from the derived wave equation that in the $(1/2, 1/2)$ representation space $[C, P] = 0$. This contrasts with $\{C, P\} = 0$ for the $(1, 0) \oplus (0, 1)$ representation space. [Notice that contrary to the canonical wisdom, the $(1, 0) \oplus (0, 1)$ representation space, when constructed in $SU(2)_R \otimes SU(2)_L$ rather than in $SL(2,C)$, supports one of the unusual Wigner classes [4]. In this representation space massive charged bosons, and antibosons, carry *opposite* relative intrinsic parities [5]. For this reason, e.g., the electroweak bosons $W^\pm$ and Z cannot be described as $(1,0) \oplus (0,1)$ particles but have to belong to $(1/2, 1/2)]$.

We are now in a position to revisit the Lorentz gauge "constraint" [2]: $\partial_\beta A^\beta = 0$, which in the momentum-space takes the form, $p_\beta A^\beta_\zeta(\vec{p}) = 0$. The *lhs* of the latter equation can be directly evaluated using the expressions for the basis vectors obtained above. We find

$$p_\beta A^\beta_\zeta(\vec{p}) = 0 \Leftrightarrow \begin{cases} c_\zeta (m^2 - p_\beta p^\beta) = 0, & \text{for } \zeta = 1, 2, 3 \\ (i/m) p_\beta p^\beta = 0, & \text{for } \zeta = 4 \end{cases}$$

where, $c_1 = ip_+$, $c_2 = -ip_z$, and $c_3 = -ip_-$. In other words, for $\zeta = 1, 2, 3$ the Lorentz condition in momentum space amounts to $p_\beta p^\beta = m^2$ and tests consistency with special relativity. As long as $A^\mu_4 = i\frac{P^\mu}{m}$, the Lorentz condition reflects the orthogonality between the $A_{1,2,3}$ and A_4. For $\zeta = 4$, one finds $p_\beta A^\beta_4 = im$, which cannot be set equal to zero unless one is considering massless particles. Therefore, the Lorentz condition restricts the four degrees of freedom that span the massive $(1/2,1/2)$ representation space to those three that satisfy Proca's equation. To state it differently, Proca's equation does not describe the full mathematical content of the massive $(1/2,1/2)$ space. In textbooks [6], the time-like degree of freedom w_4 of the (virtual) gauge field, so indispensable for the completeness relation in Eq. (8), in being associated with negative energy solutions of the Hamiltonian, is customary considered as a ghost and, though included in propagation, is excluded from the physical world by means of the well known Gupta–Bleuler mechanism.

We here undertake an attempt to soften the status of w_4 as unphysical in reporting on its intriguing property as the source of the Stückelberg term in the propagator of massive $(1/2,1/2)$ bosons.

3.   Conclusions and discussion

The massive $(1/2,1/2)$ propagator that follows from the completeness relation in Eq. (12) reads

$$S_{\mu\nu} = \frac{[A_4(\vec{p})\overline{A}_4(\vec{p})\Lambda_{00}]_{\mu\nu}}{p^2 - m^2} - \sum_{\zeta=1}^{3} \frac{[A_\zeta(\vec{p})\overline{A}_\zeta(\vec{p})\Lambda_{00}]_{\mu\nu}}{p^2 - m^2}. \qquad (14)$$

In noticing that

Table 1. Properties of the (1/2,1/2) sectors

ζ	$p^\mu(A_\zeta)_\mu$	$\widetilde{\mathcal{W}}^\mu(A_\zeta)_\mu$	$\widetilde{C}_2 A_\zeta$	Sector
1,2,3	0	$\neq 0$	$-2m^2$	Proca
4	$\neq 0$	0	0	Stückelberg

$$[A_4(\vec{p})\overline{A}_4(\vec{p})\Lambda_{00}]_{\mu\nu} = \frac{p_\mu p_\nu}{m^2}, \qquad (15)$$

$$-\sum_{\zeta=1}^{3}[A_\zeta(\vec{p})\overline{A}_\zeta(\vec{p})\Lambda_{00}]_{\mu\nu} = g_{\mu\nu} - \frac{p_\mu p_\nu}{m^2}, \qquad (16)$$

one immediately realizes that the massive (1/2,1/2) propagator contains both the Stückelberg (15) and the Proca (16) terms.

This feature of Eq. (12) appears quite appealing to us as it automatically leads to a well behaved propagator of a massive gauge boson as arising in a spontaneously broken local gauge theory. Within the context of the scenario presented above, the Proca sector is characterized by vanishing of $p^\mu A_\mu = 0$, while the Stückelberg sector is characterized by vanishing of $p^\mu \widetilde{\mathcal{W}}_\mu = 0$ (see Table 1).

Acknowledgments

This work has been supported by CONACyT project No 32067–E.

References

[1] D. V. Ahluwalia, M. Kirchbach, *Mod. Phys. Lett.* **A16** (2001) 1377.

[2] L. H. Ryder, *Quantum Field Theory*, Cambridge, (Cambridge University Press, 1996).

[3] M. Kirchbach, D. V. Ahluwalia, *Phys. Lett.* **B** (2002), in press.

[4] E. P. Wigner, in *Group Theoretical Concepts and Methods in Elementary Particle Physics, – Lectures of the Istanbul Summer School of Theoretical Physics, (July 16 - August 4, 1962)*, ed. F. Gürsey, (Gordon and Breach Sci. Publ.) (1964) pp. 37.

[5] D. V. Ahluwalia, M. B. Johnson, T. Goldman, *Phys. Lett.* **B316** (1993) 102.

[6] M. Kaku, *Quantum Field Theory*, New York-Oxford, (Oxford University Press, 1993).

INHOMOGENEOUS COSMOLOGIES WITH ADIABATIC EVOLUTION

Roberto A Sussman
Instituto de Ciencias Nucleares, UNAM,
Apartado Postal 70-543, México D.F., 04510, México.
sussman@nuclecu.unam.mx

Mustapha Ishak
Department of Physics, Queens University,
Kingston, Ontario, Canada.
ishak@hera.phy.qeensu.ca

Abstract We consider a viscous fluid source for the Lemaitre–Tolman–Bondi metrics. The isotropic pressure and matter–energy density satisfy the equation of state of a mixture of non–relativistic matter (barions and CDM) and a photon gas. The anisotropic pressure can be interpreted as shear viscosity if the mixture is interactive, or as the quadrupole moment of the photon gas if the mixture is decoupled. Suitable volume averages and suitable contrast functions are defined in order to characterize inhomogeneity in terms of density and curvature "lumps" or "voids". We show that the dynamical evolution of the mixture is qualitatively analogous to that of "exact" adiabatic perturbations on a FLRW background with the same equation of state.

Keywords: Inhomogeneous cosmological models, adiabatic perturbations.

1. Introduction

Lemaitre–Tolman–Bondi (LTB) metrics with a dust source in a comoving frame lead to very popular models of cosmological inhomogeneities of collisionless matter[1]. However, the most general momentum-energy tensor admitted by this metric form is that of a fluid with anisotropic pressure and zero energy flux. Using this tensor as the source of LTB metrics we obtain generalizations of LTB dust solutions that allow for the description of cosmological inhomogeneities with non-trivial pressure gradients, all of which could accommodate thermal motions and/or velocity dispersions of CDM. Considering this source as the sum of a perfect fluid part (the isotropic pressure, p, and the matter–energy density, ρ) plus the anisotropic tracefree pressure tensor Π_{ab}, Einstein field equations can be integrated under the assumption of various simple equations of state relating p and ρ (see [2] – [7]). In this paper we examine the case corresponding to a mixture of non–relativistic matter (barions and CDM) and a photon gas, a source with potentially good possibilities in astrophysical and cosmological applications.

In section 2 we review the process of integrating the field equations under the assumption of the equation of state described above, discussing how the resulting sources can be interpreted as an interactive, thermodynamical, mixture or as a decoupled mixture (section 3). A well posed initial value problem follows by expressing (section 4) the free parameters of the solutions in terms of initial value functions defined along an initial hypersurface of constant cosmic time (proper time of fundamental observers). Suitable volume averages and contrast functions of the initial matter and radiation densities and 3–dimensional Ricci scalar are introduced, leading to a characterization of inhomogeneity in the initial hypersurface in terms of "lumps" or "voids". These quantities have been discussed in previous work [4]–[8] and generalized in the present paper by considering (section 5) their definition along arbitrary hypersurfaces of constant t. This generalization yields interesting conservation and scaling laws involving the volume averages and contrast functions. The latter are formally analogous to "exact" perturbations with respect to the averaged densities and 3–curvature. In section 6, we identify a special class of initial conditions for which the contrast functions satisfy exactly the relations characteristic of adiabatic perturbations with respect to the averaged matter and radiation densities. Since in the limit of small contrasts, we obtain the expressions for linear adiabatic perturbations in a FLRW background [6] [7], the general case leads to exact nonlinear deviations from homogeneity, with a matter source that allows for the description of thermal effects within a formalism of irreversible causal thermodynamics. We find this result quite stimulating and will pursue its consequences in the study of the evolution of galactic and metagalactic structures in a homogeneous background.

2. Integration of the field equations

The usual form of the LTB metric is given by

$$ds^2 = -c^2\, dt^2 + \frac{Y'^2}{1+E}\, dr^2 + Y^2\left(d\theta^2 + \sin\theta\, d\phi^2\right), \tag{1}$$

where $Y = Y(t,r)$, $E = E(r)$ and $Y' = Y_{,r}$. The most general momentum energy tensor compatible with this metric in a comoving frame ($u^a = c\delta^a_t$) is

$$T_{ab} = (\rho + p) u_a u_b + p\, g_{ab} + \Pi_{ab}, \tag{2}$$

where ρ, p are matter–energy density and isotropic pressure, while Π_{ab} is the anisotropic pressure tensor ($u_a \Pi^{ab} = \Pi^a{}_a = 0$). For the metric (1), we have $\Pi^a_b = \text{diag}[0, -2P, P, P]$, where $P = P(t,r)$ is an arbitrary function. As in the dust case ($\Pi_{ab} = 0$), the 4-velocity u^a is a geodesic vector field, however now (from the momentum balance $T^{ab}{}_{;b} = 0$) the pressure gradients are exactly balanced by the divergence of Π_{ab}, hence they are nonzero [3] [4]. The field equations for (1) and (2) yield the Friedmann–like equation

$$\dot Y^2 = \frac{2M}{Y} + W\frac{Y_i^2}{Y^2} + E, \tag{3}$$

where $Y_i = Y(t_i, r)$ for $t = t_i$ arbitrary and $W = W(r)$[1]. This equation generalizes the usual evolution equation for dust sources ($W = 0$) and can be transformed into the simpler equation

$$\dot y^2 = \frac{-\kappa y^2 + 2\mu y + \omega}{y^2}, \tag{4}$$

where

$$y = \frac{Y}{Y_i}, \quad \mu \equiv \frac{M}{Y_i^3}, \quad \omega \equiv \frac{W}{Y_i^3}, \quad \kappa \equiv -\frac{E}{Y_i^2}, \tag{5}$$

Assuming a conserved particle particle number current $n^a = n u^a$ satisfying $n^a{}_{;a} = 0$, so that n is the particle number density, together with the equation of state

$$\rho = \rho^{(m)} + \rho^{(r)}, \quad 3p = \rho^{(r)}, \tag{6}$$

the field equations yield

$$\rho^{(m)} = m c^2 n = \frac{\rho_i^{(m)}}{y^3 \Gamma}, \quad \rho_i^{(m)} = m c^2 n_i \tag{7}$$

$$\rho^{(r)} = 3p = \frac{\rho_i^{(r)} \Psi}{y^4 \Gamma}, \quad \rho_i^{(r)} = a T_i^4, \tag{8}$$

$$P = \frac{\rho_i^{(r)} \Phi}{6 y^4 \Gamma}, \tag{9}$$

where m is the mass of non–relativistic particles species (barions or CDM), a is the Steffan–Boltzmann constant, T_i is the initial temperature, while the auxiliary functions Γ, Ψ, Φ, characterizing the spacial dependence of ρ, p, P, are given by

$$\Gamma = 1 + \frac{y'/y}{Y_i'/Y_i} = 1 - 3A\Delta_i^{(m)} - 3B\Delta_i^{(r)} - 3C\Delta_i^{(k)},$$

$$\Psi = \frac{1 + A\Delta_i^{(m)} + (1+B)\Delta_i^{(r)} + C\Delta_i^{(k)}}{1 + \Delta_i^{(r)}},$$

$$\Phi = \frac{4A\Delta_i^{(m)} + (1+4B)\Delta_i^{(r)} + 4C\Delta_i^{(k)}}{1 + \Delta_i^{(r)}}, \quad (10)$$

where

$$\Delta_i^{(r)} \equiv \frac{\omega'/\omega}{Y_i'/Y_i}, \quad \Delta_i^{(m)} \equiv \frac{\mu'/\mu}{Y_i'/Y_i}, \quad \Delta_i^{(k)} \equiv \frac{\kappa'/\kappa}{Y_i'/Y_i}, \quad (11)$$

while A, B, C are functions of y, μ, ω, f given by

$$A = \frac{\mu\left[\partial(Z-Z_i)/\partial\mu\right]}{y\left[\partial Z/\partial y\right]}, \quad (12)$$

$$B = \frac{\omega\left[\partial(Z-Z_i)/\partial\omega\right]}{y\left[\partial Z/\partial y\right]}, \quad (13)$$

$$C = \frac{\kappa\left[\partial(Z-Z_i)/\partial\kappa\right]}{y\left[\partial Z/\partial y\right]}, \quad (14)$$

where Z is the explicit integral of (4) and $Z_i = Z|_{y=1}$. Explicit functional forms for A, B, C are obtained in [4], [5], [6] and [7].

3. Interpretation of the solutions

The equation of state (6) can be derived from the equation of state of a mixture of non-relativistic matter and a photon gas, under the assumption that the internal energy and pressure of the former are negligible. Hence, p can be fully ascribed to the photon gas, leading to the temperature law

$$T = \frac{T_i}{y}\left(\frac{\Psi}{\Gamma}\right)^{1/4}, \quad (15)$$

We can distinguish two possible situations: (a) The mixture is interactive, hence Π^{ab} is shear viscosity, T is the common temperature and the solutions must comply with a causal thermodynamical formalism, (b) the mixture is decoupled, then T is the temperature of the photon gas and Π^{ab} can be identified with the quadrupole moment of the photon gas distribution function under a Kinetic Theory treatment [9]. Case (a) has been examined in [3]–[7], while case (b) will be studied in a future paper. The new solutions presented so far become fully determined once the Friedmann equation (4) is integrated (Z is explicitly known) for specific initial conditions provided by μ, ω and κ.

4. Initial conditions

For the metric (1) the Ricci scalar of the hypersurfaces $t = $ const. is given by

$$^{(3)}\mathcal{R} = -\frac{2(EY)'}{Y^2 Y'} \quad (16)$$

Therefore, for the initial hypersurface $t = t_i$, the initial value functions in (5) can be expressed in terms of the following volume averages

$$\mu = \frac{4\pi G}{3c^4} \langle \rho_i^{(m)} \rangle = \frac{4\pi G}{3c^4} \frac{\int \rho_i^{(m)} Y_i^2 Y_i' dr}{\int Y_i^2 Y_i' dr},$$

$$\omega = \frac{4\pi G}{3c^4} \langle \rho_i^{(r)} \rangle = \frac{4\pi G}{3c^4} \frac{\int \rho_i^{(r)} Y_i^2 Y_i' dr}{\int Y_i^2 Y_i' dr},$$

$$\kappa = \frac{1}{6} \langle ^{(3)}\mathcal{R}_i \rangle = \frac{1}{6} \frac{\int ^{(3)}\mathcal{R}_i Y_i^2 Y_i' dr}{\int Y_i^2 Y_i' dr}, \quad (17)$$

Comparing these volume averages with the initial value functions $\rho_i^{(m)}$, $\rho_i^{(r)}$ and $^{(3)}\mathcal{R}_i$. Using (5) and (11), we obtain

$$\rho_i^{(m)} = \langle \rho_i^{(m)} \rangle \left[1 + \Delta_i^{(m)} \right]$$

$$\rho_i^{(r)} = \langle \rho_i^{(r)} \rangle \left[1 + \Delta_i^{(r)} \right]$$

$$^{(3)}\mathcal{R}_i = \langle ^{(3)}\mathcal{R}_i \rangle \left[1 + \Delta_i^{(k)} \right], \quad (18)$$

so that the functions Δ_i introduced in (11) play the role of inhomogeneity gauges or "contrast functions" along $t = t_i$. The homogeneous limit of the solutions follows from choosing homogeneous initial conditions characterized by all $\Delta_i = 0$. As shown in [4], [5] and [8], the sign of the functions $\Delta_i^{(m)}$, $\Delta_i^{(r)}$ and $\Delta_i^{(k)}$ determines if there is an overdensity ("lump", negative sign) or underdensity ("void", positive sign) of the initial value functions $\rho_i^{(m)}$, $\rho_i^{(r)}$ or $^{(3)}\mathcal{R}_i$ at a symmetry center. Therefore, the characterization of initial conditions by means of these contrast functions is potentially useful for astrophysical applications.

5. Conserved quantities and scaling laws

With the help of the field equations and (5) the volume averages (17) can be generalized so that they are defined along an arbitrary hypersurface $t = \text{const.} \neq t_i$. This yields

$$\frac{4\pi G}{3c^4} \langle \rho^{(m)} \rangle \equiv \frac{4\pi G}{3c^4} \frac{\int \rho^{(m)} Y^2 Y' dr}{\int Y^2 Y' dr} = \frac{\mu}{y^3},$$

$$\frac{4\pi G}{3c^4} \langle \rho^{(r)} \rangle \equiv \frac{4\pi G}{3c^4} \frac{\int \rho^{(r)} Y^2 Y' dr}{\int Y^2 Y' dr} = \frac{\omega}{y^4},$$

$$\langle {}^{(3)}\mathcal{R} \rangle \equiv \frac{\int {}^{(3)}\mathcal{R}\, Y^2 Y' dr}{\int Y^2 Y' dr} = \frac{6\kappa}{y^2}, \qquad (19)$$

leading to the simple scaling laws

$$\langle \rho^{(m)} \rangle = \frac{\langle \rho_i^{(m)} \rangle}{y^3}, \quad \langle \rho^{(r)} \rangle = \frac{\langle \rho_i^{(r)} \rangle}{y^4}, \quad \langle {}^{(3)}\mathcal{R} \rangle = \frac{\langle {}^{(3)}\mathcal{R}_i \rangle}{y^2}, \qquad (20)$$

hence y can be identified as a scale factor so that $\langle \rho^{(m)} \rangle$, $\langle \rho^{(r)} \rangle$, $\langle {}^{(3)}\mathcal{R} \rangle$ respectively scale as y^{-3}, y^{-4}, y^{-2}. Using equations (6) to (12), the contrast functions $\Delta_i^{(m)}, \Delta_i^{(r)}, \Delta_i^{(k)}$ along $t = t_i$ can be generalized for arbitrary hypersurfaces of constant $t \neq t_i$ so that

$$\rho^{(m)} = \langle \rho^{(m)} \rangle \left[1 + \Delta^{(m)} \right],$$

$$\rho^{(r)} = \langle \rho^{(r)} \rangle \left[1 + \Delta^{(r)} \right],$$

$${}^{(3)}\mathcal{R} = \langle {}^{(3)}\mathcal{R} \rangle \left[1 + \Delta^{(k)} \right], \qquad (21)$$

and

$$P = \frac{1}{6} \langle \rho^{(r)} \rangle \Delta^{(r)}, \qquad (22)$$

with $\Delta^{(m)}, \Delta^{(r)}, \Delta^{(k)}$ are given in terms of Γ and the initial contrast functions $\Delta_i^{(m)}, \Delta_i^{(r)}, \Delta_i^{(k)}$ by

$$\left[1 + \Delta^{(m)} \right] \Gamma = 1 + \Delta_i^{(m)},$$

$$\left[1 + \frac{3}{4} \Delta^{(r)} \right] \Gamma = 1 + \frac{3}{4} \Delta_i^{(r)},$$

Inhomogeneous cosmologies

$$\left[1 + \frac{3}{2}\Delta^{(k)}\right]\Gamma = 1 + \frac{3}{2}\Delta_i^{(k)}, \qquad (23)$$

where we have eliminated Ψ in terms of Γ from (10). Since the characterization of inhomogeneity in terms of "lumps" or "voids" of the initial Δ_i functions can be applied to arbitrary hypersurfaces of constant t, we can examine the possibility that initial lumps become voids or viceversa [7].

Since the initial value functions $\langle\rho_i^{(m)}\rangle$, $\langle\rho_i^{(r)}\rangle$, $\langle^{(3)}\mathcal{R}_i\rangle$, $\Delta_i^{(m)}$, $\Delta_i^{(r)}$ $\Delta_i^{(k)}$, depend only on r, equations (19) – (23) express the conservation along the 4-velocity of the quantities like $\langle\rho^{(m)}\rangle y^3$, $\langle\rho^{(r)}\rangle y^4$, $\langle^{(3)}\mathcal{R}\rangle y^2$, as well as the quantities in the left hand side of (23). Other conserved quantities are

$$\Delta^{(s)} \equiv \left[\Delta^{(r)} - \frac{4}{3}\Delta^{(m)}\right]\Gamma = \Delta_i^{(r)} - \frac{4}{3}\Delta_i^{(m)},$$

$$\Delta^{(sk)} \equiv \left[\Delta^{(r)} - 2\Delta^{(k)}\right]\Gamma = \Delta_i^{(r)} - 2\Delta_i^{(k)}. \qquad (24)$$

6. Adiabatic initial conditions and evolution

From (24) we can readily identify a very special set of initial conditions by

$$\Delta_i^{(r)} = \frac{4}{3}\Delta_i^{(m)} = 2\Delta_i^{(k)} \quad \Leftrightarrow \quad \Delta^{(r)} = \frac{4}{3}\Delta^{(m)} = 2\Delta^{(k)}, \quad (25)$$

By comparing (21) with the forms of $\rho^{(m)}$ and $\rho^{(r)}$ given in a perturbations theory approach, it is evident that the contrast functions $\Delta^{(m)}$, $\Delta^{(r)}$, $\Delta^{(k)}$ play the role of "exact perturbations" with respect to the average values $\langle\rho^{(m)}\rangle$, $\langle\rho^{(r)}\rangle$, $\langle^{(3)}\mathcal{R}\rangle$. Since initial conditions (25) lead to $\Delta^{(m)}$ and $\Delta^{(r)}$ related for all times by a factor 4/3, we have an *exact* evolution that is formally equivalent to assuming adiabatic perturbations. Under this formal analogy, we can identify $\Delta^{(s)}$ as a radiation entropy contrast (see [4]–[7]). Therefore, we will denote (25) as "**adiabatic initial conditions**". Using (25) transforms (21) into

$$\begin{aligned}
\rho^{(m)} &= \langle\rho^{(m)}\rangle[1 + \Delta], \\
\rho^{(r)} &= \langle\rho^{(r)}\rangle\left[1 + \frac{4}{3}\Delta\right], \\
P &= \frac{2}{9}\langle\rho^{(r)}\rangle\Delta, \\
^{(3)}\mathcal{R} &= \langle^{(3)}\mathcal{R}\rangle\left[1 + \frac{2}{3}\Delta\right], \qquad (26)
\end{aligned}$$

with

$$\Delta \equiv \frac{1 - \Gamma + \Delta_i^{(m)}}{\Gamma} = \frac{\frac{3}{4}[3A + 4B + 2C]\Delta_i^{(r)}}{1 - \frac{3}{4}[3A + 4B + 2C]}, \quad (27)$$

where A, $B\,C$ are defined by (14) and computed explicitly in [4], [5] and [6], while (25) was used in order to eliminate $\Delta_i^{(m)}$ and $\Delta_i^{(k)}$ in terms of $\Delta_i^{(r)}$ in the expression for Γ in (10). The evolution of the exact adiabatic perturbation (27) yields growing and decaying modes that evolve well into the nonlinear regime that is not accessible to linear perturbations. The case of an interactive radiation–matter mixture in the context of the radiative pre-decoupling era, consistent with a thermodynamical formalism, has been examined previously [4] [5] [6]. In these papers we considered the case $\Delta \ll 1$, as well as quasi–adiabatic initial conditions ($\Delta^{(s)} \ll 1$, $\Delta^{(sk)} \ll 1$) and arrived to similar results as in the linear perturbation regime. The non linear case is currently under study [7].

Acknowledgments

The authors acknowledge financial support from grant DGAPA–PAPIIT IN–122498.

Notes

1. The subindex i will denote henceforth evaluation at $t = t_i$.

References

[1] A. Krasiński, *Inhomogeneous Cosmological Models* (CUP, Cambridge, 1997).

[2] R. Sussman, *Class. Quantum Grav.*, **15** (1998) 1759.

[3] R. Sussman and J. Triginer, *Class. Quantum Grav.*, **16** (1999) 167.

[4] R. A. Sussman and D. Pavón, *Phys. Rev.* **D60** (1999) 104023.

[5] D. Pavón and R. Sussman, *Class. Quantum Grav.* **18** (2001) 1625.

[6] R. A. Sussman and M. Ishak, "Adiabatic models of the cosmological radiative era". Submitted to *General Relativity and Gravitation*. LANL preprint gr-qc/0111010.

[7] R. A. Sussman and M. Ishak, "Exact nonlinear adibatic perturbations", in preparation.

[8] R. A. Sussman and L. Garcia Trujillo, "A new approach to initial value variables in Lemaitre-Tolman-Bondi dust solutions". Submitted to *Class. Quantum Grav.* (2002). LANL preprint gr–qc/0105081.

[9] G F R Ellis and H van Ellst, *Cosmological Models*, Cargèse Lectures, 1998. LANL e–preprints gr–qc/9812046

LEOPOLDO GARCIA–COLIN SCHERER: BRIEF BIOGRAPHY

Eduardo Piña

Leo, as he is known among some of his foreign colleagues, was born in Mexico in 1930, but with strong roots in Spain and Germany, and sharing the Spanish and English as mother languages.

He studied Chemistry and Physics simultaneously at two different faculties of the University of Mexico finishing his Bachelor in Science studies in 1954. His thesis dissertation to obtain his B.Sc. degree in Chemistry dealt with the thermodynamic properties of D_2 and HD as part of a project whose leader was one of Mexicos's foremost scientist, Alejandro Medina. The project was the construction of a nuclear reactor moderated with heavy water, which finally, was never constructed. He afterwards completed his Ph.D. work in Theoretical Physics at the University of Maryland in 1960 under Elliott Montroll's guidance. There he began his dedicated career as a teacher and researcher and started an important research program on the broad area of Statistical Mechanics, Kinetic Theory, Irreversible Thermodynamics, Critical Phenomena, Chemical Physics, and other related subjects.

By personal interest, my favorites in the hundreds of papers he has published are those on Critical Phenomena published with former students of him, others on the frontier problems of Kinetic Theory which were probably unfinished and included many years of collaboration with Melville Green and many other scientists. Also I prefer those on the Modern Foundations of Irreversible Thermodynamics. But my taste is based by personal preferences, and I do not pretend to make a fair selection.

Returning to Mexico in 1960 he faced the very hard duty of constructing from nothing, totally new research groups in the field of Sciences he cultivated, creates and loves, and he spent many of his time, energy, and insight in the development of high level Educative Institutions, and Government Laboratories with a firm scientific basis. The first task was to optimize the scientific formation of scientists, mixing the selection and preparation of students, with a program of graduate studies outside Mexico in selected places abroad, including the best universities of the United States, England, Holland, Germany, Belgium, etc. The role of Leo was essential through his personal scientific friends and friends of friends, since through their disposition of accepting students contributed to the enhancement of Statistical Physics in Mexico.

This fact is now fully acknowledged in Mexico. Prof. Garcia–Colin has been recognized as the founder of Statistical Physics in Mexico and has been honored with the highest level of National Researcher namely, he was accepted as a member of the select COLEGIO NACIONAL, hosting us today. Moreover, he was honored with the *National Prize of Sciences and Arts*, and also with the prize of the Mexican National Academy of Sciences; making him the first *Distinguished Professor* at the *Universidad Autónoma Metropolitana* of Mexico, where he has taught and done research the last twenty eight years; and appointing him *Doctor Honoris Causa* in two important universities in the country.

Along many years, beside the fields s previously mentioned, his interest in Science has had important incursions on General Relativity, Air Pollution Problems, Politics of Science, and Educational Research.

Some scientific textbooks have been authored by him, published by serious scientific editorials of Mexico for the service of the Spanish-speaking students.

This gigantic labor that can be only partially understood and appreciated in other countries, has been partaken with an intense research activity resulting in many papers published in the several International Journals of Physics and Physical Chemistry and is well known to all of those interested in current science. He has maintained a continuous correspondence with many scientists and does frequent scientific travels all around the world. He is an active member of various prestigious International Societies.

A sporting man, he is a regular amateur member of the jai alai, the basket Spanish ball game, which he inherited from his Father, an important professional of this game. But he has been also a practitioner of many other sports like racket fronton, marathon running and swimming.

He is a family man, with a beautiful family. He loves the music of the best composers and he will convince his interlocutors of this pleasure. He reads on so many matters, so many books, and so many journals.

I apologize for not being more clear on the importance of all his achievements, or the intimate collaboration with all those around him. Today I have not enough time to extend myself more on his achievements. In the future many other Mexican historians will recognize and give a fair and full explanation of my speech.

NICHOLAS G. VAN KAMPEN: BRIEF BIOGRAPHY

Rosalio F. Rodriguez

I am greatly honored and very pleased to present PROF. NICHOLAS G. VAN KAMPEN, who is the first awardee of the LEOPOLDO GARCÍA-COLÍN SCHERER MEDAL. A prize that bears the name of an outstanding Mexican scientist who started, developed and continues consolidating and expanding Statistical Physics in this country.

For more than half a century Nico van Kampen has been a pioneering force in theoretical physics and statistical mechanics. He has been an outstanding theoretical physicist ever since the publication of the results of his Ph. D. thesis in physics entitled CONTRIBUTIONS TO THE QUANTUM THEORY OF LIGHT SCATTERING, in the Proceedings of the Danish Academy of Sciences. He developed his dissertation under the advice of HANS KRAMERS in the university of Leyden and in the Niels Bohr Institute in Copenhagen, and obtained his Ph. D. degree in physics in 1951. In his thesis he showed how to overcome the singularities which arise in many quantum scattering processes. These results were essential to give the final development to Kramers's ideas which later on led to the methods of *renormalization*.

His "education" in statistical physics was accomplished in the best tradition of the Dutch School of Statistical Mechanics since his professor Kramers, was a former student of Ehrenfest whom, in turn, had been student of Boltzmann in Vienna.

In his early period of work in 1953, Nico became interested in the statistical mechanics of irreversible processes when he joined the group of PROFESSOR DE GROOT, the successor of Kramers in Leyden, and who introduced him into the field of Irreversible Processes.

In this early period in this field he addressed topics such as the quantum theory of the statistical mechanics of nonequilibrium processes, whose formulation was unsatisfactory because the familiar picture of eigenvalues and eigenstates was inappropriate for macroscopic systems.

In 1955 Nico moved to the University of Utrecht, where much experimental work on fluctuations had been done and where he developed one of his major lines of research, namely, that of the theoretical analysis of nonlinear fluctuations and of the stochastic treatment of noise in physical systems. In this context he clarified many fundamental issues concerning the use and proper place of stochastic processes in physics. He also showed how the theory of random events is spread out over the entire scientific literature from mathematics to biology. In this context, one of his most important contributions was the introduction of a systematic expansion method of the master equation. This method has clarified many misconceptions and fundamental issues in stochastic processes and has greatly expanded the applicability of Markovian processes to numerous systems in physics and chemistry.

His research on stochastic processes between the years 1955 and 1980 produced a long string of epoch making papers discussing a variety of controversial issues in the field. His powerful lines of thought, deep criticism and rigorous logical thinking addressed topics such as the *"use and abuse of the Langevin approach"*, *"the Ito–Stratonovich dilemma"*, *"the fluctuations in the Boltzmann equation"* or *"the stochastic behavior of quantum systems"*.

As is evident from the quality of his review articles on stochastic processes and other fields, he has a wonderful knowledge of the literature on this subject, and many of his views and contributions on these topics where collected in his very well known book: *"Stochastic Processes in Physics and Chemistry"*, first published in 1981 and revised and enlarged in 1992. This book is a classic in the field and has guided many graduate students and scientists through the complexities of research in different topics of fluctuations and nonequilibrium statistical mechanics.

Another field where he has made important contributions, was that of plasma physics, where in collaboration with Ubo Felderhof, he developed methods to derive the modes of the linear Vlasov equation and which are now called the *"van Kampen's modes"*.

He has been a visitor in several Universities like Aachen, in Germany, and Minneapolis, Howard's and Texas in the United States. It was precisely in the University of Texas at Austin in 1974, where he became interested on some aspects of the *foundations of quantum mechanics*, in particular, in the so–called *measurement theory of quantum mechanics*. To avoid the standard treatment of the problem which seemed to him so remote from reality, he constructed a simple model which contained all the essential elements of the problem. One of these elements was the essential relation of microscopic events to the macroscopic world which he had considered before within the context of large systems in quantum statistical mechanics. In this context, he is specially critical about those who try to endow quantum mechanics with some kind of *"mysticism"*. Argues van Kampen: *"quantum mechanics is a perfectly logical and coherent physical theory, which can be understood rationally. The mysticism is theirs."* His analysis and points of view on this issue were published in a series of papers and his main results appeared in a paper under the title *"Ten theorems about quantum mechanical measurements"* in 1988.

Among his widespread interests, and apart from his scientific writings, which stand by themselves, Nico van Kampen has also written extensively in a different category, namely, that of essays and miscellaneous writings. Recently, Paul Meijer has edited the book *"Views of a Physicist"*, which contains an extensive collection of Nico's essays on different subjects, most of which were not accessible in English. This collection of essays cover a variety of topics which range from writings for special occasions, such as invited lectures, speeches, popular–science writings, book reviews or obituaries. But it also includes several of his well known fundamental and incisive critical essays. For example, the one on the over–simplification of the statistical mechanical explanation of Ohm's law by Linear Response Theory; or the one on the usual, but erroneous, explanation of the Third Law of Thermodynamics on the basis of the non–degeneracy of the ground state.

This book also includes beautiful essays on great physicists, like Copernicus, Smoluchowski, Wigner and Kramers. Perhaps some of the most delightful essays are those about his *Recollections of Kramers*, his former teacher, and those that include his subtle, clear and first hand analysis of Kramer's work and achievements.

As Paul Meijer says in the preface of this book: *"these essays are sometimes philosophical, often critical and almost always enjoyable for their style alone. His style is incisive but not derogatory and often playful"*.

Indeed, as his scientific papers, these essays are also clear, accurate, critical and carefully written. Assets that are reflected over and over in Nico's prominent and brilliant scientific career.

As his nephew and 1999 Nobel–prize winner Gerard t'Hooft writes in the prologue of this book: *"Here comes van Kampen to show up the charlatans"*. Yes, indeed, here comes Nico, and we do hope that for many years to come! Thank you.

Index

accelerated expansion, 177, 178
Adiabatic perturbations, 286
anti–de Sitter, 23
    gravity, 21, 22
axion, 244, 248, 250, 251, 253
    abelian, 242, 243

Berezin's Quantization, 9
Bianchi models, 168–171, 174
Black hole, 137, 140, 142, 143, 145, 146, 148, 149, 151–154
Brane, 101–105
    bulk corrections, 101
    bulk interactions, 101
    effects, 106, 108
    intersections, 102
Braneworld, 102, 106, 107
    scenario, 101, 103, 108

Chern–Simons, 3–5, 13, 14, 21–23, 75
    gravity, 32
    supergravity, 14, 21
    terms, 14, 28
Collapse, 89–94, 96
    calculations, 92
    phase, 91, 96
Collinson theorem, 187, 190, 191
Conformally Flat, 188, 190, 191
    axisymmetric spacetimes, 187, 190
Cosmic Microwave Background Radiation, 99, 103, 107, 157–159, 161, 163, 164

Dark Matter, 178
Dark matter, 112–117, 119, 120, 123, 124, 133, 178
    halo, 127
de Sitter, 137, 138
Deformation, 202
Discretization, 47–49, 51–53

Eötvös, 260
    coefficient, 261, 263, 265, 272–274
electrodynamics, 241, 244

Equivalence Principle, 228, 229, 257
    satellite test, 258
    weak, 233, 257, 258, 260, 261, 266, 270, 272–274
evanescent modes, 221, 222

Field–to–particle transition, 57, 59, 63, 67
Friedmann–Robertson–Walker, 168, 170, 171, 174
    cosmology, 168, 169

Galaxy formation, 114
Gauge bosons, 277
Geon, 63, 67

Hanbury–Brown–Twiss effect, 233, 234, 238
Holographic principle, 57–59, 63, 67
Hydrodynamics, 90, 92

Inflation, 100, 101, 103, 106–108, 132, 158–160, 162–164
Inflationary
    cosmology, 106, 107
    expansion, 100
    model, 101
    models, 100
    scenario, 100, 106, 108
inflaton, 158, 159
    potential, 158, 160–162, 164
interference, 235, 236, 238
interferometer, 233
Isolated Horizon, 146–148

Kähler
    manufolds, 3–6
    polarization, 5
    potential, 7, 10
    quantization, 5
    quotient, 5, 9, 10
    quotients, 3

light cone, 245, 251

MacDowell–Mansouri, 75
matter–antimatter asymmetry, 230
metric, 243, 249–252
  pre–, 242, 244

Neutrino oscillations, 228
  flavor, 229
Newman–Penrose, 187
No–Hair
  conjeture, 82
  theorem, 81–83
non–commutative
  gravity, 31, 32, 71, 72, 76, 77
  space–times, 226
  spaces, 73
non–Newtonian, 234
  effects, 234
  gravity, 234
  situation, 237
Nonlocality
  causality, 222
  effects, 222
  evanescent modes, 221
numerical
  calculations, 96
  methods, 90
  resolution, 89
  viscosity, 90, 93

Poincaré
  group, 21
  limit, 22
Proca, 81–83, 86, 150, 282, 283

Quantum Cosmology, 33, 38–40, 43, 193
Quantum gravity, 47, 50, 53

Rarita–Schwinger
  Lagrangian, 24, 26
Rarita–Schwinger field, 18, 19, 25
rotation curves, 123, 124, 128, 132, 133

Scalar Field, 125, 126, 129, 131, 132
  complex, 131
  model, 124
  pressure, 132
  self–interaction, 124
scalar–tensor theories, 167, 168, 170, 171, 174, 193
Schwarzschild
  interior metric, 187
Seiberg–Witten, 33, 72, 74, 76, 77
  map, 32, 36, 37, 39, 43
Sigma models, 57, 60, 63, 67
skewon, 245, 246, 248–253, 255
Stückelberg, 282, 283
stars, 90, 91
  proto–, 89
super–anti–de Sitter group, 21, 23, 28
super–Poincaré group, 21, 22, 28
Superluminal
  group velocity
    signal velocity, 221
    photonic tunnelling, 222

Weyl tensor, 188, 191
  complex components, 187
Wheeler–DeWitt
  equation, 31, 38–40, 43, 193, 195, 197, 199
Witteborn–Fairbank, 263, 265
  experiment, 258, 261, 264, 267–269, 274

Yang–Mills, 32, 33, 35, 36, 71, 72, 74, 77, 150
  Einstein theory, 151
  Einstein–Higgs theory, 151, 152
  Einstein–Proca theory, 151
  Higgs theory, 152
  Proca theory, 152
  Skyrme theory, 152
Young experiment, 233, 234